SOCIAL WORK AND CHILD SERVICES

SOCIAL WORK AND CHILD SERVICES

Sharon Duca Palmer, CSW, LMSW

School Social Worker, ACLD Kramer Learning Center,
Bay Shore, New York; Certified Field Instructor,
Adelphi University School of Social Work,
Garden City, New York, U.S.A.

Apple Academic Press

Social Work and Child Services

First Published in the Canada, 2011
Apple Academic Press Inc.
3333 Mistwell Crescent
Oakville, ON L6L 0A2
Tel. : (888) 241-2035
Fax: (866) 222-9549
E-mail: info@appleacademicpress.com
www.appleacademicpress.com

The full-color tables, figures, diagrams, and images in this book may be viewed at www.appleacademicpress.com

First issued in paperback 2021

ISBN 13: 978-1-77463-248-2 (pbk)
ISBN 13: 978-1-926692-84-5 (hbk)

Sharon Duca Palmer, CSW, LMSW

Cover Design: Psqua

ataloguing in Publication Data
ry and Archives Canada

CONTENT

INTRODUCTION

Social work is a difficult field to operationally define, as it is practiced differently in many settings. It is a very diverse occupation and one that can be practiced in settings such as hospitals, clinics, welfare agencies, schools, and private practices.

The main goal of all social work practice is to assist the client to function at the best of their ability and assess what their needs are. Social workers help clients with problem-solving strategies, such as defining personal goals, focusing on what is necessary to make changes, and helping them through the process.

Social work is a demanding field and is often emotional draining. Many social workers have large caseloads, limited resources for their clients, and often work for relatively low salaries. But the personal rewards can be very satisfying.

The social work profession is committed to promoting social and economic policy though helping to improve people's lives. Research is conducted to improve social services, community development, program evaluation, and public administration. The importance of research in these areas is to examine variables that can be addressed in order to resolve issues. Research can lead to what is called "best practice". By utilizing "best practice", a social worker is engaging clients based on research that is intended to increase successful outcomes.

Social work is one of the most diverse careers available. Most social workers are employed by health care facilities and government agencies. These facilities can include hospitals, mental health clinics, nursing homes, rehabilitation centers, schools, child welfare agencies, and private practice.

Social work's interface with mental health promotion and the treatment of mental illness dates to the earliest roots of our profession. While many social workers provide mental health services in private practice settings, the majority of services are offered in community-based agencies, both public and private, and in hospitals and prisons. Social workers are the largest provider of mental health services, providing more services than all other mental health care providers combined. These workers also often provide services to those who are struggling with substance abuse.

Twenty-first century health issues are complex and multidimensional, requiring innovative responses across professions at all levels of society. Public health social workers work to promote health in hospitals, schools, government agencies and local community-based settings, making connections between prevention and intervention from the individual to the whole population.

In an ideal world, every family would be stable and supportive. Every child would be happy at home and at school. Every elderly person would have a carefree retirement. Yet in reality, many children and families face daunting challenges. For example, single parents struggle to raise kids while working. Teens may become parents before they are ready. Child social workers help kids get back on track so they can lead healthy, happy lives.

Rapid aging populations are expected worldwide. With the rapid growth of this population, social work education and training specializing in older adults and practitioners interested in working with older adults are increasingly in demand. Geriatric social workers typically provide counseling, direct services, care coordination, community planning, and advocacy in an array of organizations including in homes, neighborhoods, hospitals, senior congregate living and nursing facilities. They work with older people, their families and communities, as well as with aging-related policy, and aging research

In whatever subcategory they work, social workers help provide support services to individuals and communities by assessing their needs in order to improve the quality of life and overall well-being. This can lead to positive changes in people's environments, dignity, and self-worth. It can also lead to changes in social policy for those who are vulnerable and oppressed. Social workers change entire communities for the better.

There have been many changes emerging in the social work profession. The uses of the Internet and online counseling have been major trends. Some people are more likely to seek assistance and information first through the use of the Internet. There has also been a strong move for collaborating between professions when providing services in order to offer clients more options for success. Keeping up to date with best practice research, licensing requirements, continuing education, and professional ethics make this an exciting and challenging time to be a social worker!

— Sharon Duca Palmer, CSW, LMSW

New Estimates of the Number of Children Living with Substance Misusing Parents: Results from UK National Household Surveys

Victoria Manning, David W. Best, Nathan Faulkner and Emily Titherington

ABSTRACT

Background

The existing estimates of there being 250,000—350,000 children of problem drug users in the UK (ACMD, 2003) and 780,000—1.3 million children of adults with an alcohol problem (AHRSE, 2004) are extrapolations of treatment data alone or estimates from other countries, hence updated, local and broader estimates are needed.

Methods

The current work identifies profiles where the risk of harm to children could be increased by patterns of parental substance use and generates new estimates following secondary analysis of five UK national household surveys.

Results

The Health Survey for England (HSfE) and General Household Survey (GHS) (both 2004) generated consistent estimates—around 30% of children under-16 years (3.3—3.5 million) in the UK lived with at least one binge drinking parent, 8% with at least two binge drinkers and 4% with a lone (binge drinking) parent. The National Psychiatric Morbidity Survey (NPMS) indicated that in 2000, 22% (2.6 million) lived with a hazardous drinker and 6% (705,000) with a dependent drinker. The British Crime Survey (2004) and NPMS (2000) indicated that 8% (up to 978,000) of children lived with an adult who had used illicit drugs within that year, 2% (up to 256,000) with a class A drug user and 7% (up to 873,000) with a class C drug user. Around 335,000 children lived with a drug dependent user, 72,000 with an injecting drug user, 72,000 with a drug user in treatment and 108,000 with an adult who had overdosed. Elevated or cumulative risk of harm may have existed for the 3.6% (around 430,000) children in the UK who lived with a problem drinker who also used drugs and 4% (half a million) where problem drinking co-existed with mental health problems. Stronger indicators of harm emerged from the Scottish Crime Survey (2000), according to which 1% of children (around 12,000 children) had witnessed force being used against an adult in the household by their partner whilst drinking alcohol and 0.6% (almost 6000 children) whilst using drugs.

Conclusion

Whilst harm from parental substance use is not inevitable, the number of children living with substance misusing parents exceeds earlier estimates. Widespread patterns of binge drinking and recreational drug use may expose children to sub-optimal care and substance-using role models. Implications for policy, practice and research are discussed.

Background

Child protection cases that feature in the UK media are reminders of how babies and children can be vulnerable to harm from parents and other adults and how frequently these cases involve binge or chronic substance use. According to

ACMDs Hidden Harm report (2003) [1], parental drug use can compromise a child's health and development from conception onwards. Parental substance misuse has been associated with genetic, developmental, psychological, psycho-social, physical, environmental and social harms to children [1-5]. The unborn child may be adversely affected by direct exposure to alcohol and drugs through maternal substance use [6-8]. The risk of harm however, depends on the age of the child, the nature and patterns of substance use and contextual factors in which the substance use occurs [9]. Social deprivation and the financing of drug or alcohol consumption may restrict money allocated to meet basics needs for the child. Inadequate monitoring, early exposure to substance taking behaviours, modelling behaviour and the failure to provide a nurturing environment can result in maladaptive and dysfunctional behaviour and other poor outcomes for the child [for reviews see [1,2,10]].

The potential for harm is not likely to be limited to dependent substance use. Binge drinking or regular non-problematic drug use can affect a person's control of emotions, judgement and ability to respond to situations, particularly during periods of intoxication and withdrawal. Being under the influence of substances may affect parental responsiveness to the physical or emotional needs of a child. For example, while parents recover from a hangover, babies and young children may be under-stimulated, whilst older children may carry the burden of household responsibilities and caring roles for siblings [11].

The limited research attempting to unveil the types of harm associated with parental substance misuse is largely restricted to retrospective cohort studies. Much of this work has attempted to identify adverse childhood experiences (ACEs) in the context of parental alcohol misuse among unhealthy/addicted adult populations [12,13]. Exposure to parental alcohol abuse is highly associated with ACEs [14]. Compared to persons reporting no ACEs, the risk of heavy drinking, alcoholism and depression in adulthood is significantly increased by the presence of multiple ACEs [15]. Another study examining ten ACEs (childhood emotional, physical, and sexual abuse, witnessing domestic violence, parental separation or divorce, growing up with drug abusing, alcohol abusing, mentally ill, suicidal, or criminal household members) found that the risk of having all of these was significantly greater among adult respondents who reported parental alcohol abuse [16]. Due to its sensitive nature and parents' fear of social services involvement, it is extremely difficult to conduct research to answer these questions. We are yet to determine the effects parental heavy drinking episodes and recreational illicit drug use have on children.

The latest drug strategy document for England estimates that there are around 330,000 problem drug users in England [17]—the majority of whom are of a parenting age. The document places heavy emphasis on reducing the risk of harm to children of drug-misusers, expressing a commitment to addressing the needs of parents and children by working with whole families to prevent drug use and reduce risk. In terms of the prevention agenda, it aims to promote the sharing of information across institutions e.g. ensure children's social services are aware of drug-using parents where children could be at risk and promises to 'expand their approach so that it increasingly focuses on young children and families before problems have arisen.' Linked to this is a commitment to take a 'wider preventative view' focussing on all substances including alcohol misuse. Regarding treatment the aims are to prioritise cases causing the most harm to families, by ensuring prompt access into effective treatment, assessment of family needs and intensive parenting support. It also aims to ensure that drug-misusing parents become a target group for new parenting experts, with Family Intervention Projects for families considered to be 'at-risk.'

When it comes to estimating the number of children at risk of harm from parental substance misuse, two sources are used as the epidemiological data on which the above targets are centred. The ACMD (2003) report 'Hidden Harm' estimated there being between 250,000—350,000 children of problem drug users in the UK, representing 2-3% of all under-16 years olds and the 2004 Alcohol Harm Reduction Strategy for England [18], estimated there being 780, 000—1.3 million children living with adults with an alcohol problem. There are however limitations with both of these estimates. The number of children estimated to be living with drug- misusing parents is an extrapolation of treatment data alone, that is, records of drug users presenting for treatment until the end of 2000. There is a concern that women are less likely to access treatment [19], yet more likely to reside with the child, therefore this could be an underestimate of the true number. The estimated number of children living with adults with alcohol problems can be sourced to a 1998, EuroCare document [20]. This document indicates that the estimate is an extrapolation of data from Denmark and Finland, each using a different methodology. The same document reports UK alcohol consumption to be significantly less than most other EU countries, yet recent trends in the use and misuse of alcohol are contradictory [21]. It is unclear how alcohol problems were defined and if they relate to the UK definitions of misuse. It appears to reflect drinking at a level considered in the UK as hazardous in one of the surveys. Thus, the existing estimates used to inform current UK policy and setting of targets for the next decade are dated, not based on local epidemiological data sources and need improving and broadening to include the combination of alcohol and other co-existing problems that can lead to adversity.

In contrast to considerable policy investment in addressing the needs of children living with substance misusers and in identifying good practice, the underlying epidemiological evidence has fallen short. For policy and commissioning responses to adapt to meet the needs of both parental substance misusers and their children, we first need to understand the nature and scale of the problem. Without knowing the number of potentially at-risk families, we are unable to assist them until they come to the attention of agencies at crisis point. The current study set out to update, improve and broaden earlier estimates to include alcohol, drugs and multiple/elevated risk factors of harm e.g. concurrent mental distress and substance use. This was achieved through secondary analysis of existing national household surveys which have captured relevant data. Attempts to generate new data to answer this research question are likely to be hampered by social desirability effects, thus generating unreliable estimates. The study reports these new estimates relating to the number of children living with alcohol and drug misusing parents.

Methods

The method used was developed by one of the authors (DB) in a similar project undertaken for the Australian National Commission on Drugs [22]. A literature search identified peer-review, grey literature and key government documents on children of substance misusing parents, but the focus of the research was on identifying national databases. Household, cohort or other large-scale surveys were considered if they met the following inclusion criteria; contained information on i) domestic arrangements, ii) adult substance use and iii) number of children in the household under the age of 16 years. Permission was sought for access to the original datasets from the corresponding departments/organisations and for the data to undergo secondary analysis to address the research questions. Once granted, study questionnaires were examined to identify potentially relevant variables. Since the surveys were not designed to address this research question but were undertaken by different organisations, with different objectives, at different points in time and using different assessment tools, wide variations in the estimated methods were anticipated. The databases were examined to identify the most common and robust indicators of substance use that could be applied. Variables were created or transformed to generate consistent variables across the datasets e.g. converting daily units of alcohol into weekly units, individual illicit drugs into drug classes and continuous variables (e.g. units consumed per week into categorical data e.g. hazardous drinking). Databases containing inconsistent or incompatible variables or with large amounts of missing data on variables of interest were excluded. Surviving variables were cleaned, edited and appropriate methods for the handling of missing data were applied. Anyone under the age of 16 was considered to be a child and only those living in the same household as adult substance users constituted a case. The number of adults (>15 years) fulfilling criteria for substance use/misuse was calculated initially, followed by the presence of any children in the same household and finally the number of those children. Measures were taken to ensure that when surveys contained more than one respondent per household (e.g. both parents) each child was only counted once. The number of children reported by the adult respondents to be living in the household formed the denominator (i.e. total number of children living with the sample) and not the number of child respondents, as these could be limited to only two children per household in some surveys. Once the number of children living with parental substance misusers was calculated (as a proportion of all children living with the respondents), the figures were extrapolated to the number of under 16's in the general population at that time and in the relevant countries (e.g. England, Scotland or UK), using the Office of National Statistics interactive population pyramid http://www.statistics.gov.uk webcite. All estimates were added to a summary table and confidence intervals calculated. Regular comparisons were made across datasets and revisions made to ensure that the most consistent indicators of substance use were applied.

The five national surveys accessed and providing appropriate data were; the General Household Survey (GHS), 2004; the Household Survey for England, (HSfE) 2004; the National Psychiatric Morbidity Study (NPMS), 2000; the British Crime Survey (BCS), 2004/5 and the Scottish Crime Survey (SCS) 2000. The GHS and HSfE household surveys were conducted around the same time and used the same measures of alcohol consumption (including indicators of binge drinking), although weekly consumption could only be calculated for a sub-sample (those reporting that they drink the same amount each day). Respondents had been asked "which day in the last week did you drink the most?" and were then asked to list how many of each type of alcoholic beverage they had consumed on this day. Each recorded beverage was converted into units of alcohol and summed to provide total units consumed on that day. The UK Government definition of binge drinking was calculated for the sample [18], i.e. 6 or more units in a single drinking occasion for women and 8 or more units for men. This is above (twice) the maximum recommended daily benchmark, stating that 'regular consumption of 2-3 units a day for women and 3-4 units a day for men does not lead to significant health risk.' We adopted the governments' definition of binge drinking as an accepted UK convention—this is not to imply that there is parental risk for all drinkers meeting these criteria, nor, indeed that there is no substance-related parenting risk in those who do not reach these thresholds. The NPMS contained data on problematic (hazardous, harmful and dependent drinking). Hazardous drinking (a pattern of alcohol consumption that increases the risk of harmful consequences for the user or others) was defined as a score on the Alcohol Use disorders Identification Test [23] of 8 or more. Harmful drinking (consumption that results in consequences to physical and mental health) was defined as a score of

16 or more. The Severity of Alcohol Dependence Questionnaire [24] was used to identify alcohol dependence in this survey. The two crime surveys and the NPMS were used to examine illicit drug use, the NPMS to look specifically at cumulative risks and the SCS to look at examples of harm resulting from substance misuse in the household.

Results

Parental Alcohol Misuse

Striking similarities were observed across the surveys in the rates of parenting (both around 30%) and drinking behaviour (see Table 1). Around 81% of the population were current drinkers and around 17% had engaged in binge drinking at least once in the week before interview. Consistent rates emerged indicating that around 34% of binge drinkers had at least one child in the household. These figures were extrapolated to the population in England (for the HSfE) or UK (for the GHS) in Table 2. Estimates from the two datasets were combined to generate a single estimate which was extrapolated to the UK population.

Table 1. Key findings relating to parental alcohol misuse from two 2004 national household surveys (HfSE, GHS).

Survey	HSfE	GHS
Year	2004	2004
Sample	6,704 adults	16,715 adults
% with at least 1 child <16 in the household	30.4%	29.0%
No. of children living in the household	1990	4163
Mean age of children <16	7.7 years	7.7 years
Current drinkers	82.1%	81.4%
Binge drinkers	16.7%	17.3%
Mental Health problem[1]	12.1%	2.5%
Binge drinker with 1+ child	35.3%	33.0%
Sub-sample [2]		
Hazardous or Dangerous drinkers	5.1%	4.9%
Dangerous drinkers	1%	1%
Hazardous/Dangerous drinkers with 1+child	22.%	23.4%
Dangerous drinker with 1+child	20.0%	24.8%

[1]Defined by a score of 4+ on the GHQ in the HSfE and the presence of at least one ICD mental health problem in the GHS
[2] For a sub-sample - those reporting alcohol units consumed on heaviest day in the past week, number of drinking days in past week and reporting that they drink the same amount each day, weekly units could be calculated. Hazardous refers to those exceeding 14 and 21 units weekly for females/males respectively. Dangerous drinking refers to those exceeding 35 and 50 units weekly for females/males respectively. However this sample was too small to generate population estimates from these surveys.

Table 2. Estimates of children living in households with alcohol misusing adults based on Table 1.

Category: living with	%	CI Lower	CI Upper	Estimate	Lower Estimate	Upper Estimate
GHS						
at least one binge drinker (BD)	29.7	28.3	31.1	3,458,654	3,297,010	3,620,298
a BD who is the only adult in household	4.2	3.6	4.8	489,103	418,143	560,062
at least 2 binge drinkers	7.4	6.6	8.2	861,752	769,149	954,355
an adult with a mental health problem	2.5	2.0	3.0	291,133	235,902	346,363
a BD with a mental health problem	0.6	0.4	0.8	69,872	42,552	97,191
HSfE						
at least one binge drinker (BD)	27.8	25.8	29.8	3,237,393	3,008,163	3,466,623
a BD who is the only adult in household	3.4	2.6	4.2	395,940	303,213	488,668
at least 2 binge drinkers	9.9	8.6	11.2	1,152,885	1,000,072	1,305,698
an adult with a mental health problem	18.9	17.2	20.6	2,200,962	2,000,643	2,401,280
a BD with a mental health problem	4.5	3.6	5.4	524,039	417,970	630,107
COMBINED GHS and HfSE						
at least one binge drinker (BD)	29.1	28.0	30.2	3,388,782	3,256,612	3,520,952
a BD who is the only adult in household	3.9	3.4	4.4	458,014	401,454	514,575
at least 2 binge drinkers	8.2	7.5	8.9	957,666	877,727	1,037,606
an adult with a mental health problem	7.8	7.1	8.5	910,351	832,239	988,463
a BD with a mental health problem	1.9	1.6	2.2	221,437	181,695	261,178

Across the two surveys, 27.8—29.7% of children, representing a combined estimate of 3,388,782 (95% confidence interval 3,256,612—3,520,952) children in the UK aged under-16, lived with an adult binge drinker, 3.4—4.2% of children (representing a combined estimate of 458,014) lived in a household where the only adult was a binge drinker and 7.7—9.9% (representing a combined estimate of 957,666) lived with at least two binge drinkers (typically both parents). The combined datasets indicated that 1.9% (representing 221,437 children) lived with an adult binge drinker with concomitant psychological distress which may be exacerbated by their drinking behaviour. According to the NPMS 22.1% (representing 2,643,049) children lived with a hazardous drinker, 2.5% (298,988) with a harmful drinker and 3.7% (442,502) in households where the only adult was 'at least' a hazardous drinker. Respondents scoring more than 10 on the AUDIT then completed the SADQ, 5.9% (representing 705,611) of children lived with the 7% of drinkers who met criteria for (at least) mild alcohol dependence (see Table 3).

Table 3. Percentage of children living in households with substance using parents from the BCS, SCS and NPMS.

Survey	BCS	SCS	NPMS
Year	2004/5	2000	2000
Sample (adults)	5604[1]	2998	8580
% with at least 1 child <16 in the household	41.1%	38.1%	31.0%
No. of children living in the household	1975	2006	4783
Mean age of children <16	7.7	7.6	-
% children living with:			
an adult drug user (past year)	8.4%	4.9%	8.0%
an adult drug user (past month)	4.2%	3.3%	3.9%
an adult using class A drug (past year)	2.2%	1.4%	1.8%
an adult using class B drug (past year)	1.4%	0.4%	1.4%
an adult using class C drug (past year)	7.4%	4.2%	7.3%
an adult using class A drug (past month)	1.0%	0.9%	0.6%
an adult using class B drug (past month)	0.3%	0.2%	0.5%
an adult using class C drug(past month)	3.9%	2.7%	3.8%
a daily drug user	0.6%	0.8%	1.1%
only a drug using parent (past year)	5.7%	-	2.3%
only a drug using parent (past month)	3.3%	-	1.3%
an adult who has injected a drug	-	0.3%	0.6%
an adult who has overdosed from drugs	-	-	0.9%
an adult who is drug dependent	-	-	2.8%
an adult in drug treatment	-	-	0.6%
an adult who is a hazardous drinker (>8 on AUDIT)	-	-	22.1%
an adult who is a harmful drinker (>16 on AUDIT)	-	-	2.5%
an adult who is a dependent drinker (SADQ)	-	-	5.9%
only a hazardous or worse drinker	-	-	3.7%

[1]5604 answered the drugs section of the interview but total sample was 51,964

Parental Drug Use

The proportion of children living with illicit drugs users was relatively consistent across the two UK surveys, albeit slightly lower in the SCS (See Table 3 &4). It was not possible to collapse the datasets to form a single estimate because they were conducted in different years. According to the BCS, 8.4%

(representing 978,205 children) in 2004/5 (95% CI 835,739—1,120,671) lived with an adult who had used illicit drugs in the past year. A similar estimate of 8.0% (representing 956,760 children) in 2000 (95% CI 864,809—1-,048,711) emerged from the NPMS. The figure for Scotland in 2000 based on the SCS was 4.9% (representing 47,631 children). The UK surveys suggest

Table 4. Estimates of children living in households with substance using adults based on Table 3.

Children living with an adult who has/is......	%	CI Lower	CI Upper	Estimate	Lower Estimate	Upper Estimate
BCS						
used drugs in past year	8.4	7.2	9.6	978,205	835,739	1,120,671
used drugs in past month	4.2	3.3	5.1	489,103	386,080	592,125
used drugs class A past year	2.2	1.6	2.8	256,197	180,860	331,533
used drugs class B past year	1.4	0.9	1.9	163,034	102,691	223,377
used drugs class C past year	7.4	6.2	8.6	861,752	727,307	996,197
used drugs class A past month	1.0	0.6	1.4	116,453	65,351	167,555
used drugs class C past month	3.9	3.0	4.8	454,167	354,737	553,597
a daily drug user	0.6	0.3	0.9	69,872	30,208	109,535
used drugs in past year and is a lone parent	5.7	4.7	6.7	663,782	544,708	782,856
SCS						
used drugs in past year	4.9	4.0	5.8	47,631	38,448	56,814
used drugs in past month	3.3	2.5	4.1	32,078	24,479	39,677
used drugs class A past year	1.4	0.9	1.9	13,609	8,611	18,607
used drugs class B past year	0.4	0.1	0.7	3,888	1,203	6,573
used drugs class C past year	4.2	3.3	5.1	40,827	32,294	49,360
used drugs class A past month	0.9	0.5	1.3	8,749	4,731	12,766
used drugs class C past month	2.7	2.0	3.4	26,246	19,351	33,141
a daily drug user	0.8	0.4	1.2	7,777	3,987	11,566
an injecting drug user	0.3	0.06	0.5	2,916	590	5,243
NPMS						
used drugs in past year	8.0	7.2	8.8	956,760	864,809	1,048,711
used drugs in past month	3.9	3.4	4.5	466,420	400,804	532,0376
used drugs class A past year	1.8	1.4	2.2	215,271	170,209	260,333
used drugs class B past year	1.4	1.1	1.7	167,433	127,611	207,255
used drugs class C past year	7.3	6.6	8.0	873,043	784,874	961,213
used drugs class A past month	0.6	0.4	0.8	71,757	455,82	97,932

Table 4. *(Continued)*

used drugs class C past month	3.8	3.3	4.3	454,461	389,658	519,264
a daily drug user	1.1	0.8	1.4	131,555	96,203	166,906
an injecting drug user	0.6	0.4	0.8	71,757	45,582	97,932
overdosed	0.9	0.6	1.2	107,636	75,626	139,645
drug dependent	2.8	2.3	3.3	334,866	278,951	390,781
in drug treatment	0.6	0.4	0.8	71,757	45,582	97,932
used drugs in past year and is a lone parent	2.3	1.9	2.7	275,069	224,261	325,876
Hazardous drinker (AUDIT)	22.1	20.9	23.3	2,643,049	2,502,418	2,783,681
Harmful drinker (AUDIT)	2.5	2.1	2.9	298,988	246,071	351,904
Dependent drinker (SADQ)	5.9	5.2	6.6	705,611	625,749	785,472
Only adult is at least a hazardous drinker	3.7	3.17	4.2	442,502	378,523	506,480

that 3.9% of children in the year 2000 and 4.2% of children in 2004/5 lived with an adult who had used illicit drugs in the previous month (representing around 466,420 and 489,103 respectively). 1.8% of children (in 2000) and in 2.2% of children (in 2004/5) lived with a class A drug user (representing 215,271 and 256,197 respectively), whilst 0.6% and 1% respectively lived with someone who had used a class A drug in the previous month. 7.3% (in 2000) and 7.4% (in 2004/5) (representing 873,043 and 861,752 respectively) lived with a class C drug user, 3.8% and 3.9% respectively, with someone who had done used a class C drug in the previous month. Living in a household where the only adult was a drug user was the case for 2.3%, representing 275,069 children in 2000 and 5.7%, representing 663,782 children in 2004/5. The rates are higher for class C than class B drugs because cannabis was classified as a class C drug at the time analysis was conducted.

According to the SCS, 0.3% (representing 2,916 children in Scotland) and according to the NPMS 0.6% (representing 71,757 children in the UK) were living with an injecting drug user. The NPMS indicated that 0.6% (71,757 children) lived with a drug user in treatment, 2.8% (334,866) lived with a drug dependent user, 0.9% (107,636) lived with an adult drug user who had experienced a drug overdose and 1.1% (131,555) with a daily drug user. The number of children living with a lone drug using parent in 2004/5 had doubled at 5.7% (representing 663,782 children) from the estimated 2.3% (representing 275,069) in the UK in 2000.

Cumulative and More Severe Risks of Harm

The NPMS contained information on drinking (using standardised assessment tools such as SADQ and the AUDIT), illicit drugs and mental health status using a standardised psychiatric assessment tool (CIS-R). This enabled an estimate to be generated for the number of children exposed to multiple or cumulative risk (see Table 5 and Figure 1).

Table 5. Population estimates of children living with adults where there is cumulative risk (NPMS).

Children living with an adult who is a...	%	CI Lower	CI Upper	Estimate	Lower Estimate	Upper Estimate
problem drinker & drug user (past year)	3.6	3.1	4.1	430,542	367,402	493,683
problem drinker & drug user (past month)	1.8	1.4	2.2	215,271	170,209	260,333
problem drinker with mental health problems	4.2	3.6	4.8	502,299	434,312	570,286
drug use (past year) with mental health problems	2.6	2.2	3.1	310,947	257,010	364,884
problem drinker, drug user (past year) + mental health problems (past year)	1.0	0.7	1.3	119,595	85,871	153,319
problem drinker, drug user (past year) + mental health problems (past month)	0.6	0.4	0.8	71,757	45,582	97,932

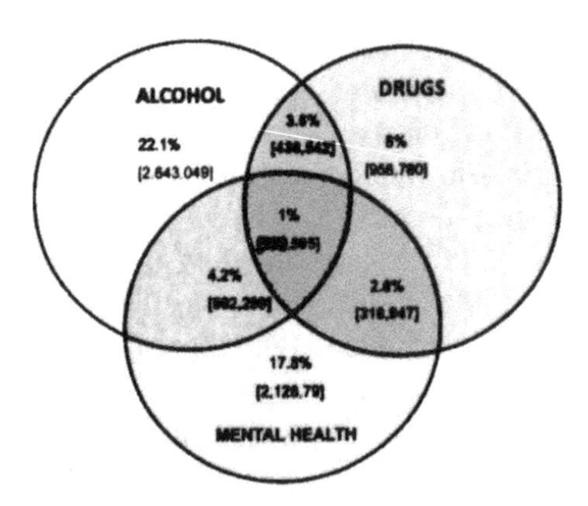

Figure 1. Cumulative risk of harm from the NPMS.

Three point six percent of children (430,542) in the UK in 2000 lived with an adult problem drinker (a score of more than 8 on the AUDIT) who had used drugs in the previous year, and 1.8% (215,271) with a problem drinker who had used drugs in the previous month. Around 4.2% (502,299) lived with a problem drinker with a concurrent mental health problem (a score of >12 on CIS-R), and 2.6% (310,947) lived with a drug user with a concurrent mental health problem. Even greater risk of harm could have existed for the 1% (119,595) of children who lived with an adult problem drinker who also used drugs and had concurrent mental health problems.

Finally, the SCS contained examples of circumstances where harm can result directly from drug or alcohol misuse (see Table 6). It recorded incidents when violence (use of force, push/shove, thrown, threatened, chocked/strangled, hit, stabbed, forced sex and other) was used on an adult (usually a parent) in the house by another adult (usually their partner) after drinking/drug use, and when these acts of violence were witnessed by children in the household. This indicated that in Scotland, 2.5%, (24,302 children) lived in a household where violent incidents occurred after the perpetrator had been drinking and 1.2% (11,665) lived in a household where violent incidents occurred after the perpetrator had been using drugs. Even greater harm may exist for the 1.2% (111,665) and 0.6% (5832) who witnessed these acts of violence occurring as a result of alcohol and drug use respectively.

Table 6. Population estimates of children living in households where violence occurs following substance use (Scottish Crime Survey, 2000)

Children living in a household where there is...	%	CI Lower	CI Upper	Estimate	Lower Estimate	Upper Estimate
violence in household when adult was drinking	2.5	1.8	3.2	24,302	17,660	30,943
violence in household when adult was using drugs	1.2	0.7	1.7	11,665	7,033	16,297
violence in household witnessed by child when adult was drinking	1.2	0.7	1.7	11,665	7,033	16,297
violence in household witnessed by child when adult was using drugs	0.6	0.3	0.9	5,832	2,547	9,118

Discussion

Overall, the figures suggest that the number of children living with at least episodic binge drinkers or illicit drug users is greater than previously thought. In 2004, 3.3—3.5 million children in the UK were living with at least one binge drinking parent. Having a non-binge drinking adult in the household offers a positive role model and can mitigate against harm caused by the problem drinking parent [25]. Therefore the near half a million children living with a lone-binge drinking parent and 957,000 with two binge-drinking parents could be more vulnerable to harm. Whilst there is no evidence to suggest that parental binge drinking is associated with harm to children, adults in this category were 'at least' binge drinking. Some would have been problem drinkers and there is literature emerging to suggest that problem drinking is associated with childhood adversity [14-16]. The estimated 2.6 million living with a hazardous drinker in the year 2000 would appear to exceed the earlier estimates of 780,000—1.3 million, although it is not clear how problem drinking was defined in the estimates from data sources in Denmark and Finland. Whilst the data does not imply that these children experience adverse consequences, the potential for exposure (assuming it occurs in the home) to modelling heavy drinking behaviour exists, as does neglect and less adequate parental responses to accidents and emergencies (child injuries, fires and other adverse events which are more likely to occur in the event of intoxicating substance use). These new estimates complement the existing estimates on treated addiction populations and add to what we know. Unfortunately, however they remain a long way from what we need to know.

Around one million children in the UK live with an adult who has used an illicit drug in the past year, and just under half a million with someone who has done so in the past month. It is not possible to compare directly with the Hidden Harm (2003) estimates since they are generated from different populations and using different methodologies. It is plausible that illicit drug use could constitute smoking cannabis when the drug user does not have responsibility for child care, thus posing no acute risk of harm. Although it could be argued that any drugs use can create a social learning model and that regular use may result in chronic effects that are more likely to compromise parenting capabilities. Equally, however it could constitute regular use of cocaine or heroin in the home environment, where the child could be exposed to drug taking behaviours, paraphernalia, dealers, and the potential to ingest or experiment with the drug. The finding that the number of children living in a household where the only adult was a drug user had more than doubled between 2000 and 2004/5, points to increasing vulnerability in single-parent families and highlights the need for child protection efforts to determine need, as well as risk. The finding that 334,000 children were estimated to be living with a dependent drug user is broadly consistent with the Hidden Harm (2003) estimate relating to treated drug users. The finding that 107,000 children lived with an adult who had experienced a drug overdose, is an indicator of the possible severity of drug misuse among this predominantly untreated population. Given that it is estimated that there are 116,809 injecting heroin or crack cocaine users in England alone in 2006/7 [26], the current estimate of only 72,000 children living with an injecting drug user in the UK is low and may reflect a reluctance to disclose injecting behaviour in the context of household surveys.

The potential for cumulative disadvantage for children living with adults with multiple problem behaviours is a particular concern as co-morbidity has been linked to less effective treatment engagement and additional difficulties in parenting [27,28]. Parental mental illness featured in one-third of 100 reviews of child deaths of abuse and neglect[29]. Parental substance misuse was a concern for 52% of families placed on child protection registers and for 62% of children subject to childcare proceedings [30,31]. Therefore the risk of harm to children of parents with co-morbid substance misuse and mental health problems is likely to be even greater. Parental experience of blunted emotions/feelings, anxiety or depression in addition to substance use may restrict the child's social and recreational activities. Finally, the observation that large numbers of children have witnessed violence occurring in the context of substance misuse is a major concern for child protection agencies and supports earlier findings [32,33].

There are some limitations with this work that are worthy of consideration. It is important to recognise that the new estimates are likely to be conservative estimates and subjected to measurement and reporting bias. Significant under-reporting of substance use is likely to have occurred, given that data were gathered in the respondents home, provided by parents and with questions relating to child welfare. Extrapolation of survey data to the UK population, poses the risk of accentuating sampling or response biases. However, the concordance of several estimates across surveys mitigate this concern to some degree. Nonetheless we recognise that there are intrinsic limitations to population surveys that make extrapolations to populations risky, and these include the under-representation of certain ethnic groups, the homeless and other vulnerable populations and the over-representation of more stable groups. The disparate rates of psychological distress reported in the GHS and HSfE are almost certainly an artefact of the different assessment tools used (GHQ-12 versus ICD-10 codes). Future research should aim to generate estimates based around different ages, since the risk and types of harms to babies and young children are likely to differ from those of teenagers. A limitation of household surveys is that they are cross-sectional in nature with respondent substance use and parental status measured at a single point in time, yet these are not stable factors but fluctuate over time. It may be more helpful to generate estimates using databases with multiple and regular assessment intervals e.g. longitudinal or cohort studies such as the Millennium Cohort Study to identify changes in exposure to risk due to factors such as child age.

The findings should be used to inform the design and content of future research aiming to explore the ways in which parenting capacity is affected by their substance use. Despite a better indication of the scale of the problem, the absence of contextual data limits the conclusions that can be drawn for both policy and practice. However, the relatively uncritical citing of the estimates used in the current UK drug and alcohol strategies should be challenged. Future research needs to examine the relative risks of harm posed to children with different substances used and patterns of use i.e. chronic patterns of relatively low-level substance use as well as those associated with binge or intoxicating use of substances. Survey items designed to capture the nature of harm must be discretely embedded within a broader research interview to prevent social desirability effects. The use of multiple methodologies will overcome the reliance on secondary analysis of surveys designed for other purposes.

While we must be cautious about the historical and opportunistic nature of the data presented here, these findings have implications for policy and practice, concerning the early identification of parental substance misuse and provision of early interventions. These can serve to minimise adverse childhood experiences and halt the escalation of recreational/binge substance use to that of problematic/dependent use more likely to cause harm. Possible indirect pathways of harm include unemployment, physical, psychological, social and financial substance-related problems and stigma [34], all of which can impede child development [1]. Whilst the bulk of the data presented here captures binge drinking and recreational drug use, universal and generic services working with parents and children should attempt to assess and be sensitive to parental substance use and possible related adversity that can serve as markers of harm. It may be time to revise the questions routinely asked when assessing parenting capacity in the Common Assessment Framework (CAF) and through Local Children's Safeguarding boards.

Research conducted on the caseloads of social services and other child welfare agencies [31,35] highlight the information, assessments and interventions required to manage risk where parental substance use exists. However, there are also implications from these findings for agencies that aim to encourage the uptake of substance use treatment and for universal educational initiatives aiming to raise awareness of the harms parental substance use can pose to their children. Training in effective management of the daily stressors associated with socioeconomic disadvantage, improving parenting practices by providing a supporting and nurturing environment for a child, and educating women about drinking during pregnancy may halt the progression from risk to actual harm.

Finally, it must be borne in mind that whilst the number of children living with parental substance misusers appears greater than earlier estimates, the picture is not universally bleak. Research findings indicate that the most high-risk drug taking behaviours tend to exist among parental substance misusers whose children live elsewhere, thus eliminating direct negative effects [36,37]. Whilst substance misuse can impair parenting capacity, harm is not inevitable and rarely exists as a consequence of substance use in isolation. Poverty, social exclusion, poor housing, a stressful environment, family tension and conflict and lack of psychosocial resources etc. collectively heighten the risk of harm [9]. This makes effective inter-agency working, involving the sharing of responsibilities, information on the potential indicators of harm and comprehensive assessments of need, harm and risk a top priority. Generating local evidence bases around this topic is as important as structured research. The 2006 'Working Together to Safeguard Children Report' [11] states that 'it is the duty of all agencies working with parents and children to make arrangements to safeguard and promote the welfare of the child,' and places a heavy emphasis on sharing and effective joint working. Further recommendations are made in the recent 'The Protection of Children in England: A Progress Report' [38], including a need to develop guidance on referral and assessment systems for children affected by parental substance misuse, using current best practice. Many substance users, particularly the stable ones in treatment, can be competent and loving parents. Nonetheless, it is important that efforts are not focussed solely on these most vulnerable families and to recognise that widespread patterns of binge drinking and recreational drug use may also disadvantage children and expose them to a poorer standard of care and safety than is acceptable.

Conclusion

Estimating the number of children living with parental substance misusers and in particular those who warrant professional intervention poses several challenges. Characterised by denial and shame, the numbers can only be inferred from available data provided by parents and the reliability of such self-report data is unknown. Nonetheless, secondary analysis of the most recent household surveys in the UK indicates that taken together, the number of children living with substance users and binge drinkers is far greater than earlier estimates would suggest. The US and emerging UK literature suggests that interventions delivered within social services and substance misuse agencies, incorporating family therapy and aiming to improve parenting skills, show much promise for families where harm already exists. However, it is the majority of those children raised in households where one or both parents are problem substance users without professional or mainstream services intervention who may experience the greatest adversity. Whilst actual harm from parental substance misuse is not inevitable, only large scale and far-reaching initiatives will likely impact on the 3.4 million children living with binge drinkers and almost 1 million living with drug users, where the potential for harm exists. In light of new findings on the scale of the problem, we urge mainstream services to reflect on their role in supporting these vulnerable families and in raising awareness of how parental substance use can elevate the risk of harm to children. Improving access to treatment, family interventions and parenting skills training, should place us in a stronger position than ever to prevent harm to children of substance-misusing parents.

Competing Interests

The authors declare that they have no competing interests.

Authors' Contributions

VM was responsible for the study design, data collection and analysis and writing the manuscript. DB was responsible for study design, analysis, development of methodology and assisting with the writing of the manuscript. Both NF and ET were responsible for identifying data sources, data collection and analysis and assisting with manuscript revisions. All authors read and approved the final manuscript.

Acknowledgements

The authors are grateful to Action on Addiction and the Wates Foundation for funding the research. The funders had no influence on the study design; in the collection, analysis, or interpretation of data; nor in the writing of the manuscript; nor in the decision to submit the manuscript for publication.

References

1. ACMD: Hidden Harm: Responding to the Needs of Children of Problem Drug Users, report of an inquiry by the Advisory Council on the Misuse of Drugs. London: Home Office; 2003.
2. ACMD: Hidden Harm: Three Years On Hidden harm: realities, challenges and opportunities. London: Home Office; 2006.
3. Barnard M, McKeganey N: The impact of parental problem drug use on children: what is the problem and what can be done to help? Addiction 2004, 99:552–559.
4. Jaudes PK, Ekwo E, Van Voorhis J: Association of drug abuse and child abuse. Child Abuse Negl 1995, 19:1065–1075.
5. Kroll B, Taylor A: Parental substance misuse and child welfare. London: Jessica Kingsley Publishers; 2003.
6. Eyler FD, Behnke M: Early development of infants exposed to drugs prenatally. Clin Perinatol 1999, 26:107–150.
7. Guerrini I, Thomson AD, Gurling HD: The importance of alcohol misuse, malnutrition and genetic susceptibility on brain growth and plasticity. Neurosci Biobehav Rev 2007, 31:212–220.
8. Huizink AC, Mulder EJ: Maternal smoking, drinking or cannabis use during pregnancy and neurobehavioral and cognitive functioning in human offspring. Neurosci Biobehav Rev 2006, 30:24–41.

9. Suchman NE, Luthar SS: Maternal addiction, child maladjustment and socio-demographic risks: implications for parenting behaviors. Addiction 2000, 95:1417–1428.
10. Johnson JL, Leff M: Children of substance abusers: overview of research findings. Pediatrics 1999, (5 Pt 2):1085–1099.
11. H.M.Government: Working Together to Safeguard Children: A guide to interagency working to safeguard and promote the welfare of children. Copyright C. London; 2006.
12. Kestila L, Martelin T, Rahkonen O, Joutsenniemi K, Pirkola S, Poikolainen K, Koskinen S: Childhood and current determinants of heavy drinking in early adulthood. Alcohol Alcohol 2008, 43(4):460–469.
13. Zlotnick C, Tam T, Robertson MJ: Adverse childhood events, substance abuse, and measures of affiliation. Addict Behav 2004, 29(6):1177–1181.
14. Dube SR, Anda RF, Felitti VJ, Croft JB, Edwards VJ, Giles WH: Growing up with parental alcohol abuse: exposure to childhood abuse, neglect, and household dysfunction. Child Abuse Negl 2001, 25(12):1627–1640.
15. Dube SR, Anda RF, Felitti VJ, Edwards VJ, Croft JB: Adverse childhood experiences and personal alcohol abuse as an adult. Addict Behav 2002, 27(5):713–725.
16. Anda RF, Whitfield CL, Felitti VJ, Chapman D, Edwards VJ, Dube SR, Williamson DF: Adverse childhood experiences, alcoholic parents, and later risk of alcoholism and depression. Psychiatr Serv 2002, 53(8):1001–1009.
17. H.M Government: Drugs: Protecting Families and Communities: The 2008 Drug Strategy London; London. 2008.
18. PMU: Alcohol Harm Reduction Strategy for England. London: Prime Minister's Strategy Unit CabinetOffice; 2004.
19. Greenfield SF, Brooks AJ, Gordon SM, Green CA, Kropp F, McHugh RK, Lincoln M, Hien D, Miele GM: Substance abuse treatment entry, retention, and outcome in women: a review of the literature. Drug Alcohol Depend 2007, 86(1):1–21.
20. Eurocare: Alcohol problems in the family—a report to the European Union. [http://www.eurocare.org/resources/eurocare_publications], Copyright European Commission edition. England 1998.
21. DHS: Binge Drinking and Europe. Hamm: Deutsche Hauptstelle für Suchtfragen; 2008.

22. Dawe S, Fry S, Best D, Moss D, Atkinson J, Evans C, Lynch M, Harnett P: Drug use in the family: impacts and implications for children. Canberra, ACT: Australian National Council on Drugs; 2007.

23. Saunders JB, Aasland OG, Babor TF, de la Fuente JR, Grant M: Development of the Alcohol Use Disorders Identification Test (AUDIT): WHO Collaborative Project on Early Detection of Persons with Harmful Alcohol Consumption–II. Addiction 1993, 88(6):791–804.

24. Stockwell T, Sitharthan T, McGrath D, Lang E: The measurement of alcohol dependence and impaired control in community samples. Addiction 1994, 89(2):167–174.

25. Foxcroft DR, Lowe G: Adolescent drinking behaviour and family socialization factors: a meta-analysis. J Adolesc 1991, 14:255–273.

26. Hay G, Gannon M, MacDougall J, Millar T, Williams K, Eastwood C, McKeganey N: National and regional estimates of the prevalence of opiate use and/or crack cocaine use 2006/07: a summary of key findings. London: Home Office; 2008.

27. Cleaver H, Unell I, Aldgate J: Children's Needs: Parenting Capacity. London: The Stationary Office; 1999.

28. Macrory F: Meeting Multiple Needs: Pregnancy, Parenting and Dual Diagnosis. In Dual Diagnosis Nursing. Edited by: Rassool PGH. Oxford: Wiley-Blackwell Ltd; 2006.

29. Falkov A: A Study of Working Together 'Part 8' Reports: Fatal child abuse and parental psychiatric disorder. In London: DOH-ACPC Series, 1. DOH-ACPC Series; 1996.

30. Forrester D: Parental Substance Misuse and Child Protection in a British Sample; A Survey of Children on the Child Protection Register in an Inner London District Office. Child Abuse Review 2000, 9:235–246.

31. Forrester D, Harwin J: Parental substance misuse and child care social work: findings from the first stage of a study of 100 families. Child and Family Social Work 2006, 11:323–355.

32. Nicholas K, Rasmussen E: Childhood Abusive and Supportive Experiences, Inter-Parental Violence, and Parental Alcohol Use: Prediction of Young Adult Depressive Symptoms and Aggression. Journal of Family Violence 2006, 21:43–61.

33. Gorin S: Understanding what children say. In Children's experiences of domestic violence, parental substance misuse and parental health problems. London: National Children's Bureau; 2004.

34. Schuckit MA: Alcohol-use disorders. Lancet 2009, 373(9662):492–501.
35. Forrester D, Harwin J: Parental Substance Misuse and Child Welfare: Outcomes for Children Two Years after Referral. British Journal of Social Work 2008, 38:1518–1535.
36. Meier PS, Donmall MC, McElduff P: Characteristics of drug users who do or do not have care of their children. Addiction 2004, 99(8):955–961.
37. Pilowsky DJ, Lyles CM, Cross SI, Celentano D, Nelson KE, Vlahov D: Characteristics of injection drug using parents who retain their children. Drug Alcohol Depend 2001, 61(2):113–122.
38. Department for Children Schools and Families D: The Protection of Children in England: Action Plan—The Government's Response to Lord Laming, [http://publications.everychildmatters.gov.uk/eOrderingDownload/HC-330.pdf] 2009.

Supporting Adolescent Emotional Health in Schools: A Mixed Methods Study of Student and Staff Views in England

Judi Kidger, Jenny L. Donovan, Lucy Biddle, Rona Campbell and David Gunnell

ABSTRACT

Background

Schools have been identified as an important place in which to support adolescent emotional health, although evidence as to which interventions are effective remains limited. Relatively little is known about student and staff views regarding current school-based emotional health provision and what they would like to see in the future, and this is what this study explored.

Methods

A random sample of 296 English secondary schools were surveyed to quantify current level of emotional health provision. Qualitative student focus groups (27 groups, 154 students aged 12-14) and staff interviews (12 interviews, 15 individuals) were conducted in eight schools, purposively sampled from the survey respondents to ensure a range of emotional health activity, free school meal eligibility and location. Data were analysed thematically, following a constant comparison approach.

Results

Emergent themes were grouped into three areas in which participants felt schools did or could intervene: emotional health in the curriculum, support for those in distress, and the physical and psychosocial environment. Little time was spent teaching about emotional health in the curriculum, and most staff and students wanted more. Opportunities to explore emotions in other curriculum subjects were valued. All schools provided some support for students experiencing emotional distress, but the type and quality varied a great deal. Students wanted an increase in school-based help sources that were confidential, available to all and sympathetic, and were concerned that accessing support should not lead to stigma. Finally, staff and students emphasised the need to consider the whole school environment in order to address sources of distress such as bullying and teacher-student relationships, but also to increase activities that enhanced emotional health.

Conclusion

Staff and students identified several ways in which schools can improve their support of adolescent emotional health, both within and outside the curriculum. However, such changes should be introduced as part of a wider consideration of how the whole school environment can be more supportive of students' emotional health. Clearer guidance at policy level, more rigorous evaluation of current interventions, and greater dissemination of good practice is necessary to ensure adolescents' emotional health needs are addressed effectively within schools.

Background

Rates of clinical and subclinical emotional health problems during adolescence [1-3], and the multiple detrimental health and social outcomes with which they are associated, such as suicide attempts, substance misuse, educational underachievement, unemployment and long-term psychiatric disorders [4-6], have been a cause of concern for some time. Many emotional disorders have their onset in adolescence [6], and a marked increase in prevalence occurs from the middle to late teenage years [7,8]. Further, many teenagers who experience emotional health problems fail to receive help from appropriate services [9,10]. Such evidence provides strong arguments for community based interventions to support emotional health in early adolescence, and prevent or reduce the onset of disorders.

Several examples exist of classroom-based programmes in schools that focus on emotional health, and these fall roughly into three groups. The first group aims to prevent or reduce emotional disorders such as depression and anxiety, by developing coping skills commonly used within psychiatry [11-13]. A number of reviews, while acknowledging the potential value of such interventions, conclude that evidence for their long-term effectiveness remains limited [14-17]. The second group comprises interventions that focus more on the improvement of knowledge and understanding regarding emotional health related issues, with a view to addressing concerns such as self-harm and suicide, negative attitudes towards emotional disorders and adolescents' unwillingness to seek help [9,18,19]. A small number of such interventions have been evaluated and found to be effective [e.g. [9,18,20]], but again evidence from suitably high quality evaluations is lacking, and there is a need to establish both long-term effectiveness, and whether changes in knowledge and understanding lead to the desired behavioural change. The third group of programmes focuses more on promoting positive emotional health, sometimes called 'emotional intelligence' or 'emotional literacy' [21]. For example the government-led initiative SEAL (Social and Emotional Aspects of Learning) in England [22], and SEL (Social and Emotional Learning) programmes developed in the USA [23], both aim to develop emotional and social skills in order to promote emotional health, improve behaviour, and support academic learning. While the evaluation of SEAL is still ongoing, SEL has been found to be effective in social and emotional skills development, increasing positive social behaviour, increasing academic attainment, and reducing emotional distress and conduct problems [23].

A concurrent development to classroom-based work has been an interest in 'whole-school' approaches, which is in keeping with a wider move towards a focus on the whole school when implementing health interventions [24]. Such programmes often aim at promoting emotional health rather than preventing emotional disorder [25], and take a holistic approach to this, considering all aspects of school life including policies and procedures, ethos, and partnerships with external agencies and parents to better support students' emotional health [25,26]. This model is not necessarily in opposition to classroom-based programmes, but is seen as providing an important backdrop against which activities such as emotional health lessons and support services for those who need them will be more effective [27]. Justification for such an approach can be found in the reported association between emotional health and factors related to school life such as academic success, peer relationships and school connectedness [28-31]. Further, there is growing evidence that whole-school approaches focusing on promoting good emotional health are more effective than stand alone classroom-based interventions that aim to prevent emotional disorder [27,32], although more evaluations, particularly beyond the elementary school years, are still needed in this area.

Within England, the significance of school settings for addressing the emotional health of all students is emphasised in key policy documents and initiatives [22,33,34]. Current thinking appears to support an emphasis on improving emotional health rather than preventing disorder, and on addressing whole-school issues such as school climate, provision of support services and the professional development of teachers, alongside curriculum based work [22,35]. A report by England's educational inspectorate Ofsted in 2005 identified several elements that characterised schools that were successfully promoting emotional health, including a caring ethos, teacher training in how to promote emotional health, and good communication with external agencies and parents [36]. However, the report concluded that very few secondary schools (age range 11-16) contained these elements or were adequately supporting emotional health, despite the national policy drive in this area [36].

Alongside the need for stronger evidence regarding what works in school-based emotional health initiatives, there is also a need to establish how far the range of potential interventions match what young people themselves say they want or need. Involving the perspectives of teenagers in the design of interventions is an important way of giving them a voice in issues that affect them, as well as helping to ensure that provision is as appropriate as possible [37]. However, interventions to date have often not taken their views into account sufficiently, resulting in a mismatch between what programmes have targeted and what young people say about their own lives [38]. Key findings from previous research on adolescents' views are that they struggle to define terms such as mental health and depression, that talking to someone is seen as an important source of support although many individuals have anxieties about seeking help from adults, and that aspects of school life such as workload and being bullied can be a significant cause of distress, but that schools are also viewed as potential sources of information and support [37,39,40]. An even smaller number of studies have explored teachers' views regarding the needs of students in relation to emotional health, and what schools could or should be doing to support those needs. Previous findings indicate that, although teachers may accept emotional health support as an integral part of their role as educators of young people, they often feel burdened by the emotional health needs of their students, and feel they lack knowledge in how to provide support, or discuss emotional health issues with them [41-43].

This paper reports findings from a study that sought to increase our understanding of both staff and students' perceptions of school-based emotional health. Specifically, the study aimed to examine the views of staff involved in emotional health work and students regarding current school-based emotional health provision, how far this meets the needs they identify and what they would like to see schools do in the future. The long term aim of this work is to develop a school-based intervention to improve the emotional health and well-being of secondary school children.

As emotional health can be addressed in a variety of ways in school, the notion of emotional health work was kept deliberately broad to ensure that all aspects that participants viewed as important were captured. It therefore included anything that had the potential to improve emotional health or reduce emotional difficulty, such as extra-curricular activities, support from counsellors or other school-based staff, and the delivery of lesson content that explicitly related to emotional health. The term 'emotional health and wellbeing' (EHWB) was used throughout the data collection as, since its introduction as a concept through the National Healthy Schools Programme [34], it has become one of the terms most commonly used within English educational settings to refer to all aspects of emotional health. The British Educational Research Association's Revised Ethical Guidelines for Educational Research 2004 were followed in the design and conduct of the study, and the study was approved by the University of Bristol's Social Sciences and Law Faculty Ethics Committee.

Methods

A mixed methods approach was used to enable us both to quantify the level of emotional health provision in English secondary schools and to use survey responses to inform the sampling of schools for in depth qualitative investigation of student and teacher views about current provision and areas for development. At the beginning of the study, a postal questionnaire enquiring about EHWB activities was sent to a random sample of 296 secondary schools in England. The questionnaire was sent to the headteacher, with the request that they pass it on to the member of staff best placed to give an overview of EHWB provision within and outside the curriculum. Questions asked how high a priority EHWB was in the school (choice of five responses from very low to very high), whether a list of EHWB topics were taught in lessons, and if a range of other EHWB related activities or support were provided such as on-site counselling, links with external agencies, relevant policies and teacher training (respondents were asked to tick yes/no in each case).

Qualitative methods were used to address the main study question, that is staff and students' views regarding current and potential emotional health provision in schools. This approach was taken due to its emphasis on participants' perspectives and interpretation of the world [44], and its ability to allow in-depth and unanticipated findings to emerge, which was deemed advantageous due to the under-explored nature of this area.

Sample Selection

The questionnaire sample was obtained from EduBase—a register of all educational establishments in England and Wales, http://www.edubase.gov.uk—who provided a random sample of schools from a sampling frame of all non-fee paying secondary schools in England. The sample was made up of two schools from every Local Education Authority area (n = 296), and stratified for deprivation of catchment area (using percentage of students eligible for free school meals as a proxy measure). This stage of the process required a random sample, as the aim was to gain a representative picture of what was happening in schools, and to create a representative sampling frame for the qualitative study.

Schools were selected for the qualitative study purposively, to ensure those that were included covered the range of amount and types of emotional health support seen in the national picture. Three factors were taken into account in sampling schools: a) extensiveness of EHWB curriculum coverage and support (as reported in the questionnaire) b) geographical position of the school c) percentage of pupils eligible for free school meals (an indicator of parental income) relative to the national average. The aim was for the final sample to represent different combinations of these three factors, as it was hypothesised that they may have a bearing on student and staff opinions about EHWB need and provision [45].

When approached, three of the schools declined to take part due to lack of time or in one case, impending closure, and they were replaced by schools that matched them as far as possible on level of activity, free school meal eligibility and location. A summary of the final qualitative sample is given in table 1.

Table 1. Summary of schools that participated in the study

School	Region	Free School Meal Eligibility (%)	EHWB Activity	Focus Groups Undertaken		Staff Interviewed
				Year 8 12/13 yrs	Year 9 13/14 yrs	
1	SW	22.9	Low	1 male 1 female		PSHE coordinator
2	SW	6.0	High	1 male 1 female		Head of year
3	SW	27.6	High	1 male 1 female	1 male 1 female	Assistant Principal Learning Support Manager
4	SW	4.2	High	1 male 1 female	1 male 1 female	Head Key Stage 3
5	NW	3.5	Low	1 male 1 female	1 male 1 female	Head Key Stage 3
6	London	17.5	Low	1 male 1 female	1 male 1 female	Learning Support Manager Psychologist 3 Teaching assistants
7	Midlands	21.3	Low	1 male	1 male 1 female	2 Learning Mentors
8	NE	26.4	High	1 male 1 female	1 male 1 female	SEN Coordinator Head of Year

Notes
1. Percentage of pupils eligible for free school meals was taken as a proxy measure of deprivation of school catchment area. National average for England is 14%.
2. See table 2 for fuller description of staff roles.

Qualitative Data Collection

Staff Interviews

Initial contact with each of the eight schools was made through a deputy or assistant head whose job included responsibility for pastoral care, and this was the individual who agreed to the school taking part. They then identified another member of staff who was responsible for some aspect of EHWB work in the school to liaise with the study team. The nature of this contact's role varied, but included teachers responsible for coordinating the personal, social and health education (PSHE) programme within the curriculum, staff heading up a support unit for students at risk of exclusion, and staff who managed the school's special educational needs provision. Each contact was invited to be interviewed, and to suggest one or two colleagues also involved in delivering work that supported EHWB, who might be willing to be interviewed. As there is a great deal of variety from school to school regarding which staff are involved with EHWB support, this method of identifying two or three key staff members to approach was deemed most appropriate, and generated a wide range of interviewees (see table 2), which helped build up a picture of the different ways in which schools support emotional health. The interviews were held on school grounds during the school day, either in private offices or empty classrooms. Written information was given to each potential interviewee about the aims of the project, and the procedures to ensure confidentiality and anonymity that would be adhered to, and written consent obtained from those who agreed to take part.

Table 2. Description of Staff Interviewees

Job Title	Number of Participants	Job Description
Teaching Staff		
Head of Key Stage 3 (11-14 year olds)	2	Senior teacher. Role involves teaching but also managing other teachers and coordinating teaching and support for first three years (grades) of secondary school.
Head of Year	2	Senior teacher. Role involves teaching but also managing other teachers and coordinating teaching and support for particular year (grade).
PSHE Coordinator	1	Role involves teaching, but also coordinating the Personal Social and Health Education curriculum, which is where most teaching about health takes place.
Teaching Assistant	3	Role involves supporting teachers within the classroom, often by offering one to one support or support to small groups of students with particular needs.
Other Positions		
Assistant Principal	1	Member of senior management team. No longer involved in teaching, but manages particular aspects of the school. This interviewee was responsible for the pastoral care in the school among other things.
Learning Mentor	2	Employed to provide one to one support to students to improve attainment and reduce exclusions. One of these interviewees had previously been a teacher.
Learning Support Manager	2	Responsible for the Learning Support Unit, which provides short term teaching and support to vulnerable students and those at risk of exclusion. One of these interviewees had previously been a teacher.
Special Educational Needs Coordinator	1	Responsible for the support offered to students with special educational needs, for example those that arise in relation to learning difficulties or mental health problems.

Student Focus Groups

In each of the eight schools, focus groups were conducted with students in years 8 (12-13 year olds) and 9 (13-14 year olds). There were two main reasons for using this method. Kitzinger writes that "group processes can help people to explore and clarify their views in ways that would be less easily accessible in a one to one interview" [[46] p300], therefore it was hoped that a group discussion might enable participants to develop ideas and co-construct knowledge around issues that they may not spend a lot of time reflecting on in their day to day lives. Secondly, the stigma that can surround emotional distress and disclosure of emotions makes discussion of emotional health a potentially sensitive topic [47]. Although discussing sensitive issues in groups can create difficulties in terms of certain views being stigmatised or remaining unexpressed, it was felt that the power differential in a one to one interview between an adult researcher and a teenager might make such discussions even less comfortable for the participants. Therefore the focus group was chosen in the hope that it would help break the ice for the more inhibited participants, and provide a supportive setting in which they could express their views [46]. Because of this concern about the potentially sensitive nature of the subject matter, the topic guides only asked about the needs of teenagers in general, they did not ask about personal experiences, which are likely to be more appropriately explored within an interview setting. In addition, the groups were made up of friendship pairs, that is half of the focus group participants were selected by form teachers or heads of year, and then each chosen participant was invited to bring a friend with them, in an attempt to ensure a supportive atmosphere, a technique which has been used elsewhere [48].

Most writers suggest that homogeneity in focus groups can be an important way to reduce inhibitions and ensure participants feel comfortable enough to share their views and feelings [49,50]. Evidence that boys and girls may have differences in emotional health needs and favoured coping strategies [2,51], and that both boys and girls can find it difficult to speak openly in front of the other gender [52] led to a decision to make the focus groups single sex. Single-sex focus groups to explore health related topics where gender differences exist has been used elsewhere [48].

The staff selecting the participants were asked to provide a range of students in terms of confidence and academic ability. Information leaflets were provided to each student and information letters were sent to their parents or guardians, both of which outlined the purpose of the study, what would be done with the data, and how confidentiality and anonymity would be assured. Written consent to participate was gained from both student and parent or guardian. The focus groups were held on school grounds during lesson time, in empty classrooms or offices.

In both the focus groups and the interviews, a list of topics was followed, but with sufficient flexibility to allow participants to raise and focus on the issues that they identified as important. The staff interview topic guide is shown in table 3, and the focus group topic guide is shown in table 4.

Data Analysis

All interviews and focus groups were audio-recorded and transcribed. The focus groups and interviews were analysed separately, using constant comparison techniques common to qualitative research [53]. Transcripts were scrutinised for emergent themes, and relevant sections of text were coded and grouped together according to those themes. Where the original theme identified was quite broad, it was broken down into sub themes, as further relevant sections of text were identified. After the first transcript had undergone analysis in this way, the identified themes were compiled into a coding frame. Each new transcript was then analysed for themes, which were compared against this initial coding frame. Those themes that did not fit the existing frame were either added as new themes, or were used to expand and modify existing themes, until all data had been accounted for. As this process continued, conceptual themes were developed from the initial descriptive themes, and previously analysed transcripts were then re-examined to identify any other sections of text relevant to these new themes. To explore similarities and differences across groups on key themes, matrix displays were used [54], in which the relevant sections of text from each focus group were drawn together in a table format for easy comparison. Descriptive accounts of each focus group were constructed, to ensure that any within group differences regarding the themes identified were acknowledged and included in the analysis.

Table 3.

Staff Interview Topic Guide
A. GENERAL
Their role in relation to EHWB provision
What they understand by the term EHWB
Students' main needs in this area
Teachers' main needs in this area
Is it important for schools to deliver EHWB?
B. WITHIN THE CURRICULUM
Anything that is covered now, how and by whom
What other things they would like to see covered, how and by whom
How they feel about teaching EHWB and related topics
Barriers and facilitators to the school covering EHWB in the curriculum
C. OUTSIDE THE CURRICULUM
Anything the school provides to enhance the emotional health and wellbeing of pupils
What they would like to see that is not currently provided
Anything this school provides to enhance the emotional health and wellbeing of teachers
What they would like to see for teachers that is not currently provided
What services are provided for students experiencing emotional difficulty?
Are there things that could be provided to help students deal with emotional difficulties that are not currently?
Barriers and facilitators to the school supporting EHWB outside the curriculum
Links with external agencies
D. FINAL COMMENTS
What is their understanding of a whole-school approach to addressing EHWB?
Do they think this school takes a whole-school approach?
What would help this school address EHWB more effectively?
Anything else they want to say about EHWB in this school or in schools generally

Table 4. Student Focus Group Topic Guide

A. OUTSIDE THE CURRICULUM
What makes young people feel good or happy?
Are there things in school that help students feel good or happy?
What else could be done or changed in school to make students feel good or happy?
What sort of problems do people your age have?
What difficult feelings do people your age have?
Do you think teachers are aware of the sorts of problems that young people face?
Are there things this school does or provides that help students who are having problems or difficult feelings?
What do young people do if they have a problem/difficult feelings?
Are there people in school you can go to if you have those problems/feelings?
Are there things that could be changed/set up in schools that would help people who were having problems/difficult feelings?
B. IN THE CURRICULUM
Do you have lessons on emotional health and wellbeing?
IF YES: how many, topics, who teaches, what they do, what they like/dislike about the lessons, whether useful or not
IF NO: would you like lessons on this?
Does the subject emotional health and wellbeing come up in other lessons?
Is there anything else you want to say about what this school could do, to make it a better place for students to be?

Initial focus group and interview transcripts were analysed independently by two members of the research team, to check the reliability of the coding frame [55]. Inter-rater reliability was good, with no disagreements arising over the definition of themes that emerged from the text, or regarding the themes that each section of text related to. Some differences between raters arose as to whether the themes they identified were new or were subthemes of existing themes. For example one member of the team included depression as a subtheme of negative feelings experienced by teenagers, whereas another identified depression as a new main theme. Where such differences arose, the theme in question was included as a main theme, but was cross referenced with the existing relevant theme.

Results

Questionnaire Survey

Questionnaires were returned from 75 schools (25.3%). This low response rate may have been due the questionnaire not reaching the right member of staff—all questionnaires were sent to the headteacher with the request to pass it to the most appropriate person. A further explanation may have been the feeling among schools that they receive too many surveys, which has led to a decline in responses to research surveys and government audits in recent years, with the majority conducted since 2000 having a response rate of less than 50% [56]. Responders and non-responders were compared using chi-square tests on dimensions that were hypothesised to affect emotional health provision in schools: size, free school meal eligibility, SATS results (national examinations taken at age 14) and religious affiliation. Differences found on these dimensions were slight (table 5). A phone survey of a subset of the questions was conducted with 17 non-responding schools. It revealed that they were less likely to view EHWB as of high importance (47.1%) compared to those who had responded (68.6%), and were less likely to provide an onsite counsellor (47.1% compared to 80%) and a drop-in health service (68.8% compared to 84.5%). Therefore it may be that the schools that did respond were those who were most interested in EHWB provision.

Table 5. Comparison of responders and non-responders on key variables

Key Variables	Responders N (%)	Non-responders N (%)
Below average size	30 (42.25)	118 (52.7)
Below average free school meal eligibility	38 (53.5)	114 (50.9)
Above average SATS results	37 (52.1)	105 (46.9)
Religious affiliation	26 (36.6)	71 (31.7)

Notes

1. Although all state schools receive local authority funding, follow the national curriculum, and are inspected by the national childcare inspectorate Ofsted, two types (voluntary-controlled and voluntary-aided) work in partnership with at least one charitable foundation, which is often a religious organisation. This organisation typically owns the school buildings, and may appoint some members of the governing body, as well as having some influence over the ethos and religious teaching in the school.

Respondents to the survey were generally deputy headteachers responsible for pastoral care, Personal, Social and Health Education (PSHE) coordinators, staff in charge of a whole year (grade) or staff responsible for student support. The proportions of responding schools that provided key emotional health and wellbeing activities are shown in table 6. While most respondents covered bullying and improving emotional health in lessons, only around one third covered self-harm or depression. At least 80% provided an on-site counsellor, a drop-in health service, a peer support service and support groups for vulnerable students. Less than half provided any training for staff members involved in delivering EHWB work, and only 26.5% provided emotional health training for teachers in general.

Table 6. Proportion of schools providing key EHWB activities

Provision/coverage	N (%)
EHWB topics covered in the curriculum	
Improving emotional health	51 (78.5)
Bullying	65 (97.0)
Strategies for dealing with emotional distress	44 (66.7)
Reducing the stigma of emotional distress	33 (44)
Causes, symptoms and treatment of depression	22 (35.5)
Self-harm	21 (32.8)
Extro-curricula EHWB activities	
On-site counsellor	56 (80.0)
Drop-in health service	60 (84.5)
Peer support service	58 (82.9)
Support groups for vulnerable students	64 (90.1)
Training for teachers delivering EHWB lessons and activities	34 (47.8)
Training for teachers in general about emotional health	18 (26.5)
School policy that focuses on emotional health	33 (47.8)
Regular contact with relevant health services e.g. CAMHS	56 (82.4)

Focus Groups and Interviews

Twenty seven focus groups (154 students) ranging from four to eight students in size, and 12 individual or paired staff interviews (15 individuals) were conducted in total (table 1). The same number of focus groups and staff interviews were not completed in all eight schools, due to the amount of time each school was able to give to the study. Participants' perceptions of the emotional health provision within their schools led to identification of three overarching categories: (i) EHWB in lessons (ii) support for students experiencing emotional difficulty, and (iii) wider aspects of the school environment. The key themes that emerged with regard to each of these categories are presented, with illustrative quotes.

EHWB in Lessons

The main themes within this category were a lack of EHWB material in lessons, the inadequacy of the EHWB teaching that did exist, a desire for external experts to teach EHWB topics, concerns regarding the stigma of discussing emotions or of experiencing emotional difficulty and the potential for learning about EHWB through other, non-health lessons.

The general feeling among students was that they did not receive much teaching that focused overtly on EHWB, regardless of whether a school had been identified as relatively high in EHWB provision or not. Several participants indicated that, where health related lessons did occur—delivered as part of the (non-statutory) PSHE curriculum—these tended to focus on other things:

Interviewer (I): do you have lessons on emotional health?

P1: ummm

P2: there's PSHE, it's usually drugs and that

P3: it's usually all the same thing about drugs or alcohol

P4: we've done cigarettes twice as well

P3: sometimes we do sex education, watching the same videos again and again

School 4 Boys 12/13 Years Old

The staff interviews revealed a similar picture of a lack of EHWB lessons, although several individuals felt that they did cover emotional literacy. Five of the schools delivered PSHE as regular lessons, but in three cases all health education was limited to PSHE days or half days held a few times a year. In both scenarios, other topics tended to be prioritised:

When I think of the things we've been doing a lot of it is careers work, ummm, Industry Day, interviews, there's global awareness, fair-trade. There's not a massive amount on health and emotion. Oh, there's a friendship day in year 7. In year 9 we do an AIDS and relations game, in the past we've done 'no smoking' workshops.

School 2 Head of Year

A second key theme from the student focus groups, was the inadequacy of EHWB lessons where they had been delivered, with participants feeling they did not cover useful topics and were not taught effectively:

P1: in citizenship sometimes they do stuff on bullying but it's not really that useful

P2: and it's just the same stuff over and over again

P1: yeah so maybe it would be good if there's like for a day or something someone comes in and we go in like small groups with your friends and they'd talk to you about how to stop bullies and stuff like that rather *than just watching a tape or reading a book.*

School 6 Boys 12/13 Years Old

Variations between but also within schools emerged as to how useful existing EHWB lessons were:
P1: some of them do actually teach you what you need to know but some of the others will be like
P2: just treat it like a free lesson
P3: others just read out what the lesson's about and then sit and do it. Most of them just do something else because they've no idea what to do

School 8 Girls 13/14 Years Old

A possible reason for this emerged from the staff interviews. In six of the eight schools PSHE lessons were delivered by teachers or form tutors (a teacher assigned to each class who oversees its welfare, and monitors attendance, behaviour and support needs), who had mixed feelings about this role:

In this school it's delivered by tutors and for some people it's not what they would choose to be doing, it's not what they find easy to be doing. So in some instances people will prepare and take it seriously and teach it to the best of their ability and they might have an interest in it and they'll have some experience in various aspects and then for other people it's just a burden which they're not interested in so it's very last minute and the actual quality of children's experiences of those lessons will be quite variable.

School 2 Head of Year 8

A small number of students felt tutors were appropriate people to deliver EHWB lessons, but a key theme from the majority was that neither tutors nor teachers should teach this topic. This was either because of perceptions that teachers did not have the desire or expertise to do so, or because students felt uncomfortable discussing sensitive issues with them:

If you talk in front of a teacher they'll sort of always like be, probably every time they look at you you'll be thinking "oh no are they thinking about this, what I've talked about" and it'll be kind of awkward.

School 8 Girls 12/13 Years Old

Instead, students preferred the idea of having outside speakers with relevant training or experiences to lead EHWB sessions.

A majority of students felt that having more and improved EHWB lessons would be useful. They wanted such lessons to include activities, discussions, and IT, and for them to be conducted in a relaxed and informal way. Two groups suggested using an anonymous emotions box in which students placed an emotion or question relating to emotional health, which then formed the basis of a lesson. Some participants raised concerns that students may find speaking openly about their feelings difficult due to fears of being laughed at:

Some people would just laugh, like if someone had been bullied or the class doesn't like them and they get up and say something everybody would just laugh at them.

School 3 Girls 13/14 Years Old

For this reason many groups felt that small groups rather than whole classes would be the most appropriate format, and four of the 13 girls' groups suggested that EHWB sessions would best be conducted in single sex groups, due to concerns that boys would take the lessons less seriously, and would engage in teasing or disruptive behaviour.

The topics that students suggested should be covered were understanding feelings and how to express them appropriately, coping strategies for dealing with emotional distress, solving problems such as bullying and peer pressure, and accessing sources of help:

I think people don't know enough about that really like how to stop yourself feeling depressed or sad or something or how to make yourself feel happy.

School 6 boys 12/13 Years old

A few students suggested depression and self-harm as potentially useful topics to be covered, but others disagreed. A small minority were not in favour of any EHWB lessons, because they felt that it cannot be taught, they did not believe it to be relevant or they were concerned about the adverse effects of learning about such things:

P1: I don't want to think about people like cutting themselves and things

I: right, but do you think you don't need to know about it then?

P1: no, we already know

School 5 Boys 13/14 Years Old

All staff interviewees also wanted to see more EHWB teaching, with lessons exploring coping strategies, understanding feelings, bullying, self-harm and bereavement mentioned in particular. However, three individuals raised the point that EHWB should not be seen as exclusively or even mainly a PSHE subject, believing it should pervade all aspects of teaching and school life, a suggestion which is returned to below:

I think it would be a complete and utter disaster if schools saw the word health in emotional health and thought oh that's PSHE isn't it, and so it gets stuffed over there. So I much prefer it to be a whole school thing... it's about the way you respond to and deal with people in your classroom, that's what the emotional curriculum should really be about.

School 1 PSHE Coordinator

Underpinning much of the student discussion about best formats for learning about EHWB within lesson time was a concern about stigma, both the stigma of experiencing emotional difficulty, but also the stigma attached to talking about emotions. This explained the concerns voiced by several students that they might be laughed at or treated differently if they contributed to EHWB lessons delivered by teachers or tutors, in whole class, mixed gender situations. Concerns over stigma led some students to comment on the value of other, non-health lessons such as English, Art and Sport, as potential places to learn about positive and negative emotional states more obliquely:

P1: you can build up anger in the day something might have happened and say you've got a rugby match after school you can let al.l your anger out then

P2: and you feel good about it

Voices: yeah

P3: it's like relaxation, even though you're getting munched

School 2 Boys 12/13 Years Old

P1: the whole creative side like English, Art, Dance, Drama 'cause you can all express yourself through that sort of thing, but it doesn't necessarily have to be about you and people can't guess if it's about you or not so it's like expressing yourself without others knowing, that's what I find really good about it.

School 8 Girls 12/13 Years Old

Non-health related lessons therefore gave students the opportunity to discuss emotions impersonally, or to express feelings in anonymous ways, which avoided the possibility of being stigmatised.

Support for Students in Distress

The main themes that emerged in relation to schools supporting students experiencing distress were the importance of having someone to talk to, a lack of confidential, accessible and sympathetic help sources within schools, the importance of having easily accessible safe spaces within school to work through emotions, and the need to find non-stigmatising ways of providing support.

A majority of students discussed the importance of having someone to talk to when coping with emotional difficulty. In addition to teachers, a range of personnel within the schools existed who might be considered potential help sources, such as teaching assistants, who provide support to teachers within the classroom (in all 8 schools), learning mentors, who work with individual students to raise attainment (in 7 schools), a counsellor (in 6 schools), a chaplain (in 4 schools), a school nurse (in all schools) and a peer support service (in 7 schools). However, there was a great deal of variation between the schools in whether and when such people were available. Even in schools that appeared to provide a good amount of support, the students often did not know very much about the options available:

I: do you know much about the counsellor?

P1: I wouldn't know where to go, I think you go up to the Personal Guidance Centre I think

P2: who is the counsellor?

P1: I don't know much about it

School 4 Boys 12/13 Years Old

Further, through the students' discussions of the various help sources, it became clear that they looked for particular attributes when evaluating potential help sources. Where students could approach help sources directly without anyone else knowing, and where they trusted them to maintain confidentiality, these tended to be the most popular. Other characteristics that emerged as important were having good availability and accessibility, being good at listening, being respectful of teenagers and their problems, and being able to provide useful information or advice. Just under one third of the girls' groups raised the gender of the help source as an issue, indicating that they would prefer help sources to be the same gender, although no boys' groups commented on this. Several students liked the idea of a help source being the same age because they were perceived to be more likely to understand and more trustworthy, but others argued that adults were more likely to have the information or knowledge necessary to help.

Conversely, several barriers to seeking help from these various sources emerged. Some of these related to characteristics that were attributed to the help source, in particular being difficult to access, being unable or unwilling to help or being unlikely to be trustworthy. Other barriers related to perceived consequences of seeking help, such as not being taken seriously, being treated differently in the future, or exacerbating the problem. Many of the students' comments about the negative consequences of help-seeking related to fear of being stigmatised, which was partly why students emphasised the need for help sources to be confidential, and able to be accessed without anyone else knowing. These concerns also led to some students feeling they would prefer a help source to be a relative stranger, echoing some of the concerns regarding EHWB lessons:

And sometimes talking to the head of year isn't really good because you see your head of year every day so sometimes you just want to talk to somebody you don't see every day because [pause] it's like you don't want to be treated differently, you want to talk about it but you want to be treated the same anyway. You want to be normal basically, everybody wants to be normal.

School 6 Girls 13/14 Years Old

Of all possible school-based help sources, the chaplains where they existed appeared to meet most of the desired criteria most of the time, and therefore received the most favourable reaction:

P1: you can just go up to the desk and ask for them [the chaplains] and also it's better than going to teachers 'cause if it's a teacher it might sometimes affect your lessons and you hardly ever see them anywhere and you know it's not going to affect the rest of your school life

P2: they don't just sit there taking notes and have a glazed look over their eyes 'cause they're not actually listening

P1: yeah and they don't go "and how do you feel about that" they just talk to you like you're a normal person

School 8 Girls 12/13 Years Old

Learning mentors, counsellors and teachers had a mixed reaction, with some students regarding them as helpful, whereas others expressed concerns that they would not keep problems confidential, that they were not be understanding, or that they were not available to everyone enough of the time. School nurses were not regarded positively by any of the seven groups that suggested them as a possible help source, due to perceptions that they were unavailable, or were only interested in physical illness. Similarly, the peer support schemes, when mentioned, engendered a negative response from most students in the schools where they existed, due to concerns that the service would not be confidential, that the peer supporters were only the clever or popular students who would not understand, that they might laugh at the person asking for help, and that they were unlikely to be able to help:

P1: you talk to them and then they probably go off and laugh about you and have a discussion

P2: they don't know anything do they? They're just children

P3: they get no training whatsoever, they just give it to the clever people

School 5 Boys 13/14 Years Old

Students felt that a priority for schools should be having more help sources with the right attributes available. Although staff were generally more positive about the help available, some agreed that more was needed:

Before we were saying our job is learning support you know and we're supposed to be in the lesson but 90% of my time is listening to a child it could be something going on at home, at school, anything...there is just not enough of us, you know, I mean I'm on dinner duties as well a couple of times and they could literally be queuing you know to talk to you, um because you're there and you're a face and they just need someone to talk to.

School 6 Teaching Assistant

A few students also suggested that schools could do more to assist them in accessing external help sources, for example by providing leaflets and posters. All the schools displayed various posters relating to mental health, but one student pointed out this was not always helpful, due to concerns of being seen by others standing in front of a poster trying to memorise a phone number. One school avoided this problem by putting helpline numbers in the back of homework diaries, which all students had to keep with them.

As well as having access to appropriate help sources, some students felt strongly that there was a need for safe spaces, where students could relax away from difficult situations, or be left alone to process emotions such as anger, without other students or staff attempting to intervene:

P1: I don't know how to say it but have like a room where you can let your anger out

P2: somewhere like you can let off steam

P1: yeah

School 7 Boys 13/14 Years Old

In some of the schools, staff did discuss separate areas or buildings that were intended to function as a safe haven for those feeling vulnerable, and a place to access help sources. However, the student participants did not always know much about these places where they existed, and access was generally monitored and controlled by the staff involved, therefore they did not provide the more informal and anonymous 'chill out' spaces that the students said they wanted.

The other main support mechanism identified by staff and students, provided in all eight schools, were support groups for 'vulnerable' individuals. On the whole, students felt that these were a good idea, although there were some concerns expressed that the selective nature of such groups could lead to some students missing out, or those attending being stigmatised:

P1: people will use it against you and like cuss us or something and say "ah you have to go to your stupid emotional health thing"

P2: you'll feel like you're special needs and they'll say like "well at least I'm not dumb and at least I'm not special needs"

School 6 Boys 12/13 Years Old

One staff member described a support group in her school that avoided these problems by being open to everyone, and that while originally set up as an exam support group, had successfully evolved into one that dealt with any emotional health issues that were raised. Two student groups suggested the idea of student-led groups, following a self-help model, but this was not currently provided by any of the schools.

The School Environment

Much of the discussion in the staff interviews and student focus groups highlighted the importance of the school environment, both physical and psychosocial, in impacting on student emotional health. Key themes that emerged were the sources of distress that existed in the school context—specifically bullying from peers, difficult relationships with teachers and academic work—and the potential of schools to bolster student emotional health through whole-school changes such as providing a supportive culture and improving the physical environment.

All of the focus groups discussed ways in which schools had a negative impact on emotional health, with three key sources of distress that were raised directly relating to school-life. Bullying was identified as a major cause of emotional distress by 23 focus groups, with the majority of participants feeling that schools did not do enough to deal with it effectively:

P1: if you do tell the teachers they usually say they'll do something about it and then nothing happens

P2: or they go up to them and have a word with them and then it just carries on it has no effect at all

P1: they don't really follow anything through

School 8 Girls 12/13 Years Old

Several staff members also saw bullying as a significant problem, and outlined various ways in which their schools had attempted to address this, such as poster competitions, a focus on bullying in lessons and assemblies, and an ongoing anonymous e-mail and texting service to report incidents. However, only one school appeared to have been successful in the students' eyes, with participants at this school less likely to see bullying as a problem than those in the other schools.

A second main source of emotional distress, identified by 25 of the 27 student focus groups, was problematic relationships with certain teachers, with students describing unfair, unhelpful or even humiliating treatment:

P1: she [teacher] says like "I'll pick you up by your sidies [sideburns] and throw you out of the window"

I: how did that make you feel?

P2: like an idiot

P1: yeah

P3: ashamed

P1: yeah

School 5 Boys 12/13 Years Old

Even where teachers were not the source of emotional distress, some students noted that their lack of sympathy to students' difficulties could make matters worse:

P1: sometimes if things are happening outside of school that teachers don't understand and then they have a go at you anyway it just makes it worse

P2: Yeah that's true, things will happen at home with me and then I come to school and the teachers will shout at you and you feel so frustrated and you tell them what's happened at home and they still wouldn't, oh that's no excuse

School 6 Girls 13/14 Years Old

Staff interviewees did not express staff-student relationships quite so negatively, but some agreed that colleagues could contribute to students' emotional distress:

You know a lot of problems are caused by just a throw away remark, and it's not meant to be malicious or anything like that but that child takes it away with them and it makes them feel as though they're not wanted and that can be done by saying "oh it's good of you to drop in" or something like that you know? The teacher doesn't mean it nastily, but if that's someone that's found it difficult to get into the classroom"

School 3 Learning Support Manager

Finally, being overburdened with academic work and exams were mentioned by more than half the student focus groups as a source of emotional difficulty. Some staff agreed:

I think some of our youngsters are coming in to school dealing with all sorts of issues which I as a child I never had to deal with and they're really burdened with things, some are young carers for example, and then we're expecting them to work quite a long day and under a great deal of pressure and I do think we need in our curriculum time to give them some creative time or relaxation time.

School 3 Assistant Principal

Conversely, the importance of school as a place where emotional health can be enhanced, or at least emotional distress relieved, was a theme that emerged from several staff and students' discussions:

I feel one of our selling points is that we offer very much a caring environment, we might not have the 49 sports halls but we know every student that walks through the door

School 8 Head of Year 7

But some people bring it to school 'cause they can't bring it anywhere else, they have so much anger it's like they can't take it out anywhere else.

School 3 Girls 12/13 Years Old

Students made several suggestions about changes that could be made, in order to achieve a physical and psychosocial environment that supported emotional health in this way. Key ideas were increasing the amount of rewards and praise for good behaviour, making lessons more enjoyable through activities and use of IT, having more extra-curricular activities, improving the physical environment, and being less strict about what students regarded as trivial matters such as having the correct uniform. Suggestions from staff echoed some of these views, in particular having more encouragement and attention for students who behave well, providing a wider range of activities during breaks, and improving the physical environment:

We're looking at students being safe and enjoying the school, we need picnic tables so they can sit down and eat their sandwiches in the summer when it's nice, in the shade. We need some basketball nets and netball nets out in the courts, they need safe areas, an activities club for the lower school where students who aren't sporty who want to sit down and play chess games who just want to do some crafts can go, it's at lunchtime so they're safe and they know it's somewhere they can go.

School 5 Head of Key Stage 3

Discussion

Main Findings

The school is frequently identified as an important setting for addressing emotional health [15,26,34]. The participants of this study echoed this view, through their identification of a number of ways in which this could be achieved. However, several gaps were revealed between what students and staff perceive as important and what is currently provided.

All participants agreed that very little lesson time was spent on EHWB, with comments indicating that other topics, notably sex and drugs, were more commonly taught. It could be argued that emotional health/distress underpins much of the behaviour and decision-making relating to drug use and sexual activity, and indeed the survey indicated that more work was being done on emotional health in lesson time than the respondents appeared aware of, which may be due to these links with other curriculum content. Nevertheless, it is noteworthy that many participants wanted more lessons that dealt overtly with emotional health, covering topics such as understanding feelings, coping with emotional distress and accessing help sources. A minority of students did have reservations about this idea, and several voiced concerns about the stigma attached to experiencing or even discussing difficult emotions, suggesting instead that emotional health lessons should be delivered in small groups, by outside experts rather than teachers. An equally important finding is therefore the way that other, non-health related curriculum subjects could provide students with a safer context in which to explore and learn about EHWB.

Students and staff also felt that schools should provide better support for those in distress, for example through someone to talk to, support groups and safe spaces for relieving emotions and receiving help. Although the survey data indicate that a majority of schools currently provide some form of support for students in need, the qualitative data revealed great variety in the number and quality of those services. Students did not always have a favourable opinion of the support that existed, particularly if it was not perceived to be confidential, available to all or sympathetic to the needs of teenagers. A key concern was stigma, and fear of being seen or treated differently, which could create a significant barrier to a help source being used.

Finally, several participants emphasised the importance of the school environment in supporting or damaging students' emotional health. A clear message was therefore that any EHWB work had to include consideration of the physical and psychosocial environment. Suggestions for changes in this regard included reducing bullying, improving teacher—student relationships, increasing rewards and recognition of good behaviour, and developing the range and number of extra-curricular activities available.

Strengths and Limitations

This is the largest study to have qualitatively explored staff and student views of EHWB provision in secondary schools. The background survey data provided a useful indication of the sorts of provision that is currently delivered in schools, but the in-depth qualitative findings made a much more detailed and contextualised picture available regarding the variability, acceptability and usefulness of those activities from the perspective of those at whom they are targeted. In fact few differences between schools designated as providing high levels of EHWB activity and those labelled as low level emerged in the qualitative data, which may be explained by the different levels of detail that each data collection method allowed. For example, the survey asked whether a school provided an onsite counsellor or not, to which 80% of respondents answered yes. However, the qualitative data revealed several problems with this provision from the viewpoint of participants, such as the counsellor not being in the school for many hours, students not being able to self-refer, and students not knowing about the service or how to access it. Such findings highlight the value of gathering qualitative data when assessing what schools can do to support emotional health, in order to explore the quality and appropriateness of the provision from the perspective of the users.

The fact that the focus of the study was on students' and staffs' views means the findings may not be a comprehensive reflection of all the ways in which schools do support emotional health. For example emotional health may be covered more obliquely in the PSHE curriculum as it relates to topics such as drugs and sex and relationships. However, by focusing on the views of those involved, this study was able to identify discrepancies between what students were aware they were receiving in terms of emotional health support and education, and how far they felt this measured up to what they would find useful. A further strength was the ability to compare students' perceptions with those of staff involved in delivering EHWB provision—although there were some disparities regarding what staff said was available and what students were aware of, there were striking similarities in terms of the improvements that both groups suggested could be made.

One limitation of the study is that because teachers selected half the participants for the focus groups, it is possible that an element of "gatekeeping" went on. For example it is possible that students were more likely to have been selected if they were expected to portray the school favourably, or if they were known to have experienced emotional distress or made use of certain services. A second potential limitation is that by conducting student focus groups rather than individual interviews, an element of group conformity may have been introduced, for example students who had more positive views of provision in their school may have felt unable to express this. The groups were single sex and made up of friendship pairs in an attempt to encourage students to speak openly, and in fact the many examples of students disagreeing and contradicting each other, suggested this group effect was limited. Further, using focus groups allowed students to comment on and add to each other's views, creating richer and more detailed accounts than might have been gathered in a one to one situation.

Implications for Policy and Future Research

The fact that very little lesson time is spent on EHWB means that key areas relating to emotional health such as understanding feelings, depression and self-harm are neglected by a majority of schools. Evidence that young people downplay symptoms of emotional distress and delay accessing formal help sources [47,57] indicates that topics such as understanding and coping with difficult feelings, and knowing when and how to seek help could usefully be addressed within lesson time. Although there is some promising evidence that both preventative and educational classroom interventions might be effective at improving knowledge and skills related to emotional health and distress [17,23], studies with more robust designs are needed to establish the most useful ways to approach this long term. Such studies must also explore the potential for harm to which explicit teaching about EHWB might lead. A minority of participants did not want such lessons to be introduced, not least because of the stigma associated with disclosing emotions and emotional distress. Further, emotional literacy programmes such as SEAL in England have been criticised for imposing a particular way of experiencing and expressing emotions on all young people, and disempowering individuals by positioning them as emotionally vulnerable in the first place [58]. Given such concerns, the point raised by several students in this study that creating greater opportunities for exploring emotions more anonymously in other lessons such as English and Drama, should also be acknowledged and supported in policy documents.

The students and staff made several insightful comments regarding the best format for EHWB lessons, which should be explored in future research, for example whether small and informal groups would result in more valuable EHWB lessons. The question of whether such lessons should be single-sex should also be examined, both in terms of whether the views expressed here by some female participants are shared by other students, particularly males and other age groups, and also in terms of any evidence that single-sex or mixed-sex groupings are more effective. The point made by some girls here that boys would be disruptive echoes a similar debate within the sex education literature. Here, it has been suggested that single-sex lessons are desired by many students due to embarrassment of both girls and boys in talking openly in mixed classes, reports of disruptive and demeaning behaviour of boys towards girls, and differing needs in terms of content [52,59], although others have challenged these points [60,61].

Another key question raised by the students' preferences reported here is whether using outside experts is more effective than using teachers, and whether this is sustainable long term. There is some evidence from other areas of health education that involving outside speakers is something many students want [61], and is something that can be effective as a supplement to a school's ongoing educational programme [62]. However, concerns have been raised that there can be practical difficulties in organising this, and that 'cultural clashes' can arise in terms of key messages [63]. Two published studies have examined the involvement of external contributors in mental health interventions in English secondary schools and, although showing promising results, neither had control groups with which to compare the results [20,64]. Collaboration between schools and external agencies would help identify outside experts within the local context who might be willing to contribute to lessons [65]. Indeed where schools have counsellors or other emotional health support staff based in schools, they too could play a role. However, the practical and financial implications of involving external experts makes it likely that teachers would still play a role in delivery, meaning better EHWB training, and specialist teaching teams in every school, are essential.

Several extra-curricular changes also need to be considered. The finding that many students consider having someone to talk to within school to be an important source of support resonates with previous research indicating that school-based counselling services are effective and regarded positively by students and staff [66]. However, the schools visited here varied a great deal in the amount and type of help available. A key issue that emerged was the need to provide help sources that fulfilled certain criteria, in particular that were confidential, easily available, good at listening, respectful of teenagers and their problems, and able to provide useful information or advice. These features resonate with previous findings regarding students' views of school-based help sources [67,68], as well as with the views of adolescents regarding previous experiences with 'helping professionals' more generally [69]. Such findings underline the importance of ascertaining students' views of services, to ensure they have a correct understanding of what they provide, but also to identify areas where provision can be improved to better meet their needs. The peer support schemes were a striking example of this; such interventions have become very popular among practitioners over the past decade [70]—indeed the survey revealed that over 80% of schools provided such a service—but this favourable view was not shared by the students in this study. Conversely, where chaplains existed, they did appear to match the criteria that were important to students. Generally it is only schools with a religious affiliation that would have regular access to a chaplain, however more exploration of this role—what it is about the way chaplains operate or behave that enables them to fulfil the desired criteria—may help identify a blueprint useful for all schools in providing the most effective forms of support possible.

A key concern raised by students was that some of the support offered—for example support groups for 'vulnerable' students, and non-confidential help sources—could lead to stigma and students in need missing out on help. The relative value of targeted versus universal approaches in the field of adolescent emotional health is yet to be clearly established [71], but it is noteworthy that the student views reported here emphasise an approach that allows confidential access to interventions for all individuals. One such intervention that might be easily introduced and that appeared popular with participants was a 'chill out' area, in which students could be left alone to work through their emotions.

In addition to the introduction of specific educational and supportive interventions, students and staff felt strongly that changes to the whole school environment needed to be considered, due to the potential of schools to contribute to emotional distress or to bolster emotional health. This view resonates with the emphasis on whole-school approaches to tackling emotional health in some studies, and in most policy documents in England [25-27,34]. The three aspects of the school context that students identified as causing the most emotional distress—

bullying, difficult relationships with teachers and academic stress—have all been found to be detrimentally associated with mental health [30,72,73], and would be useful areas for schools to target, for example through developing relationship policies, or providing workshops focused on coping with exams. Conversely, staff and students identified several areas for change that might enhance emotional health, including greater recognition for good work and behaviour, a pleasanter physical environment, and a greater range of extra-curricular activities. The survey revealed that only just over a quarter of the responding schools provided any EHWB training in general for teachers and this would need to be increased, if teachers are to understand the school-based causes of emotional distress, and ways in which emotional health can be better supported through day to day school life and interactions. These findings indicate the need to avoid introducing additional EHWB lessons or support services in isolation, but to take a more holistic approach, in which consideration is given to current policies and practices throughout the school, and ways in which these can be made more supportive of student emotional health [74].

Conclusion

Although all the schools within this study were delivering some form of EHWB work, students and staff agreed that much more needs to be done in terms of teaching about emotional health within lessons, provision of appropriate support, and improving key aspects of the physical and psychosocial environment. Such changes require greater clarity at policy level in terms of what schools, in collaboration with external agencies, should provide in this area. Further, more rigorous evaluation is needed to determine what initiatives are effective, so examples of effective practice can be implemented more widely. Improving emotional health provision in schools is vital for supporting the emotional health of all teenagers, and for intervening early to reduce emotional disorder and associated detrimental outcomes among this age group.

Competing Interests

The authors declare that they have no competing interests.

Authors' Contributions

JK conceived of the study, participated in its design, conducted the data collection and analysis, and wrote the paper. JD contributed to the study design, data analysis and interpretation, and revision of the manuscript. LB contributed to the data collection, analysis and interpretation, and revision of the manuscript. RC contributed to the data analysis and interpretation, and revision of the manuscript. DG contributed to the study conception and design, data analysis and interpretation, and revision of the manuscript. All authors have read and approved the final manuscript.

Acknowledgements

Funding for this study came from the UK Economic and Social Research Council (RES-000-22-1462). We would like to thank the staff and students who gave up their time to participate in this study. Thanks are also due to Elise Whitley for help with the quantitative analysis, and to Angela Afonso and Belle Harris for secretarial support.

References

1. Costello JE, Mustillo S, Erkanli A, Keeler G, Angold A: Prevalence and development of psychiatric disorders in childhood and adolescence. Archives of General Psychiatry 2003, 60:837–844.
2. Green H, McGinnity A, Meltzer H, Ford T, Goodman R: Mental health of children and young people in Britain, 2004. Hamshire: Palgrave MacMillan; 2005.
3. Harrington R, Clark A: Prevention and early intervention for depression in adolescence and early adult life. European Archives of Psychiatry and Clinical Neuroscience 1998, 248:32–45.
4. Fombonne E, Wostear G, Cooper V, Harrington R, Rutter M: The Maudsley long-term follow-up of child and adolescent depression. British Journal of Psychiatry 2001, 179:210–217.
5. Fergusson DM, Woodward LJ: Mental health, educational and social role outcomes of adolescents with depression. Archives of General Psychiatry 2002, 59:225–231.
6. Kessler R, Bergland P, Demler O, Jin R, Walters EE: Lifetime prevalence and age-of-onset distributions of DSM-1V disorders in the National Comorbidity Survey Replication. Archives of General Psychiatry 2005, 62:593–602.
7. Fergusson DM, Horwood J: The Christchurch health and development study: review of findings on child and adolescent mental health. Australian and New Zealand Journal of Psychiatry 2001, 35:287–296.

8. Hankin BL, Abramson LY, Moffit TE, Silva PA, McGee R, Angell KE: Development of depression from preadolescence to young adulthood: emerging gender differences in a 10-year longitudinal study. Journal of Abnormal Psychology 1998, 107:128–140.
9. Naylor PB, Cowie HA, Walters SJ, Talamelli L, Dawkins J: Impact of a mental health teaching programme on adolescents. The British Journal of Psychiatry 2009, 194:365–370.
10. Ford T, Goodman R, Meltzer M: Service use over 18 months among a nationally representative sample of British children with psychiatric disorder. Clinical Child Psychology and Psychiatry 8:37–51.
11. Barrett PM, Lock S, Farrell LJ: Developmental differences in universal prevention intervention for child anxiety. Clinical Child Psychology and Psychiatry 2005, 10:539–555.
12. Merry S, McDowall H, Wild C, Bir J, Cunliffe R: A Randomized placebo-controlled trial of a school-based depression prevention programme. Child and Adolescent Psychiatry 2004, 43:538–547.
13. Clarke GN, Hawkins W, Murphy N, Sheeber LB, Lewisohn PM, Seeley JR: Targeted prevention of unipolar depressive disorder in an at-risk sample of high school adolescents: a randomized trial of a group cognitive intervention. Journal of American Academy of Child and Adolescent Psychiatry 1995, 34:312–321.
14. Neil AL, Christensen H: Australian school-based prevention and early intervention programs for anxiety and depression: a systematic review. Medical Journal of Australia 2007, 186:305–308.
15. Spence SH, Shortt AL: Research review: can we justify the widespread dissemination of universal, school-based interventions for the prevention of depression among children and adolescents? Journal of Child Psychology and Psychiatry 2007, 48:526–542.
16. Sutton JM: Prevention of depression in youth: a qualitative study and future suggestions. Clinical Psychology Review 2007, 27:552–571.
17. Merry S, McDowall H, Hetrick S, Bir J, Muller N: Psychological and/or educational interventions for the prevention of depression in children and adolescents (Cochrane Review). Chichester, UK: John Wiley & Sons, Ltd; 2004.
18. Santor DA, Poulin C, LeBlanc JC, Kusamakar V: Facilitating help seeking behaviour and referrals for mental health difficulties in school aged boys and girls: a school-based intervention. Journal of Youth and Adolescence 2007, 36:741–752.

19. Hess SG, Cox TS, Gonzales LC, Kastelic EA, Mink SP, Rose LE, Swartz KL: A Survey of adolescents' knowledge about depression. Archives of Psychiatric Nursing 2004, 18:228–234.

20. Pinfold V, Toulmin H, Thornicroft G, Huxley P, Farmer P, Graham T: Reducing psychiatric stigma and discrimination: evaluation of educational interventions in UK secondary schools. British Journal of Psychiatry 2003, 182:342–346.

21. Weare K, Gray G: What works in developing children's emotional and social competence and wellbeing?. London, UK: Department for Education and Skills; 2003.

22. Department for Education and Skills: Social and Emotional Aspects of Learning (Secondary Schools). Nottingham: DfES Publications; 2007.

23. Payton J, Weissberg RP, Durlak JA, Dymnicki AB, Taylor RD, Schellinger KB, Pachan M: The positive impact of social and emotional learning for kindergarten to eighth-grade students: Findings from three scientific reviews. Chicago, IL: Collaborative for Academic, Social, and Emotional Learning; 2008.

24. Jenson BB, Simovska V, eds: Models of Health Promoting Schools in Europe. Copenhagen, Denmark: European Network of Health Promoting Schools; 2002.

25. Wyn J, Cahill H, Holdsworth R, Rowling L, Carson S: MindMatters, a whole-school approach promoting mental health and wellbeing. Australian and New Zealand Journal of Psychiatry 2000, 34:594–601.

26. Bond L, Patton G, Glover S, Carlin JB, Butler H, Thomas L, Bowes G: The Gatehouse Project: can a multilevel school intervention affect emotional wellbeing and health risk behaviours? Journal of Epidemiology and Community Health 2004, 58:997–1003.

27. Weare K, Markham W: What do we know about promoting mental health through schools? Promotion and Education 2005, 12:118–122.

28. Fergusson DM, Beautrais AL, Horwood LJ: Vulnerability and resiliency to suicidal behaviours in young people. Psychological Medicine 2003, 33:61–73.

29. Hawton K, Hall S, Simkin S, Bale L, Bond A, Codd S, Stewart A: Deliberate self-harm in adolescents: a study of characteristics and trends in Oxford, 1990-2000. Journal of Child Psychology and Psychiatry 2003, 44:1191–1198.

30. Bond L, Carlin JB, Thomas L, Rubin K, Patton G: Does bullying case emotional problems? A prospective study of young teenagers. BMJ 2001, 323:480–484.

31. Resnik MD, Bearman PS, Blum RW, Bauman KE, Harris KM, Jones J, Tabor J, Beuhring T, Sieving RE, Shaw M, Ireland M, Bearinger LH, Udry JR: Pro-

tecting adolescents from harm. Findings from the National Longitudinal Study on Adolescent Health. The Journal of the American Medical Association 1997, 278:823–832.

32. Wells J, Barlow J, Stewart-Brown S: A systematic review of universal approaches to mental health promotion in schools. Health Education 2003, 4:197–220.
33. Department for Education and Skills: Promoting Children's Mental Health Within Early Years and School Settings. Nottingham: DfEE Publications; 2001.
34. Department of Health: National Healthy Schools Status: a guide for schools. London: DH Publications; 2005.
35. Department for Children, Schools and Families: Whole-School Approach to the National Healthy Schools Programme. London: DCSF Publications; 2007.
36. Ofsted: Healthy Minds: promoting emotional health and well-being in schools. London. 2005.
37. Harden A, Rees R, Shepherd J, Brunton G, Oliver S, Oakley A: Young people and mental health: a systematic review of research on barriers and facilitators. London: EPPI-Centre Publications; 2001.
38. Oliver S, Harden A, Rees R, Shepherd J, Brunton G, Oakley A: Young people and mental health: novel methods for systematic review of research on barriers and facilitators. Health Education Research 2008, 23:770–790.
39. Fortune S, Sinclair J, Hawton K: Adolescents' views on preventing self-harm. Social Psychiatry and Psychiatric Epidemiology 2008, 43:96–104.
40. Randhawa G, Stein S: An exploratory study examining attitudes towards mental health and mental health services among young South Asians in the United Kingdom. Journal of Muslim Mental Health 2007, 2:21–37.
41. Roeser RW, Midgley C: Teachers' views of issues involving students' mental health. The Elementary School Journal 1997, 98:115–133.
42. Cohall AT, Cohall R, Dye B, Dini S, Vaughan RD, Coots S: Overheard in the halls: what adolescents are saying, and what teachers are hearing, about health issues. Journal of School Health 2007, 77:344–350.
43. Walter HJ, Gouze K, Lim KG: Teachers' beliefs about mental health needs in inner city elementary schools. Journal of the American Academy of Child and Adolescent Psychiatry 2006, 45:61–68.
44. Mason J: Qualitative Researching. 2nd edition. London: Sage Publications; 2002.

45. Goodman E, Huang B, Wade TJ, Kahn RS: A multilevel analysis of the relation of socioeconomic status to adolescent depressive symptoms: does the school context matter? The Journal of Pediatrics 2003, 143:451–456.

46. Kitzinger J: Qualitative research: Introducing focus groups. BMJ 2005, 311:299–302.

47. Biddle L, Donovan J, Sharp D, Gunnell D: Explaining non-help-seeking amongst young adults with mental distress: a dynamic interpretive model of illness behaviour. Sociology of Health and Illness 2007, 29:983–1002.

48. Milton B, Woods SE, Dugdill L, Porcellato L, Springett RJ: Starting young? Children's experiences of smoking during pre-adolescence. Health Education Research 2008, 23:298–309.

49. Krueger RA, Casey MA: Focus Groups: A Practical Guide for Applied Research. 3rd edition. California: Sage Publications; 2000.

50. Morgan DL: Focus Groups as Qualitative Research. 2nd edition. California: Sage Publications; 1997.

51. Biddle L, Gunnell D, Sharp D, Donovan J: Factors influencing help-seeking in mentally distressed young adults: a cross-sectional survey. British Journal of General Practice 2004, 54:248–253.

52. Strange V, Oakley A, Forrest S: Mixed-sex or single-sex education: how would young people like their sex education and why? Gender and Education 2003, 15:201–214.

53. Donovan J, Sanders C: Key issues in the analysis of qualitative data in health services research. In Handbook of Health Research Methods. Edited by: Bowling A, Ebrahim S. Berkshire: Open University Press; 2005:515–532.

54. Miles MB, Huberman AM: Qualitative Data Analysis: An Expanded Source Book. 2nd edition. Thousand Oaks: Sage Publications; 1994.

55. Barbour R: Checklists for improving rigour in qualitative research. BMJ 2001, 322:1115–1117.

56. Sturgis P, Smith P, Hughes G: A Study of Suitable Methods for Raising Response Rates in School Surveys. Nottingham, England: DfES Publications; 2006.

57. Fortune S, Sinclair J, Hawton K: Help-seeking before and after episodes of self-harm: A descriptive study in school pupils in England. BMC Public Health 2008, 8:369.

58. Craig C: The Potential Dangers of a Systematic, Explicit Approach to Teaching Social and Emotional Skills (SEAL). Glasgow: Centre for Confidence and Wellbeing; 2007.

59. Measor L, Tiffin C, Miller K: Gender and sex education: a study of adolescent responses. Gender and Education 2006, 8:275–288.

60. Allen L: Poles apart? Gender differences in proposals for sexuality education content. Gender and Education 2008, 20:435–450.

61. Mellanby AR, Phelps FA, Crichton NJ, Tripp JH: School sex education, a process for evaluation: methodology and results. Health Education Research 1996, 11:205–214.

62. Buckley EJ, White DG: Systematic review of the role of external contributors in school substance use education. Health Education 2007, 107:42–62.

63. Crosswaite C, Tooby J, Cyster R: SPICED: Evaluation of a drug education project in Kirklees primary schools. Health Education Journal 2004, 63:61–69.

64. Essler V, Arthur A, Stickley T: Using a school-based intervention to challenge stigmatizing attitudes and promote mental health in teenagers. Journal of Mental Health 2006, 15:243–250.

65. Weist MD, Axelrod Lowie J, Flaherty LT, Pruitt D: Collaboration among the education, mental health, and public health systems to promote youth mental health. Psychiatric Services 2001, 52:1348–1351.

66. Cooper M: Counselling in Schools Project Phase II: Evaluation Report. Glasgow: Counselling Unit, University of Strathclyde, Scotland; 2004.

67. Aseltine RH, DeMartino R: An outcome evaluation of the SOS suicide prevention program. American Journal of Public Health 2004, 94:446–451.

68. Fox CL, Butler I: 'If you don't want to tell anyone else you can tell her': young people's views on school counselling. Journal of Guidance and Counselling 2007, 35:97–114.

69. Freake H, Barley V, Kent G: Adolescents' views of helping professionals: A review of the literature. Journal of Adolescence 2007, 30:639–653.

70. Cowie H, Naylor P, Talamelli L, Chauhan P, Smith PK: Knowledge, use of and attitudes towards peer support: a two-year follow-up to the Prince's Trust survey. Journal of Adolescence 2002, 25:453–467.

71. Sheffield JK, Spence SH, Rapee RM, Kowalenko N, Wignall A, Davis A, McLone J: Evaluation of universal, indicated and combined cognitive-behavioural approaches to the prevention of depression among adolescents. Journal of Consulting Clinical Psychology 2006, 74:66–79.

72. Delfabbro P, Winefield T, Trainor S, Dollard M, Anderson Sarah, Metzer J, Hammarstrom A: Peer and teacher bullying/victimization of South Australian

secondary school students: Prevalence and psychosocial profiles. British Journal of Educational Psychology 2006, 76:71–90.

73. Chen X, Rubin KH, Li B: Depressed mood in Chinese children: relations with school performance and family environment. Journal of Consulting and Clinical Psychology 1995, 63:938–947.

74. Spratt J, Shucksmith J, Philip K, Watson C: 'Part of who we are as a school should include responsibility for well-being': links between the school environment, mental health and behaviour. Pastoral Care in Education 2006, 24:14–21.

Depressive Symptoms from Kindergarten to Early School Age: Longitudinal Associations with Social Skills Deficits and Peer Victimization

Sonja Perren and Françoise D. Alsaker

ABSTRACT

Background

Depressive symptoms in children are associated with social skills deficits and problems with peers. We propose a model which suggests different mechanisms for the impact of deficits in self-oriented social skills (assertiveness and social participation) and other-oriented social skills (pro-social, cooperative and non-aggressive behaviors) on children's depressive symptoms. We hypothesized that deficits in self-oriented social skills have a direct impact on children's depressive symptoms because these children have non-rewarding interactions with peers, whereas the impact of deficits in other-oriented social skills on depressive symptoms is mediated through negative reactions from peers such as peer victimization.

Method

378 kindergarten children (163 girls) participated at two assessments (Age at T1: M = 5.8, T2: M = 7.4). Teachers completed questionnaires on children's social skills at T1. Teacher reports on peer victimization and depressive symptoms were assessed at both assessment points.

Results

Our study partially confirmed the suggested conceptual model. Deficits in self-oriented social skills significantly predicted depressive symptoms, whereas deficits in other-oriented social skills were more strongly associated with peer victimization. Longitudinal associations between other-oriented social skills and depressive symptoms were mediated through peer victimization.

Conclusion

The study emphasizes the role of deficits in self-oriented social skills and peer victimization for the development of internalizing disorders.

Background

Decades of research have shown that depressive symptoms in children are associated with social skills deficits and problems with peers [1-4]. Nevertheless, not much is known about the mechanisms underlying these associations. In the current paper a conceptual model is tested which tries to explain the interdependent effects of social skills and peer relations on the development of depressive symptoms.

Generally speaking social skills have been defined as behaviors that affect interpersonal relations [5]. In our work, we conceive of social competence as the ability to use social interactions to satisfy one's own goals and needs while at the same time considering the needs and goals of others. We differentiate between two dimensions: (a) self-oriented social skills which are aimed at satisfying one's own needs (e.g. assertiveness and social participation) and other-oriented social skills which are aimed at satisfying another's goals and needs (e.g. pro-social, co-operative and non-aggressive behavior) [6]. We assume that these dimensions operate through different mechanisms on depressive symptoms. First, we suggest that deficits in self-oriented social skills are directly associated with children's depressive symptoms because children's social needs and goals remain unsatisfied as a consequence of their inability to initiate social contacts, to express their needs, to assert themselves or to set limits to others' demands. These children therefore experience non-rewarding social interactions. Second, we propose that the impact of deficits in other-oriented social skills on depressive symptoms is mediated through negative peer relations, such as peer victimization.

Social Skills and Depressive Symptoms

Interpersonal theories of depression have tried to explain the observed associations between depression and relationship problems and deficits in social skills in children, adolescents and adults [7]. For example Lewinsohn [8] has suggested that a lack of certain social skills in depressed persons reduces their experience of positive reinforcement by others because they do not engage in behavior that leads to rewarding consequences. Coyne's theory [9] proposes that over the long term, a vicious circle develops: Depressive symptoms trigger negative reactions from others (e.g. aggressive behavior), which hinder recovery from depression. Moreover, intervention studies have shown that social skills training is an effective means to reduce depressive symptoms [4]. These latter results may even suggest that social skills play a causal role in the development of depression.

There is broad agreement that deficits in self-oriented social skills are associated with depressive symptoms. Withdrawal is considered as one of the main behavioral precursors of depressive symptoms in children [10] and it may exacerbate internalizing problems [11]. The empirical data on the association between deficits in other-oriented social skills and depressive symptoms are somewhat contradictory. Interestingly, aggressiveness—considered a deficit in other-oriented social skills—is positively associated with self-oriented social skills, i.e. aggressive children are more assertive and more prone to engage in social interactions [12-14]. Therefore, aggressiveness might even protect children from depressive symptoms. However, the specific role of pro-social behavior in the development of psychopathology has received only limited attention. It has been suggested that low levels of pro-social behavior may place children at risk for externalizing problems, whereas high levels of pro-social behavior might be a risk factor for internalizing problems [15-17]. It might be that a lack of pro-social behavior could be considered an indicator of a deficit in other-oriented skills, whereas high levels of pro-social behavior could also reflect a lack of self-oriented skills because these children might be too considerate of the needs of others and neglect their own feelings and needs [18]. In fact, cross-sectional and longitudinal studies have found that children who are overly concerned for the welfare of others, are highly cooperative or over-friendly, have elevated levels of emotional symptoms [19-21]. Perren and collaborators showed that at kindergarten age pro-social behavior predicted increases in emotional symptoms, but only in children who already had emotional problems at the first assessment point [18]. On the other hand, some studies found pro-social behavior to be negatively associated with emotional symptoms [22,23]. Pro-social behavior was also shown to be a protective factor in terms of peer acceptance in children with emotional symptoms [24]. Therefore, considering simultaneously the impact of self- and other-oriented social skills on depressive symptoms might give further insights into these apparently controversial results.

Social Skills and Peer Victimization

Deficits in certain social skills may lead to negative reactions by peers, such as rejection or victimization. Peer rejection may be a precursor of peer victimization [25] and may play a crucial role in stabilizing a child's victim role [26]. In fact Ladd and Troop-Gordon [27] reported peer rejection to be predictive of later victimization and victimization of later rejection.

Peer rejection has consistently been associated with aggressive and withdrawn behavior [28], i.e. with deficits in other- and self-oriented social skills. Likewise, in bully/victim research two different pathways to victimization are suggested and the need to differentiate between two types of victims has been emphasized: (1) children who are aggressive and victimized (aggressive victims) and (2) children who are victimized without being aggressive (passive victims) [12,14,29,30]. Accordingly, aggressive as well as withdrawn-submissive behavior patterns are related to peer victimization.

Submissiveness has been discussed as a hallmark of victimization. One explanation for this association is that bullies are looking for easy targets for their assaults [31]. Several studies revealed that passive victims have problems defending themselves [14,32] and that they are less assertive, for example using fewer persuasion attempts [33]. However, Perren et al. [6] could not confirm submissiveness as being an overall predictor of peer victimization in kindergarten children and the study by Fox and Boulton [34] reported submissive behavior in school children to be longitudinally predictive of social exclusion only. Withdrawal behavior (also reflecting a deficit in self-oriented skills) may be associated with victimization by (1) suggesting vulnerability, (2) suggesting low risk of retaliation, and (3) hindering children to find supporting and protecting friends in the class [35]. Also, withdrawn children are not salient and socially less rewarding for their peers. This, in itself, might lessen the chance that peers would help them when they become victimized.

However, in younger children, aggressive behavior seems to be a stronger predictor of victimization and rejection than withdrawn-submissive behavior

[36-38]. Nevertheless, not all aggressive children are at risk for becoming victimized and different findings suggest that the most important difference between non-victimized aggressive children (bullies) and aggressive victims consists in their respective ability or inability to control their physical aggression. These uncontrolled aggressive children are fairly disturbing in the class and it seems rather obvious that peers could easily be influenced to assist the bullies [35].

In sum, deficits in self- and other-oriented social skills are associated with rejection and victimization and deficits in other-oriented social skills have been found to be stronger predictors than deficits in self-oriented social skills.

Peer Victimization and Depressive Symptoms

Children with depressive symptoms have generally been reported to have poorer peer relations in terms of popularity, rejection or victimization [22,39,40]. Hawker and Boulton's [41] meta-analysis of cross-sectional associations between peer victimization and psychosocial maladjustment showed that victimization is most strongly related to depression and least strongly to anxiety. Peer victimization and exclusion may also increase children's depressive symptoms [11,42-44] or even be causally related to the development of self-derogation and depressive problems [45,46]. Peer victimization is also associated with health problems, suicidality, and poor school adjustment [47-50].

In sum, empirical findings consistently show that depressive symptoms are associated with negative peer relations and that peer rejection and victimization may play a causal role in the development of depressive symptoms.

The Interplay of Social Skills, Peer Victimization, and Depressive Symptoms

As shown above, social skills deficits are not only associated with depressive symptoms but are a strong predictor of peer victimization. Therefore, we assume that these variables interact in systematic ways, especially when we differentiate between self-oriented and other-oriented skills.

As outlined by Bukowski and Adams [51], peer relations have been discussed as markers, mediators, or moderators for maladjustment in children and adolescents. Similarly, Ladd [52] suggested different "Child by Environment Models" which take into account the interplay between child behavior, peer relations, and the development of internalizing (and externalizing) disorders. Empirical results mainly support additive or mediation models. For example a four-year longitudinal study by Ladd [10] has provided support for the additive model. Withdrawn behavior and peer rejection were shown to be overlapping risk factors for the development of children's internalizing problems. A study by Dill and collaborators demonstrated that peer rejection and victimization mediate between children's withdrawn/shy behavior and negative affect [53]. Similarly, a study among young adults showed interpersonal relationships to mediate the impact of social skills on well-being [54]. The conceptual model to be tested in the current paper was partly tested in a cross-sectional study with 198 kindergarten children [6]. The study confirmed the distinct contribution of self- and other-oriented social skills on children's peer victimization and emotional well-being. Deficits in self-oriented social skills predicted higher levels of emotional symptoms, whereas deficits in other-oriented social skills predicted higher levels of peer victimization. The suggested mediating role of peer victimization was partly confirmed.

In the present paper we aim to replicate the findings in a larger sample and most importantly, to include longitudinal data. We hypothesize that deficits in self- and other-oriented social skills are associated with both peer victimization and depressive symptoms. Furthermore, we hypothesize that the impact of deficits in other-oriented social skills on depressive symptoms is mediated through peer victimization whereas deficits in self-oriented social skills are directly associated with depressive symptoms. We also hypothesize that peer victimization is associated with an increase in depressive symptoms over time. We will examine whether the associations are moderated by the child's gender.

In our study we are adopting a dimensional approach of assessing depressive symptoms [55], i.e. we do not use clinical diagnoses of depression.

Method

Study Design and Sample

The data come from a large longitudinal study of pathways to peer victimization in a representative sample of kindergarten and elementary school children in the German-speaking part of Switzerland [56]. Written parental consent was obtained for all participants. Children themselves gave oral assent prior to the first interview and knew they could withdraw from the study at any time.

Due to organizational reasons the original recruiting was lagged (T1 assessment at two consecutive school years) but the follow-up was conducted in the same school year. Therefore, the time interval between the two assessments ranges between one and two years; and children of subsample 1 are older than those of subsample 2 at Time 2. Thus, we control for subsample membership in all analyses. Some kindergarten groups also participated in a prevention program against bullying. Thus, we reran all analyses reported in this paper, including intervention participation as a control variable. None of the reported results showed any change. Thus, in the final analyses this variable was not included. Only children with valid data at both assessment points are included in the current paper.

378 children (163 girls) participated (Age_T1: M = 5.8, SD = 0.6, Age_T2: M = 7.4, SD = 0.8). At T1 35% children were in the first year and 65% in their second year of kindergarten (age-mixed groups). At T2, most children were in the first grade of primary school (54%), 24% in second grade and 22% were in their second year of kindergarten.

Children were primarily white and German speaking, but there is large proportion of children with an immigrant background: 26% of participating children have both parents and 16% one parent with a non-Swiss country of origin. These percentages correspond with general population statistics.

Instruments

Teachers completed questionnaires on a range of dimensions covering children's psychosocial adjustment and social behavior. Teachers rated children's peer victimization and depressive symptoms at Time 1 and Time 2, social skills were only assessed at Time 1. Children's behaviors were rated by different teachers at Time 1 and at Time 2. Prior to data collection at T1, teachers were offered a workshop during which they received in-depth information about bullying and victimization (each class was represented by at least one teacher). Teachers who completed the questionnaires at T2 did not participate in this workshop.

Peer Victimization

Teachers rated each child on four victimization items (physical, verbal, object-related, exclusion; e.g. 'child is victimized verbally, i.e., laughed at, called names, teased.') on a 5-point rating-scale (never, seldom, once or several times a month, once a week, or several times a week [14,57]). The four items yielded an adequate reliability coefficient (T1: α = .81; T2: α = .85). For the purpose of the present research question, a mean score of the four items was used.

Depressive Symptoms

The scale consists of three items. Two items were selected from the German version of the Child Behavior Checklist/TRF [58]("He/she seems to be unhappy, sad"; "He/she speaks pejoratively about him/herself") and one item was added that was very similar to the first item ("He/she looks a little sad") in order to increase the scale's reliability. This item was taken from the list of typical symptoms for children aged 3-6 years according to the German guidelines for the diagnosis of psychological disturbances in infants, children and adolescents [59,60] All three items were rated on a four-point scale (0 = not at all true to 4 = definitely true) and yielded adequate reliability (T1: α = .71, T2: α = .76).

Social Skills

Teachers completed a questionnaire on children's social behavior (SOCOMP; [61]) The two dimensions self- versus other-oriented social skills were built using a combination of six subscales of social behavior patterns All items were rated on a four-point scale (0 = not at all true to 4 = definitely true). First, six different social skills scales were computed. The cooperative behavior subscale consists of five items (e.g. "Compromises in conflicts with peers"; α = .79). The pro-social behavior subscale consists of five items, covering helping, comforting, and sharing behavior (e.g. "Frequently helps other children"; α = .86). The overt aggression subscale (physical and verbal) consists of 4 items (e.g. "Kicks, bites or hits other children"; α = .88). The social participation subscale, covering propensity to participate in social interactions, consists of 4 items (e.g. "Converses with peers easily," α = .82). The leadership subscale consists of three items ("organizes, suggests play activities to peers"; α = .82). The setting limits subscale also consists of three items (e.g. "Refuses unreasonable requests from others"; α = .76).

Validation of the Two-Dimensionality of Social Skills

A principal component analysis was conducted to validate the suggested dimensions of self- and other-reported social skills. The six social behavior subscales were entered in the analysis (PCA with Varimax-Rotation, Eigenvalue > 1 is used as criterion). In accordance with the hypothesized dimensions the analysis yielded two factors. The first factor (explained variance 41%) had high factor loadings for the subscales cooperative behavior (.90), pro-social behavior (.81) and overt aggression (-.85). The second factor (explained variance 34%) had high factor loading for the subscales sociability (.85), leadership (.91) and setting limits (.81). All cross-loadings were <.33. Based on this analysis, two dimensions of social skills were computed. The scale other-oriented social skills is the mean of the z-standardized subscales cooperative behavior, pro-social behavior and overt aggression (reversed). The scale self-oriented social skills is the mean of the z-standardized subscales social participation, leadership and setting limits.

Overview of the Statistical Analyses

For the bivariate associations Pearson correlations were used, for the multivariate analyses several multiple regression analyses were computed (all predictors were entered simultaneously and remained in the final model).

Mediation

To test for mediation, we adopted the commonly used approach by Baron and Kenny [62]. This mediation analysis consists of three steps. In a first step, we investigated the impact of social skills on peer victimization. In step 2, we investigated the impact of social skills on depressive symptoms. In step 3, we analyzed whether peer victimization mediates the associations between social skills and depressive symptoms. A variable is then considered to mediate when the effect of the first predictor drops to zero, when controlling for the mediator variable.

Changes Over Time

For each of the above mentioned steps, two regression models were computed. First the effect of the control variables and social skills at T1 was tested for T2 outcomes. In a second step the score of the outcome measures at T1 was entered as additional independent variable in order to predict changes in the outcome variable over time [63].

Gender as Moderator

As we were interested whether gender moderates the associations we also included the interaction effects. Gender was dummy coded (boys = 0, girls = 1). All analyses were first computed without the gender interaction and than including the gender interaction. As the results remained largely the same, only the results including the interactions are presented below.

Results

Bivariate correlations between all study variables are shown in Table 1. The two dimensions of social skills were not significantly associated. Deficits in other-oriented skills were significantly associated with depressive symptoms and victimization at both assessment points. Deficits in self-oriented skills were significantly associated with depressive symptoms (T1 and T2), and victimization (T1 only).

Depressive symptoms and victimization were significantly associated at both assessment points. Furthermore, depressive symptoms and peer victimization were moderately stable from T1 to T2.

Table 1. Bivariate associations between main study variables (Pearson correlations)

	SOS	Dep T1	Dep T2	Vict T1	Vict T2	Age	Gender (girls = 1)
Other-oriented social skills (T1)	.09	-.25**	-.18**	-.46**	-.30**	.15**	.23**
Self-oriented social skills (T1)		-.48**	-.23**	-.18**	.01	.24**	.14**
Depressive symptoms (T1)			.35**	.29**	.12*	.01	-.05
Depressive symptoms (T2)				.15**	.30**	.01	-.13*
Victimization (T1)					.26**	.02	-.21**
Victimization (T2)						-.03	-.22**

** $p < .01$, *$p < .05$ (two-tailed)

Age was positively associated with both dimensions of social skills. In addition, girls showed higher levels of other- and self-oriented social skills and lower levels of victimization (T1 and T2) and depressive symptoms (T2) than boys.

Social Skills Predicting Peer Victimization (Step 1)

In a first set of regression analyses we investigated the impact of social skills on victimization at T2. Gender, age and subsample membership (see method section) were entered as control variables. In a second step victimization at T1 was entered as additional independent variable in order to predict changes in victimization over time. The analyses yielded significant effects of gender and other-oriented social skills on level of victimization at T2 and changes in victimization (see Table 2).

Table 2. Results for the regression analyses predicting peer victimization and depressive symptoms at T2 (betas).

	STEP 1		STEP 2		STEP 3	
	Victimization (T2)	Victimization (Change[b])	Depressive symptoms (T2)	Depressive symptoms (Change[b])	Depressive symptoms (T2)	Depressive symptoms (Change[b])
Gender[a]	-.162**	-.153**	-.059	-.078	-.024	-.052
Age (T1)	.006	.019	.090	.044	.082	.042
Sample	-.070	-.053	-.012	.001	-.001	.005
Other-oriented social skills (T1)	-.304***	-.240***	-.209**	-.148*	-.117	-.083
Self-oriented social skills (T1)	.053	.075	-.248**	-.104	-.251***	-.121
Gender*OOS[a]	.064	.049	.079	.089	.044	.060
Gender*SOS[a]	-.013	-.017	.023	.023	.007	.008
Victimization (T1)	--	.125**	--	--	.047	.011
Victimization (T2)			--	--	.217***	.202**
Gender*VIC (T1)[a]			--	---	-.202	-.196
Gender*VIC (T2)[a]			--	--	.202	.200
Depressive symptoms (T1)				.285***		.268***
Model: R^2	.126	.137	.094	.131	.167	.190

*** $p < .001$ ** $p < .01$, *$p < .05$

[a] Dummy coded variables (girls = 1, boys = 0)

[b] The outcome at T2 were used as dependent variable but controlled for its level at T1, thus we predict changes from T1 to T2 (positive scores = increases).

Boys were more frequently victimized than girls, and also showed a larger increase in victimization over time. The lower the level of other-oriented social skills, the higher the level of peer victimization, and also the greater the increases in peer victimization over time.

Social Skills Predicting Depressive Symptoms (Step 2)

Following the same procedure as presented above, we examined the impact of social skills on depressive symptoms. Results are shown in Table 2.

The control variables gender, age and sample had no significant effect on level and change of depressive symptoms. No significant interactions with gender emerged. The analyses yielded significant effects of both dimensions of social skills on depressive symptoms. Children with lower levels of other-oriented social skills and self-oriented social skills at T1 had higher levels of depressive symptoms at T2. Deficits in other-oriented social skills also predicted increases in depressive symptoms.

Peer Victimization as a Potential Mediator (Step 3)

To analyze the potential mediating effects of peer victimization, the variable was entered at the third step of the regression analyses [62]. In this analysis, peer victimization at T1 and T2 were entered as independent variables (including gender interactions). Victimization at T2 was a significant predictor of depressive symptoms at T2 and also predicted increases in depressive symptoms over time (see Table 2). When controlling for peer victimization, the effect of self-oriented social skills remained unchanged. That is, self-oriented social skills predicted depressive symptoms at T2, but no increases in depressive symptoms. In contrast, when controlling for peer victimization at T2, the significant associations between other-oriented social skills and depressive symptoms at T2 disappeared. The same effect was shown in regards changes in depressive symptoms.

Discussion

Our study has shown that self- and other-oriented social skills are independent dimensions, and that both are directly or indirectly associated with depressive symptoms.

The Role of Self-Oriented Social Skills

We hypothesized that deficits in self-oriented skills could make a child an easy target for bullies. The hypothesis received some support, but the association was relatively weak, in the multivariate analyses the significant effect even disappeared. This finding confirms that in preschool and kindergarten age submissive-withdrawn behaviors are less strong predictors for peer victimization than aggressive behaviors [6,38]. Victimization in itself was also a strong predictor of depressive symptoms but it did not act as a mediator regarding self-oriented social skills. That is, both deficits in self-oriented skills and victimization predict depressive symptoms in young children. This is in line with previous studies. The association between victimization and depressive symptoms has been repeatedly documented, both in school-age and kindergarten children [64,65], and the effect of victimization on depressive symptoms has been shown to persist into early adulthood, long after victimization had stopped [66].

It still remains to elucidate which processes might link self-oriented social skills with depressive tendencies. It might be that children who have difficulties in setting limits or asserting themselves seldom experience the satisfaction of their own wishes and needs. They might therefore experience relatively little reward in social situations and see themselves as incompetent in social contexts. Knowing that their chances to achieve their social goals are modest, they might gradually take less initiative in the peer group, and thus also become less attractive to their peers. The quality of their peer relations might therefore become highly dependent on the personality and behavior of their peers. In other words, they might perceive themselves as low in self-efficacy and highly dependent on their peers' moods. Knowing how important positive peer relations are to children's well-being [67], it seems plausible that deficits that prevent children from developing rewarding relations with their peers might lead at least to sadness and possibly to other depressive symptoms. There might also be a reciprocal relationship between lack of social skills and depressive symptoms [2]. On the one hand, being sad and depressed or being victimized may result in further withdrawal from peers, i.e. unsociability may turn to peer avoidance [68]. A lack of experience with peers may in turn decrease children's social skills as they miss important learning opportunities. In fact, it has been shown that depressive symptoms lead to a decrease in self-evaluated competence [69].

The Role of Other-Oriented Social Skills

Considering deficits in the second dimension of social skills examined in our study, other-oriented skills, we found a mediating effect of victimization. Other-oriented skills were operationalized in terms of pro-social and cooperative behavior and non-aggressiveness. In other words, children low in other-oriented skills were typically high in aggressive behavior and low in behaviors such as sharing, helping and being cooperative. These are all characteristics of aggressive victims [35]. In fact, our results show that deficits in other-oriented social skills were associated with victimization. Most importantly, the association between depressive symptoms and this type of deficit disappeared when victimization was entered in the regression analysis. This is perfectly in line with earlier studies on bully/victim problems that have demonstrated that bullies (who are defined as being aggressive and not victimized by their peers) do not suffer from internalizing problems. That is, deficits in other-oriented social skills, that are typical both for bullies and aggressive victims, do not automatically lead to depressive symptoms. They only do so when the children are experiencing victimization by their peers.

Gender Differences

The study yielded several gender differences, but gender did not moderate the associations between social skills, victimization and depressive symptoms. Boys were more frequently victimized than girls. This goes in line with studies showing that boys are more frequently aggressive victims (but not passive victims) than girls [14]. In correspondence with other studies [15], girls were more prosocial-cooperative and less aggressive than boys.

In our study we did not differentiate between different forms of victimization. However, some studies have shown that boys and girls may not be victims of the same types of aggression, e.g. preschool boys were more frequently victimized physically than girls [70]. Further studies have to show whether different types of social skills deficits are associated with different forms of victimization.

Strengths and Limitations of the Study

Results are based on data from a rather large representative sample of kindergarten and primary school children. In the current paper, only data from children who participated also in the follow-up were included. Attrition was mainly due to the unwillingness of school principals or teachers who had not participated in the first wave of data collection, i.e. there was no systematic attrition with respect to children's or families' characteristics. Therefore, we do not expect systematic biases due to attrition. However, our results may be somewhat biased because we did not control for other potentially important variables such as family background or children's cognitive or verbal abilities [71,72].

One important limitation of our study is that we only used a three-item measure to assess children's depressive symptoms. Due to the small number of items, other potential symptoms for depression (e.g. lack of energy, hopelessness, psychosomatic complaints, preoccupation with death) were not assessed. Further studies have to show whether lack of assertiveness and sociability and peer victimization are merely associated with low emotional well-being or depressed mood or in fact are predictive for more severe cases of childhood depression.

In our conceptual model we assume that deficits in social skills precede children's depressive symptoms and peer victimization. However, social skills were only assessed at the first measurement point. Therefore, we could not analyze whether depressive symptoms or peer victimization predict changes in social skills.

In the current study we only included teacher reports; therefore we have shared method variance in our data. Regarding the validity of teacher reports on depressive symptoms, we relied on the strong tradition of the Achenbach questionnaires which reliably can be completed by parents or teachers. Studies have shown a moderate agreement between teacher, parent and self reports regarding children's depressive symptoms [40,73,74] indicating that teachers can provide valid reports on children's depressive symptoms. Nevertheless, the inclusion of other informants might give even more reliable and valid data [75]. The inclusion of children's self-perceptions, e.g. of depressive symptoms, may be particularly important as children have their own meaningful perspective on their experience [40]. As to victimization, an earlier study revealed a high degree of correspondence between teacher reports and peer nominations in kindergarten [14].

Clinical Implications

Our model suggests different intervention strategies considering both improvements in children's social skills and their peer relations. Until now, most (universal) prevention and intervention strategies in children focus on the promotion of empathy, pro-social behavior, and reduction in aggressive behavior [76]. Our results suggest that prevention and intervention efforts in children should also consider the promotion of self-oriented social skills, i.e. efforts to improve children's social participation and assertiveness. Training specifically aimed at improving children's self-oriented social skills, i.e. their social initiative and assertiveness, might even be considered as a psychotherapeutic approach to childhood depression. Shifting the focus toward self-oriented social skills is also important, as studies have shown that for some children being highly pro-social or empathic may even be negative for their own emotional well-being [18]. Prevention approaches should aim at establishing a balance between both dimensions of social skills, i.e. to promote children in their ability to use social interactions to satisfy their own goals and needs while at the same time considering the needs and goals of others.

Conclusion

Our study confirmed the suggested conceptual model. Deficits in self-oriented social skills significantly predicted depressive symptoms, whereas deficits in other-oriented social skills were more strongly associated with peer victimization. Longitudinal associations between other-oriented social skills and depressive symptoms were mediated through peer victimization. We suggested that the direct link between deficits in self-oriented social skills and depressive symptoms might partly be mediated by non-rewarding interactions with peers. More specific research is needed on these potential mediating mechanisms, for example in terms of socio-cognitive or socio-emotional processes. The study emphasizes the role of deficits in self-oriented social skills and peer victimization for the development of internalizing disorders.

Competing Interests

The authors declare that they have no competing interests.

Authors' Contributions

SP was responsible for the conceptual background of the paper, analyzed and interpreted the data and drafted the manuscript. FA is grant-holder, conceived and directed the study, and was actively involved in writing up the manuscript. Both authors read and approved the final manuscript.

Acknowledgements

This study was supported by a grant from the Swiss National Science Foundation (National Research Programme 52, Grant-No 4052-69011) to the second author.

References

1. Deater-Deckard K: Annotation: Recent research examining the role of peer relationships in the development of psychopathology. Journal of Child Psychology and Psychiatry and Allied Disciplines 2001, 42:565–579.
2. Hay DF, Payne A, Chadwick A: Peer relations in childhood. Journal of Child Psychology and Psychiatry and Allied Disciplines 2004, 45(1):84–108.

3. Masten AS: Peer Relationships and Psychopathology in Developmental Perspective: Reflections on Progress and Promise. Journal of Clinical Child and Adolescent Psychology 2005, 34(1):87–92.
4. Segrin C: Social skills deficits associated with depression. Clinical Psychology Review 2000, 20(3):379–403.
5. Elliott SN, Sheridan SM, Gresham FM: Assessing and treating social skills deficits: A case study for the scientist-practioner. Journal of School Psychology 1989, 27:197–222.
6. Perren S, Groeben M, Stadelmann S, von Klitzing K: Selbst- und fremdbezogene soziale Kompetenzen: Auswirkungen auf das emotionale Befinden. In Soziale Kompetenz bei Kindern und Jugendlichen: Entwicklungsprozesse und Förderungsmöglichkeiten. Edited by: Malti T, Perren S. Stuttgart: Kohlhammer; 2008.
7. Lara ME, Klein DN: Psychosocial processes underlying the maintenance and persistence of depression: Implications for understanding chronic depression. Clinical Psychology Review 1999, 19(5):553–570.
8. Lewinsohn PM: A behavioral approach to depression. In The psychology of depression: Contemporary theory and research. Edited by: Friedman RJ, Katz MM. New York: Wiley; 1974:157–185.
9. Coyne JC: Toward an interactional description of depression. Psychiatry: Journal for the Study of Interpersonal Processes 1976, 39(1):28–40.
10. Ladd GW: Peer Rejection, Aggressive or Withdrawn Behavior, and Psychological Maladjustment from Ages 5 to 12: An Examination of Four Predictive Models. Child Development 2006, 77(4):822–846.
11. Gazelle H, Ladd GW: Anxious solitude and peer exclusion: A diathesis-stress model of internalizing trajectories in childhood. Child Development 2003, 74(1):257–278.
12. Pellegrini AD, Bartini M, Brooks F: School bullies, victims, and aggressive victims: Factors relating to group affiliation and victimization in early adolescence. Journal of Educational Psychology 1999, 91(2):216–224.
13. Pepler DJ, Craig WM, Roberts WL: Observations of aggressive and nonaggressive children on the school playground. Merrill Palmer Quarterly 1998, 44(1):55–76.
14. Perren S, Alsaker FD: Social behavior and peer relationships of victims, bully-victims, and bullies in kindergarten. Journal of Child Psychology and Psychiatry 2006, 47(1):45–57.

15. Eisenberg N, Fabes RA: Prosocial development. In Handbook of child psychology. Volume 3. 5th edition. Edited by: Damon W. New York: Wiley; 1998:701–778.

16. Hastings PD, Zahn-Waxler C, Robinson J, Usher B, Bridges D: The development of concern for others in children with behavioral problems. Developmental Psychology 2000, 36:531–546.

17. Hay DF: Prosocial development. Journal of Child Psychology and Psychiatry 1994, 35(1):29–71.

18. Perren S, Stadelmann S, von Wyl A, von Klitzing K: Developmental pathways of emotional/behavioural symptoms in kindergarten children: What is the role of pro-social behaviour? Eur Child Adolesc Psychiatry 2007, 16(4):209–214.

19. Bohlin G, Bengtsgard K, Andersson K: Social inhibition and overfriendliness as related to socioemotional functioning in 7- and 8-year-old children. Journal of Clinical Child Psychology 2000, 29(3):414–423.

20. Gjerde P, Block J: Preadolescent antecedents of depressive symptomatology at age 18: A prospective study. Journal of Youth and Adolescence 1991, 20:217–232.

21. Hay DF, Pawlby S: Prosocial development in relation to children's and mothers' psychological problems. Child Development 2003, 74(5):1314–1327.

22. Rudolph KD, Clark AG: Conceptions of relationships in children with depressive and aggressive symptoms: Social-cognitive distortion or reality? Journal of Abnormal Child Psychology 2001, 29(1):41–56.

23. Wentzel KR, McNamara CC: Interpersonal relationships, emotional distress, and prosocial behavior in middle school. Journal of Early Adolescence 1999, 19(1):114–125.

24. Henricsson L, Rydell AM: Children with Behaviour Problems: The Influence of Social Competence and Social Relations on Problem Stability, School Achievement and Peer Acceptance Across the First Six Years of School. Infant and Child Development 2006, 15(4):347–366.

25. Hodges EVE, Malone MJ, Perry DG: Individual risk and social risk as interacting determinants of victimization in the peer group. Developmental Psychology 1997, 33(6):1032–1039.

26. Hanish LD, Guerra NG: Aggressive victims, passive victims, and bullies: Developmental continuity and developmental change? Merrill-Palmer Quarterly 2004, 50(1):17–38.

27. Ladd GW, Troop Gordon W: The role of chronic peer difficulties in the development of children's psychological adjustment problems. Child Development 2003, 74(5):1344–1367.

28. Newcomb AF, Bukowski WM, Pattee L: Children's peer relations: A meta-analytic review of popular, rejected, neglected, controversial, and average sociometric status. Psychological Bulletin 1993, 113(1):99–128.

29. Olweus D: Aggression in the schools bullies and whipping boys. Washington (D.C.): Hemisphere Publ; 1978.

30. Veenstra R, Lindenberg S, Oldehinkel AJ, De Winter AF, Verhulst FC, Ormel J: Bullying and Victimization in Elementary Schools: A Comparison of Bullies, Victims, Bully/Victims, and Uninvolved Preadolescents. Developmental Psychology 2005, 41(4):672–682.

31. Perry DG, Perry LC, Boldizar JP: Learning of aggression. In Handbook of developmental psychopathology. Edited by: Lewis M, Miller S. New York: Plenum Press; 1990:135–146.

32. Perry DG, Kusel SJ, Perry LC: Victims of peer aggression. Developmental Psychology 1988, 24(6):807–814.

33. Schwartz D, Dodge KA, Coie JD: The emergence of chronic peer vicitimization in boys' play groups. Child Development 1993, 64(6):1755–1772.

34. Fox CL, Boulton MJ: Longitudinal associations between submissive/nonassertive social behavior and different types of peer victimization. Violence and Victims 2006, 21:383–400.

35. Alsaker FD, Gutzwiller-Helfenfinger E: Social behavior and peer relationships of victims, bully-victims, and bullies in kindergarten. In The Handbook of School Bullying An International Perspective. Edited by: Jimerson SR, Swearer SM, Espelage DL. Mahwah, New Jersey: Lawrence Erlbaum Associates; 2010:87–99.

36. Asher SR: Recent advances in the study of peer rejection. In Peer rejection in childhood. Edited by: Asher SR, Coie JD. Cambrigde: University; 1990:3–14.

37. Coie JD, Dodge KA, Kupersmith JB: Peer group behavior and social status. In Peer rejection in childhood. Edited by: Asher SR, Coie JD. Cambrigde: University; 1990:17–57.

38. Hanish LD, Eisenberg N, Fabes RA, Spinrad TL, Ryan P, Schmidt S: The expression and regulation of negative emotions: Risk factors for young children's peer victimization. Development and Psychopathology 2004, 16:335–353.

39. Henricsson L, Rydell AM: Elementary school children with behavior problems: Teacher-child relations and self-perception. A prospective study. Merrill Palmer Quarterly 2004, 50(2):111–138.
40. Perren S, Von Wyl A, Stadelmann S, Burgin D, von Klitzing K: Associations between behavioral/emotional difficulties in kindergarten children and the quality of their peer relationships. Journal of the American Academy of Child and Adolescent Psychiatry 2006, 45(7):867–876.
41. Hawker DSJ, Boulton M: Twenty years' research on peer victimization and psychosocial maladjustment: A meta-analytic review of cross-sectional studies. Journal of Child Psychology and Psychiatry and Allied Disciplines 2000, 41:441–455.
42. Goodman MR, Stormshak EA, Dishion TJ: The significance of peer victimization at two points in development. Journal of Applied Developmental Psychology 2001, 22(5):507–526.
43. Hanish LD, Guerra NG: A longitudinal analysis of patterns of adjustment following peer victimization. Development and Psychopathology 2002, 14(1):69–89.
44. Hodges EVE, Perry DG: Personal and interpersonal antecedents and consequences of victimization by peers. Journal of Personality and Social Psychology 1999, 76(4):677–685.
45. Alsaker FD, Olweus D: Stability and change in global self-esteem and self-related affect. In Understanding early adolescent self and identity: Applications and interventions SUNY series, studying the self. Edited by: Brinthaupt TM, Lipka RP. Albany, NY: State University of New York Press; 2002:193–223.
46. Arseneault L, Milne BJ, Taylor A, Adams F, Delgado K, Caspi A, Moffit TE: Being bullied as an environmentally mediated contributing factor to children's internalizing problems. Archives of Paediatric and Adolecent Medicine 2008, 162:145–150.
47. Graham S, Bellmore AD, Mize J: Peer Victimization, Aggression, and Their Co-Occurrence in Middle School: Pathways to Adjustment Problems. Journal of Abnormal Child Psychology 2006, 34(3):363–378.
48. Nishina A, Juvonen J, Witkow MR: Sticks and Stones May Break My Bones, but Names Will Make Me Feel Sick: The Psychosocial, Somatic, and Scholastic Consequences of Peer Harassment. Journal of Clinical Child and Adolescent Psychology 2005, 34(1):37–48.
49. Rigby K: Health consequences of bullying and its prevention in schools. In Peer harassment in school: The plight of the vulnerable and victimized. Edited by: Juvonen J, Graham S. New York, NY: Guilford Press; 2001:310–331.

50. Storch EA, Phil M, Nock MK, Masia Warner C, Barlas ME: Peer Victimization and Social-Psychological Adjustment in Hispanic and African-American Children. Journal of Child and Family Studies 2003, 12(4):439–452.

51. Bukowski WM, Adams R: Peer Relationships and Psychopathology: Markers, Moderators, Mediators, Mechanisms, and Meanings. Journal of Clinical Child and Adolescent Psychology 2005, 34(1):3–10.

52. Ladd G: Children's peer relations and social competence. A century of progress. New Haven: Yale University Press; 2005.

53. Dill EJ, Vernberg EM, Fonagy P, Twemlow SW, Gamm BK: Negative Affect in Victimized Children: The Roles of Social Withdrawal, Peer Rejection, and Attitudes Toward Bullying. Journal of Abnormal Child Psychology 2004, 32(2):159–173.

54. Segrin C, Taylor M: Positive interpersonal relationships mediate the association between social skills and psychological well-being. Personality and Individual Differences 2007, 43(4):637–646.

55. Kraemer H, Noda A, O'Hara R: Categorical versus dimensional approaches to diagnosis: methodological challenges. Journal of Psychiatric Research 2004, 38(1):17–25.

56. Alsaker F, Nägele C: Bullying in kindergarten and prevention. In An International Perspective on Understanding and Addressing Bullying. Edited by: Craig WM, Pepler DJ. Kingston, Canada: PREVNet; 2008:230–252.

57. Alsaker F, Valkanover S: Early diagnosis and prevention of victimization in kindergarten. In Peer harassment in school: The plight of the vulnerable and victimized. Edited by: Juvonen J, Graham S. New York: Guilford Press; 2001:175–195.

58. Döpfner M, Melchers P, Fegert J, Lehmkuhl G, Lehmkuhl U, Schmeck K, et al.: Deutschsprachige Konsensus Versionen der Child Behavior Checklist (CBCL 4-18), der Teacher Report Form (TRF) und der Youth Self Report Form (YSR). Kindheit und Entwicklung 1994, 3:54–59.

59. DGKJP , ed: Leitlinien zur Diagnostik und Therapie von psychischen Störungen im Säuglings-, Kindes- und Jugendalter [Guidelines for the diagnosis and therapy of psychological disturbances in infants, children and adolescents]. Köln: Deutscher Ärzte Verlag; 2002.

60. Groen G, Petermann F: Depressive Kinder und Jugendliche [Depressive children and adolescents]. Göttingen: Hogrefe; 2002.

61. Perren S: SOCOMP. Ein Fragebogen zur Erfassung von selbst- und fremdorientierten verhaltensbezogenen sozialen Kompetenzen (Manual). Zürich: Jacobs Center for Productive Youth Development, Universität Zürich; 2007.

62. Baron RM, Kenny DA: The moderator-mediator variable distinction in social psychological research: Conceptual, strategic, and statistical considerations. Journal of Personality and Social Psychology 1986, 51(6):1173–1182.
63. Twisk JWR: Applied longitudinal data analysis for epidemiology. A practical guide. Cambridge: University Press; 2003.
64. Alsaker F: Psychische Folgen von Mobbing. In Schule und psychische Störungen. Edited by: Steinhausen H-C. Stuttgart: Kohlhammer; 2006:35–47.
65. Rigby K: Consequences of bullying in schools. Canadian Journal of Psychiatry 2003, 48(9):583–590.
66. Olweus D: Bullying at school: Long-term outcomes for the victims and an effective school-based intervention program. In Aggressive behavior: Current perspectives Plenum series in social/clinical psychology. Edited by: Huesmann LR. New York, NY: Plenum Press; 1994:97–130.
67. Hartup WW: Social relationships and their developmental significance. American Psychologist 1989, 44(2):120–126.
68. Asendorpf JB: Beyond social withdrawal: Shyness, unsociability, and peer avoidance. Human Development 1990, 33(4-5):250–259.
69. Cole DA, Martin JM, Powers B: A competency-based model of child depression: A longitudinal study of peer, parent, teacher, and self-evaluations. Journal of Child Psychology and Psychiatry 1997, 38(5):505–514.
70. Hanish LD, Kochenderfer-Ladd B, Fabes RA, Martin CL, Denning D: Bullying among young children: The influence of peers and teachers. In Bullying in American schools: A social ecological perspective on prevention and intervention. Edited by: Espelage DL, Swearer S. Erlbaum Publishers; 2004:141–160.
71. von Grünigen R, Perren S, Nägele C, Alsaker F: Immigrant children's peer acceptance and victimization in kindergarten: the role of local language competence. British Journal of Developmental Psychology 2009, in press.
72. Perren S, Stadelmann S, von Klitzing K: Child and family characteristics as risk factors for peer victimization in kindergarten. Schweizerische Zeitschrift für Bildungswissenschaften 2009, 31(1):13–32.
73. Clarizio HF: Assessment of depression in children and adolescents by parents, teachers, and peers. In Handbook of depression in children and adolescents. Edited by: Reynolds WM, Johnston HF. New York: Plenum Press; 1994:235–248.
74. Epkins CC, Meyers AW: Assessment of childhood depression, anxiety, and aggression: Convergent and discriminant validity of self-, parent-, teacher-, and peer-report measures. Journal of Personality Assessment 1994, 62:364–381.

75. Kraemer H, Measelle JR, Ablow JC, Essex MJ, Boyce WT, Kupfer DJ: A new approach to integrating data from multiple informants in psychiatric assessment and research: Mixing and matching contexts and perspectives. American Journal of Psychiatry 2003, 160:1566–1577.
76. Perren S, Malti T: Soziale Kompetenz entwickeln: Synthese und Ausblick. In Soziale Kompetenz bei Kindern und Jugendlichen: Entwicklungsprozesse und Förderungsmöglichkeiten. Edited by: Malti T, Perren S. Stuttgart: Kohlhammer; 2008.

Exploring the Impact of the Baby-Friendly Hospital Initiative on Trends in Exclusive Breastfeeding

Sheryl W. Abrahams and Miriam H. Labbok

ABSTRACT

Background

The Baby-Friendly Hospital Initiative (BFHI) seeks to support breastfeeding initiation in maternity services. This study uses country-level data to examine the relationship between BFHI programming and trends in exclusive breastfeeding (EBF) in 14 developing countries.

Methods

Demographic and Health Surveys and UNICEF BFHI Reports provided EBF and BFHI data. Because country programs were initiated in different years, data points were realigned to the year that the first Baby-Friendly hospital was certified in that country. Pre-and post-implementation time periods were analyzed using fixed effects models to account for grouping of data by country, and compared to assess differences in trends.

Results

Statistically significant upward trends in EBF under two months and under six months, as assessed by whether fitted trends had slopes significantly different from 0, were observed only during the period following BFHI implementation, and not before. BFHI implementation was associated with average annual increases of 1.54 percentage points in the rate of EBF of infants under two months ($p < 0.001$) and 1.11-percentage points in the rate of EBF of infants under six months ($p < 0.001$); however, these rates were not statistically different from pre-BFHI trends.

Conclusion

BFHI implementation was associated with a statistically significant annual increase in rates of EBF in the countries under study; however, small sample sizes may have contributed to the fact that results do not demonstrate a significant difference from pre-BFHI trends. Further research is needed to consider trends according to the percentages of Baby-Friendly facilities, percent of all births occurring in these facilities, and continued compliance with the program.

Background

Breastfeeding, especially exclusive breastfeeding (EBF), is one of the most effective preventive health measures available to reduce child morbidity and mortality [1]. The international Baby-Friendly Hospital Initiative (BFHI) was launched in 1991 by UNICEF and WHO to promote and protect maternal and child health by ensuring support for breastfeeding in maternity care facilities [2]. Since that time, more than 20,000 health care facilities in more than 150 countries around the world have achieved Baby-Friendly certification from their national certifying body (Labbok M, personal communication from global query carried out in 2006, [3]) by implementing the Ten Steps to Successful Breastfeeding and ending the practice of distributing free or low-cost breast milk substitutes [4,5].

Evidence from developed and developing countries indicates that the BFHI has had a direct impact on breastfeeding rates at the hospital level [6-11]. In a randomized controlled trial in Belarus, Kramer et al. noted improved rates of any and exclusive breastfeeding at 3 and 6 months and any breastfeeding at 12 months, in infants of mothers giving birth at hospitals randomized to follow BFHI policies, compared to those delivering at control hospitals [7]. A 2003 analysis of data from Swiss mothers demonstrated that rates of EBF for infants 0 to 5 months was significantly higher among those delivered in Baby-Friendly hospitals than in the general sample, and that average breastfeeding duration was longer for infants born in Baby-Friendly hospitals that had maintained good compliance with the Ten Steps [8]. Analysis of data from 57 hospitals in Oregon, United States, show that breastfeeding rates at two days, and two weeks postpartum increased with the institution's implementation of the Ten Steps [9]. Similarly, results of the United States Infant Feeding Practices II Study indicate that mothers who experienced no Baby-Friendly practices in-hospital were 13 times more likely to stop breastfeeding before six weeks than mothers who experienced six specific Baby-Friendly practices [10]. Widespread implementation of the BFHI has also been associated with increased rates of breastfeeding and exclusive breastfeeding at the regional and national levels [8,12,13].

Although global trends in breastfeeding initiation, duration and exclusivity have generally increased during the years since the introduction of the UNICEF/WHO BFHI, few studies have examined these trends specifically within the context of BFHI activities [13-15]. The purpose of this study is to investigate the contribution of the BFHI to trends in EBF in a group of selected developing countries, through analysis of trends before and after their implementation of the BFHI. We sought to test the hypothesis that overall trends in EBF in countries that implemented the BFHI had increased significantly during the time period after BFHI implementation, compared to the time period prior to the program's launch, and also to estimate the program's contribution to upward trends in EBF in the countries under study.

Methods

Data Selection

Data were taken from the Demographic and Health Surveys (DHS) from 1986-2006. These nationally representative sample surveys captured population, health and nutrition-related indicators in 72 developing countries (DHS), including prevalence of EBF according to child's age [16]. Use of the DHS limited our analysis to developing countries, but was thought to provide the single best source of consistent data on EBF rates.

First, we selected countries with a minimum of two DHS surveys within the given time frame (the minimum number of data points necessary to establish a trend). A total of 45 countries met this criterion, of which all but three were found to have implemented the BFHI. Due to the lack of "non-BFHI" countries in the data set, we elected to compare trends in EBF before and after implementation of the BFHI, rather than comparing trends between "BFHI" and "non-BFHI" countries.

To allow countries to serve as their own comparisons, we selected those with a minimum of two surveys prior to their country's year of initial BFHI implementation, and two after. Few countries had data on EBF available from the year of initial implementation (the "zero" year). In order to capture trends immediately before and after program launch in countries with no "zero" year data, any data point available within two years of BFHI implementation (i.e., within the range -2 to 2) was included in both the before and after data sets for trend. Therefore some countries with only three data points were included in the sample (a decision reflected in the overlapping trend lines seen in Figure 1).

Outcome Variables

The primary outcome variables were the percent of living children under the age of two months and under the age of six months who were exclusively breastfed at the time of survey. EBF as defined by DHS refers to the practice of giving no food or drink other than human milk, measured by 24-hour recall [17]. Although this definition has remained constant, DHS surveys have changed the number of possible responses over time, adding additional categories of "food and drink." These changes occurred in DHS modules used across countries over the years. Adding these additional prompts likely reduces the number of infants who would be considered "exclusively breastfed" over time in all countries. Since this change occurred across all countries, comparison over time remains valid, albeit reducing potential observed increases in EBF rates [18].

Independent Variable

An independent variable was created, "years from BFHI," by subtracting the year of BFHI implementation in the country from the year in which the EBF rate was collected. Negative and positive values denoted data points collected before and after the start of BFHI activities, respectively. The year of BFHI implementation was defined individually for each country as the year in which the first in-country hospital achieved Baby-Friendly status from its national certifying body, as determined from queries to UNICEF country offices and review of UNICEF BFHI reporting data from 1994-2006 [3]. This variable allowed all data points to be considered in relation to their distance from the time of initial implementation, despite countries having implemented the program in different years.

Statistical Analysis

To examine EBF trends, we fitted fixed-effects models in STATA 9.1 (StataCorp, College Station, TX) for the time periods prior to and following program launch. Fixed effects models were chosen to account for the grouping of data points by country and to control for observed and unobserved between-country differences that were fixed over the time period (such as the starting prevalence of EBF immediately prior to program implementation). Linear models were selected based on the positive trends observed in global rates and the relatively small number of data points available.

To compare pre-BFHI and post-BFHI trends in the countries under study, we assessed whether trends were statistically significant during either time period, i.e., whether the slopes of the observed trend pre-BFHI, or the observed trend post-BFHI, was significantly different from zero, and whether the slopes of these observed trends were statistically significantly different from one another.

Results

Characteristics of Countries in the Sample

A total of 14 countries were included in the final sample (Table 1). The latest year of initial BFHI implementation in the sample was found to be 1997. The percentage of in-country maternity hospitals ever certified as Baby-Friendly by 2006 ranged from 3% to 69%, with a median of 17% [3]. The percentage of institutional births (using a 2000-2006 composite measure) ranged from 17% to 97%, with a median of 61% [19].

Table 1. Characteristics of countries included in the analysis

Country (n = 14)	Year of BFHI launch	Percentage institutional deliveries, 2000-2006 [19]	Percentage of in-country maternity hospitals ever-certified as Baby-Friendly*
Bolivia	1992	57	20
Brazil	1992	97	10
Colombia	1993	92	37
Dominican Republic	1994	95	3
Egypt	1993	65	3
Ghana	1996	49	13
Indonesia	1992	40	5
Jordan	1997	97	4
Kenya	1992	40	69
Mali	1995	38	33
Niger	1996	17	49
Peru	1994	70	66
Uganda	1994	41	3
Zimbabwe	1993	68	23

* Percentage of hospitals ever-certified as Baby-Friendly calculated by dividing the total number of hospitals ever certified as Baby-Friendly as determined by 2006 UNICEF reporting records by the total number of in-country maternity hospitals as determined by 1999 UNICEF reporting records [3,4].

Trends in the Rate of EBF Pre-BFHI Implementation and Post-BFHI Implementation

When rates of EBF for children less than two months were examined, statistically significant upward trends were observed post-BFHI implementation. Prior to the initiation of BFHI, there was a slow upward trend in EBF, but the slope did not achieve statistical significance (Table 2; Figure 1). Implementation of the BFHI was followed by an average annual increase of 1.54 percentage points in the rate of EBF of infants under two months of age ($p < 0.001$). Pre-BFHI, the rate of increase was only 0.88 percentage points annually, a difference of 0.66 percentage points. The difference between the slopes of the pre-BFHI and post-BFHI trend lines was not itself statistically significant (95% CI for difference: -0.82, 2.14; $p = 0.384$).

Table 2. Trends in exclusive breastfeeding before and after implementation of the Baby-Friendly Hospital Initiative

Estimated annual change in the percent of children under two months of age exclusively breastfed (in percentage points)		
	Parameter estimate	p-value for trend
Pre-BFHI	0.88	0.14
Post-BFHI	1.54	< 0.01*
Estimated annual change in the percent of children under six months of age exclusively breastfed (in percentage points)		
Pre-BFHI	0.20	0.67
Post-BFHI	1.11	< 0.01*

*Statistically significant at the $p < 0.01$ level

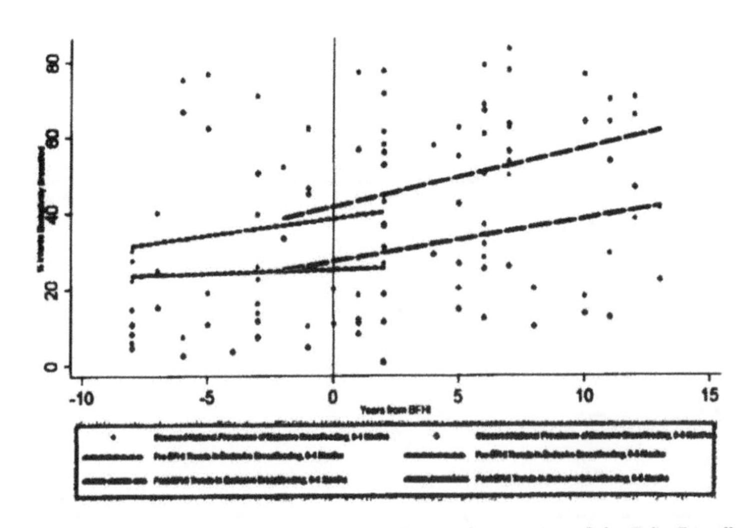

Figure 1. Trends in exclusive breastfeeding before and after implementation of the Baby-Friendly Hospital Initiative (BFHI).

Results were similar using EBF for children less than six months as the outcome. Statistically significant upward trends were observed only during the period post-BFHI implementation (Table 2; Figure 1). BFHI implementation was associated with a 1.11- percentage point annual increase in the rate of exclusive breastfeeding in the first six months ($p < 0.001$), 0.91 points greater than the rate of increase estimated for the period prior to BFHI launch. Again, the difference between the slopes of the pre-BFHI and post-BFHI trend lines was not itself statistically significant (95% CI for difference: -0.22, 2.09; $p = 0.131$).

Discussion

Program Impact Implications

There is little debate as to the importance of exclusive breastfeeding [1]; however, the effectiveness of programs such as the BFHI has been questioned and there has been a reduction in international support for this program (Labbok M, personal communication from global query carried out in 2006, [3]). Despite the many studies demonstrating the impact of BFHI on breastfeeding rates at the hospital level [6-11], and those that show its impact at the individual country or sub-regional level [8,12,13], no previous study has utilized a multi-country construct based on the actual year of national BFHI implementation, with exclusive breastfeeding as the outcome variable.

Results suggest that among the countries under study, there were no significant upward trends in EBF rates in the years prior to BFHI implementation, but that BFHI implementation was associated with a statistically significant annual increase in rates of EBF in the first two, as well as during the first six, months. The two month rate of increase was higher than the six month rate, as might be expected with an immediate postpartum intervention.

According to the models, a country that implemented the BFHI would experience, on average, a 7.7- and 5.5-percentage point increase in the first two, and first six, months of EBF respectively, over a subsequent period of five years. If improvements in EBF practices are sustained over time, such an increase could provide a significant improvement in child health outcomes. One can estimate the impact of such an increase as follows: based on the accepted estimate that a 51% increase in EBF is needed to reduce child mortality by 13% (i.e., from the 2006 estimate of a 39% prevalence to the 90% prevalence used for calculation of the 13% reduction in child mortality [1]), we estimate that a 5.5% increase in EBF in the first six months has directly reduced annual child mortality by about 1.4%, or prevented about 140,000 deaths. The fact that the slopes of these trend lines did not differ significantly from one another is a call for caution in interpreting these findings. However, the fact that the definitions of EBF became more conservative over time may have blunted the slopes in the later data, reducing the likelihood of achieving significance even if a true increase in positive breastfeeding patterns had occurred.

One strength of our analysis is the use of the "zero" year to re-center all data to the time of country-specific BFHI implementation. This allowed for countries to serve as their own comparisons over time, and for cross-national trends to be considered in relation to the start of BFHI programming, adding strength to the argument that observed trends are derived from BFHI activities. This adds to our understanding of the impact of the BFHI as it was implemented.

Limitations

The limitations of this study include the reality that the 14 countries analyzed represent a small portion of all developing nations that have implemented the BFHI, and exhibit relatively low rates of hospital certification. In addition, we had no measure of the level of ongoing adherence to the Ten Steps or of the general quality of BFHI implementation over time in this sample. Our results, therefore, are not necessarily reflective of the program's potential to improve breastfeeding rates if implemented on a national or global level.

A serious limitation of any effort to evaluate the effectiveness of the BFHI on cross-national breastfeeding trends is the lack of data collected specifically for this purpose. In this study, we have had to rely on a relatively small number of data points. As such, our study was not powered to detect small differences in trends between pre- and post-BFHI time periods. The limited number of data points also hindered examination of non-linear models that may have provided more insight into the behavior of these trends over time. Our use of overlapping trend lines to compensate for a lack of "zero" year data points, and incomplete information on Baby-Friendly changes that may have been instituted prior to actual certification, further impaired our ability to detect differences between pre- and post-BFHI time periods. We cannot fully predict how access to additional EBF measurements, data from additional countries or information about possible pre-existing Baby-Friendly practices may have changed our results.

The use of fixed effects models allowed us to control for the presence of measured and unmeasured confounders that were fixed over the time period studied, but did not control for factors that were variable over the time period, such as demographic changes, shifts in maternal employment patterns, or other breastfeeding promotion programs implemented concurrently with the BFHI. We lacked sufficient information to control for these variables appropriately. With the exception of concurrent public health programming, we would expect most changes over this broad time period, including increased urbanization and women's employment, to have negatively impacted EBF [20]. For this reason, we feel that the observed trends may represent a conservative estimate of the program's potential.

Future Research

If and when sample size and available data permit, additional analyses are needed to consider trends taking into account the percentages of maternity facilities ever-certified as Baby-Friendly, the percent of all births that occur in these facilities, and continued compliance with and investment in the program. Such analyses would help to determine whether a dose-response relationship exists between the level of BFHI programming and trends in EBF over time. Further research is also needed to investigate the existence and impact of other local and national breastfeeding promotion and support programs implemented concurrently with the BFHI.

Conclusion

Implementation of the international BFHI was associated with a statistically significant annual increase in rates of EBF among infants 0 to 2 months of age and among infants 0 to 6 months of age in the 14 countries studied. Further research is needed to explore fully the impact of the BFHI on cross-national breastfeeding trends, including studies that could better control for individual country's rates of BFHI certification, for whether the practices were maintained, and for the proportion of all births that occurred in Baby-Friendly facilities.

In sum, although the trends following BFHI introduction were not statistically significantly increased from the pre-BFHI trends, and although we were unable to control for all possible confounders, our findings indicate that implementation of the BFHI is associated with positive changes in EBF at a level that would result in improved child health and survival outcomes.

Competing Interests

SA declares no competing interests. ML has served as a consultant to UNICEF and WHO, the implementing agencies of the international Baby-Friendly Hospital Initiative.

Authors' Contributions

SA was responsible for the preparation and analysis of data, wrote the paper with inputs from ML, and contributed to study design. ML was responsible for study concept, and contributed to study design and writing and editing of the final paper.

Acknowledgements

This research was supported by the Carolina Global Breastfeeding Institute at the University of North Carolina at Chapel Hill, with partial support from UNICEF/NYHQ. The authors would like to thank Dr. E. Michael Foster for his input on methodology. The authors are solely responsible for the content and any remaining errors.

References

1. Jones G, Steketee RW, Black RE, Bhutta ZA, Morris SS, Bellagio Child Survival Study Group: How many child deaths can we prevent this year? Lancet 2003, 362(9377):65–71.
2. Naylor AJ, Baby-Friendly Hospital Initiative: Protecting, promoting, and supporting breastfeeding in the twenty-first century. Pediatr Clin North Am 2001, 48(2):475–483.
3. UNICEF 1990-2005: Celebrating the Innocenti Declaration on the Protection, Promotion and Support of Breastfeeding: Past Achievements, Present Challenges and Priority Action for Infant and Young Child Feeding. 2nd edition. UNICEF Innocenti Research Centre, Florence, Italy; 2006.
4. UNICEF and WHO: Protecting, Promoting and Supporting Breastfeeding: The Special Role of Maternity Services. World Health Organization, Geneva; 1989.
5. UNICEF: The Baby-Friendly Hospital Initiative, [http://www.unicef.org/programme/breastfeeding/baby.htm].
6. Britton C, McCormick FM, Renfrew MJ, Wade A, King SE: Support for breastfeeding mothers (Review). Cochrane Database Syst Rev 2007, 1:CD001141.
7. Kramer MS, Chalmers B, Hodnett ED, Sevkovskaya Z, Dzikovich I, Shapiro S, Collet JP, Vanilovich I, Mezen I, Ducruet T, Shishko G, Zubovich V, Mknuik D, Gluchanina E, Dombrovskiy V, Ustinovitch A, Kot T, Bogdanovich N, Ovchinikova L, Helsing E: PROBIT Study Group: Promotion of Breastfeeding

Intervention Trial (PROBIT): a randomized trial in the Republic of Belarus. JAMA 2001, 285(4):413–420.

8. Merten S, Dratva J, Ackermann-Liebrich U: Do Baby-Friendly hospitals influence breastfeeding duration on a national level? Pediatrics 2005, 116(5):e702–708.
9. Rosenberg KD, Stull JD, Adler MR, Kasehagen LJ, Crivelli-Kovach A: Impact of hospital policies on breastfeeding outcomes. Breastfeed Med 2008, 3(2):110–116.
10. DiGirolamo AM, Grummer-Strawn LM, Fein SB: Effect of maternity-care practices on breastfeeding. Pediatrics 2008, 122(Suppl 2):S43-49.
11. Cattaneo A, Buzzetti R: Effect on rates of breast feeding of training for the baby friendly hospital initiative. BMJ 2001, 323(7325):1358–62.
12. Caldeira AP, Goncalves E: Assessment of the impact of implementing the Baby Friendly Hospital Initiative. J Pediatr (Rio J) 2007, 83(2):127–132.
13. Labbok MH, Wardlaw T, Blanc A, Clark D, Terreri N: Trends in exclusive breastfeeding: findings from the 1990s. J Hum Lact 2006, 22(3):272–276.
14. Pérez-Escamilla R: Evidence based breast-feeding promotion: the Baby-Friendly Hospital Initiative. J Nutr 2007, 137(2):484–487.
15. Hofvander Y: Breastfeeding and the Baby Friendly Hospitals Initiative (BFHI): organization, response and outcome in Sweden and other countries. Acta Paediatr 2005, 94(8):1012–1016.
16. Measure DHS: Demographic and Health Surveys Overview, [http://www.measuredhs.com/aboutsurveys/dhs/start.cfm].
17. Rutstein SO, Rojas G: MEASURE DHS: Online guide to DHS statistics, [http://www.measuredhs.com/help/Datasets/index.htm].
18. Measure DHS: DHS Model Questionnaires with Commentary, [http:/ / www.measuredhs.com/ pubs/ search/ search_results.cfm?Type=35&srchTp=type&newSrch=1].
19. UNICEF: State of the World's Children 2008: Child Survival, [http:/ / www.unicef.org/ publications/ files/ The_State_of_the_Worlds_Children_2008.pdf].
20. Grummer-Strawn LM: The effect of changes in population characteristics on breastfeeding trends in fifteen developing countries. Int J Epidemiol 1996, 25(1):94–102.

Perception of Neighborhood Safety and Reported Childhood Lifetime Asthma in the United States (U.S.): A Study Based on a National Survey

S. V. Subramanian and Malinda H. Kennedy

ABSTRACT

Background

Recent studies have emphasized the role of psychosocial stressors as a determinant of asthma, and neighborhoods can be a potential source of such stressors. We investigated the association between parental perception of neighborhood safety and reported lifetime asthma among children.

Methodology/Principal Findings

Data for the study came from the 2003–04 National Survey of Children Health (NSCH); a nationally representative cross-sectional sample of children aged 0–17 years. Demographic, socioeconomic and behavioral covariates were included in the study. Models were estimated after taking account of weighting and complex survey design. Parental report of whether the child has ever been diagnosed with asthma by a physician was used to define the outcome. Parental report of perception of neighborhood safety was the main exposure. In unadjusted models, the odds ratio (OR) for reporting asthma associated with living in neighborhoods that were perceived to be sometimes or never safe was 1.36 (95% confidence intervals [CI] 1.21, 1.53) compared to living in neighborhoods that were perceived to be always safe. Adjusting for covariates including exposure to second hand tobacco smoke, mother's self-rated health, child's physical activity and television viewing attenuated this association (OR 1.25, 95% CI 1.08, 1.43). In adjusted models, the increased odds ratio for reporting asthma was also higher among those who perceived neighborhoods as being usually safe (OR 1.15 95% CI 1.06, 1.26), as compared to always safe, suggestive of a dose-response relationship, with the differentials for usually safe and never safe being statistically significant (p = 0.009).

Conclusion

Psychosocial stressors may be important risk factors that may impact the pathogenesis of asthma and/or contribute to asthma morbidity by triggering exacerbations through neuroimmunologic mechanisms, as well as social mechanisms.

Introduction

According to the most recent data an average annual 20 million children and adults in the U.S. had asthma, with a current asthma prevalence of 7.2% [1]. Research on the determinants of asthma has largely investigated the role of environmental exposures, such as exposure to aeroallergens, indoor and outdoor air pollution, endotoxin, smoking, and viral infections [2], [3]. More recently, the role of psychosocial stressors as a determinant of asthma has been proposed [4], [5], [6], [7], [8]. Laboratory as well as prospective population-based studies have shown associations between stress experiences and asthma expression [9], [10], potentially mediated through physiologic pathways resulting in enhanced IgE expression, enhanced allergen-specific lymphocyte proliferation and differential cytokine expression in children [11], [12], [13]. A potential source of psychosocial stress is the type of neighborhood in which people reside, which increasingly has been given important consideration in asthma disparities[14], [15], [16], [17] and epidemiologic research [18], [19], [20], [21], [22]. One hypothesized mechanism through which a neighborhood can induce stress among its residents is through exposure to violence. Residents chronically feel unsafe in these neighborhoods, [23] which leads to increased stress. There are few studies of the association between exposure to neighborhood violence and asthma, [24], [25], [26] with most studies relying on small community samples. We investigated the association between parental perception of neighborhood safety and reported lifetime asthma in a large, nationally representative sample of children aged 5–17 years in the United States (U.S.)

Methods

Study Population and Design

Data for the study came from the National Survey of Children's Health (NSCH) which was a telephone survey conducted in 2003–2004 that gathered information from parents on the physical, emotional and behavioral health of a nationally representative sample of 102,353 children aged 0–17. A full description of the survey can be found at www.cdc.gov/nchs/about/major/slaits/nsch.htm. A complex sample design involving clustering of children within households and stratification of households within states based on the National Immunization Survey sampling plan was used to collect the data. Random digit dialing was used to create the sample in each state. One child per household was randomly selected for enrollment. The survey is representative of non-institutionalized children ages 0–17 at both the national and state levels. Approximately 2000 surveys were collected per state (1,483–2,241) using the SLAITS (State and Local Area Integrated Telephone Survey) program. The SLAITS methodology recognizes the potential for bias resulting from the non-coverage of households without telephones and applies direct adjustments in an attempt to minimize this potential bias [27]. Specifically, in order to adjust for potential bias of under-representation of households without telephones, post-stratification adjustments were made to the NSCH based on evidence which suggests that households with interrupted telephone service are more similar to households without telephones than to households with telephone service. Using data from other national surveys about the proportion of children in households without telephones in each state, weights were increased for completed NSCH interviews in households with interrupted telephone service (see http://www.cdc.gov/nchs/data/series-/sr_01/sr01_043.pdf).

The response rate for the survey was 55.3%. This response rate was calculated as the product of the interview completion rate (68.8%), the screener completion rate ("the proportion of known households where a resident reported whether a child lived in the household") (87.8%), and the resolution rate ("the proportion of telephone numbers that could be positively identified as residential or nonresidential") (91.6%) [28]. The survey was conducted in English and Spanish. A total of 102,353 children were enrolled in the survey. Of these, 74,264 children between the ages 5 and 17 were eligible for this study. Those with missing information on asthma (n = 124), exposure (n = 3,834) and covariates (n = 18,270) were excluded, yielding a final analytic sample of 54,002 children, constituting 73% of the surveyed population. In the NSCH survey, approximately 9% of respondents were missing poverty data. To account for this, the National Center for Health Statistics provided five publically available datasets containing imputations for the missing poverty values [29]. These data were used for the analysis.

Outcome

Lifetime asthma was measured using respondent (primary caregiver) reports in the NSCH survey of whether "a doctor or health professional ever told you that the child has asthma."

Exposure

Perceived neighborhood safety was measured through one survey question "How often do you feel (Selected Child) is safe in your community or neighborhood?" with the possible responses of never, sometimes, usually, and always. For analysis, the "never" and "sometimes" responses were combined because of low numbers.

Covariates

Covariates for the study included child's age, gender, a detailed race/ethnicity variable, a measure of household income/poverty (<100% of poverty level, 100-<200% of poverty level, 200-<400% of poverty level, and ≥400% of poverty level), highest level of education obtained by anyone in household (< high school, high school degree, > high school), child weight status, child physical activity, child television viewing, and mother's health status and household smoking. Child physical activity was measured by two questions. The first measured the number of days in the past week that the child exercised or participated in physical activity lasting at least 20 minutes that made him/her sweat and breathe hard. The second question determined whether the child participated in sports lessons or on a sports team in the past year. Television viewing was defined as the number of hours on an average day that a child watches television, videos or plays video games. Mother's health was included as a series of dichotomous dummy variables (poor health, fair health, good health, or very good health with excellent health as the referent). Household smoking was defined as a dichotomous variable indicating whether "anyone in the household uses cigarettes, cigars, or pipe tobacco."

In general the rationale for including the above covariates was to adjust for previously identified risk factors for asthma and potential confounders. However, given the cross-sectional design of the survey, it is impossible to establish the temporal relationship of these factors. Consequently, we report the unadjusted and fully adjusted models to give a perspective on the potential confounding due to observed covariates, and build models in sequential steps from being most to least parsimonious.

Analysis

All analyses used the provided sampling weights developed to provide nationally representative estimates [28]. Final multivariable analyses were conducted using STATA to ensure control of the study design effects, proper analysis of the five imputed datasets and correct estimates of standard errors [29], [30], [31]. We first estimated the unadjusted effect of neighborhood safety on the log odds of reporting lifetime asthma and then sequentially adjusted the effect estimate for socio-demographic variables (sex, age, race/ethnicity, SES and child weight status), followed by household variables related to household smoking and mother's health and finally child behavioral variables associated with physical activity and TV viewing.

Ethics

Verbal consent for participation in the study was obtained from NSCH respondents. Respondents were informed about the voluntary nature of the survey, the authorizing legislation, and confidentiality of data collected. In addition, the informed consent script provided information about the content of the survey and the expected duration. The informed consent was verbal since the survey was via telephone. The study was reviewed by Harvard School of Public Health Institutional Review Board and was considered as exempt from full review as the study was based on an anonymous public use data set with no identifiable information on the survey participants.

Results

The gender distribution of the study population was relatively equal. A majority of the children were recorded as non-Hispanic white (67.5%), followed by non-Hispanic black (16.1%), White or Black Hispanic (8.0%), and other non-Hispanic (4.6%). The majority of children (65.8%) were in households where at least one resident had more than a high school education. Seventeen percent of children lived in households that were less than 100% of the poverty level. Approximately 15.5% were at risk for overweight (BMI≥85th percentile and <95th percentile) and 21% were overweight (BMI≥95th percentile). Almost half of parents were minimally concerned about neighborhood safety (48.6%), while 16.1% were most concerned (Table 1).

Table 1. Sample distribution across exposure and different covariates along with prevalence of reported lifetime asthma, National Survey of Children's Health (Children ages 5–17).

Variables	Frequency (%)	Number (%) of total ever diagnosed with asthma
Total population	74,264 (100)	10,271 (14.0)
Neighborhood Safety		
Neighborhood sometimes or never safe	8,809 (16.1)	1,492 (18.9)
Neighborhood usually safe	27,257 (35.4)	3,913 (37.1)
Neighborhood always safe	37,299 (48.6)	4,745 (43.9)
Child demographics		
Gender		
Female	36,039 (48.9)	4,339 (11.8)
Male	38,156 (51.1)	5,925 (16.2)
Race/ethnicity		
White Hispanic	4,022 (7.3)	522 (12.1)
Black Hispanic	352 (0.7)	67 (17.1)
Multiple race Hispanic	289 (0.3)	54 (15.3)
Other race Hispanic	358 (0.5)	67 (18.1)
White non-Hispanic	51,820 (67.5)	6,704 (13.3)
Black non-Hispanic	7,176 (16.1)	1,394 (19.1)
Multiple race non-Hispanic	2,608 (3.1)	465 (17.2)
Other race non-Hispanic	2,742 (4.6)	413 (14.3)
Weight Status of Child		
Underweight	4,475 (6.8)	498 (12.0)
Normal Weight	40,637 ([illegible])	5,250 (13.2)
At-Risk for Overweight	10,672 (15.5)	1,633 (14.9)
Overweight	13,216 (21.0)	2,354 (18.6)
Child on sports team or took sports lessons in past year	42,926 (57.6)	5,879 (14.1)
Household characteristics		
Derived Poverty level (* range based on imputed datasets)		
Less than 100% of the poverty level	8,335–8,396 (16.8)	1,397–1,415 (15.4)
100% to <200% of the poverty level	14,756–14,797 (22.7)	2,101–2,111 (14.1)
200% to <400% of the poverty level	27,213–27,334 (33.6)	3,592–3,621 (13.7)
≥400% of the poverty level	23,796–23,925 (26.9)	3,146–3,173 (13.4)
Education		
Highest Household education is < high school	3,173 (7.6)	386 (11.8)
Highest Household education is = high school	15,584 (26.6)	2,271 (14.0)
Highest Household education is > high school	55,235 (65.8)	7,583 (14.3)
Smoking		
Anyone in household use cigarettes, cigars, or pipe tobacco	22,334 (30.3)	3,377 (15.3)
Mother's health status		
Mother in poor health	1,340 (2.3)	335 (23.6)
Mother in fair health	5,225 ([illegible])	1,021 (18.2)
Mother in good health	15,082 (24.6)	2,400 (15.3)
Mother in very good health	23,566 (32.1)	3,235 (14.1)
Mother in excellent health	22,829 (32.1)	2,408 (10.8)

unweighted numbers, weighted percents.

There was an inverse association between perceptions of neighborhood safety and the odds of reporting asthma among children (Table 2). In unadjusted models, children living in neighborhoods that were perceived to be sometimes or never

Table 2. Odds ratios (OR) and 95% confidence intervals (CI) for reporting lifetime asthma among children aged 5–17 for exposure and covariates.

Variable	Model 1[a] (Unadjusted)	Model 2[b] (Adjusted for sex, age, race/ethnicity, poverty, education and child weight status)	Model 3[c] (Model 2 additionally adjusted for mother's health status and household smoking)	Model 4[d] (Model 3 additionally adjusted for children's TV viewing and physical activity)
[illegible]				
Always safe	Reference	Reference	Reference	Reference
[illegible] safe	1.19 (1.10–1.28)	1.19 (1.10–1.30)	1.14 (1.05–1.24)	1.15 (1.06–1.26)
Sometimes or never safe	1.36 (1.21–1.53)	1.27 (1.11–1.45)	1.23 (1.08–1.41)	1.25 (1.08–1.43)
[illegible]				
Female		0.69 (0.64–0.75)	0.67 (0.62–0.73)	0.68 (0.62–0.74)
[illegible]		Reference	Reference	Reference
Child age		1.02 (1.01–1.03)	1.02 (1.01–1.03)	1.02 (1.00–1.03)
[illegible]				
White non-Hispanic		Reference	Reference	Reference
[illegible] Hispanic		0.99 (0.82–1.20)	1.01 (0.83–1.24)	1.03 (0.84–1.27)
Black Hispanic		1.29 (0.79–2.09)	1.00 (0.58–1.73)	1.10 (0.62–1.94)
[illegible] non-Hispanic		1.40 (1.25–1.57)	1.36 (1.20–1.53)	1.38 (1.22–1.57)
Multi race Hispanic		1.20 (0.69–2.09)	0.99 (0.54–1.84)	0.88 (0.49–1.57)
[illegible] race non-Hispanic		1.30 (1.07–1.59)	1.19 (0.98–1.44)	1.18 (0.97–1.45)
Other race Hispanic		1.71 (0.92–3.20)	1.83 (0.94–3.53)	1.81 (0.91–3.62)
[illegible] race non-Hispanic		0.97 (0.75–1.27)	1.07 (0.82–1.41)	1.05 (0.81–1.37)
Body mass index				
[illegible]		0.95 (0.79–1.13)	0.99 (0.82–1.19)	1.00 (0.82–1.22)
Normal Weight		Reference	Reference	Reference
[illegible] risk for overweight		1.15 (1.03–1.28)	1.13 (1.01–1.26)	1.15 (1.03–1.29)
Overweight		1.45 (1.30–1.60)	1.38 (1.24–1.54)	1.38 (1.24–1.55)
[illegible] participates in sports				0.97 (0.89–1.07)
Child physical activity				1.00 (0.98–1.02)
[illegible] viewing				0.98 (0.95–1.00)
Household poverty				
< 100% of the poverty level		Reference	Reference	Reference
100-<200% of poverty level		0.86 (0.74–1.00)	0.92 (0.79–1.08)	0.93 (0.79–1.09)
[illegible]< 400% of poverty level		0.80 (0.70–0.92)	0.91 (0.79–1.05)	0.93 (0.80–1.08)
≥400% of poverty level		0.80 (0.69–0.92)	0.96 (0.83–1.12)	0.97 (0.83–1.14)
[illegible]				
Household education less than highschool		0.84 (0.65–1.09)	0.71 (0.55–0.93)	0.72 (0.55–0.95)
[illegible] education equal to [illegible] school		0.89 (0.80–0.98)	0.84 (0.76–0.94)	0.85 (0.76–95)
Household education more than high school		Reference	Reference	Reference
[illegible]				
Presence of Household smoking			1.09 (0.99–1.19)	1.09 (1.00–1.20)
[illegible] self-rated health				
Mother's health excellent			Reference	Reference
[illegible] health very good			1.33 (1.21–1.48)	1.34 (1.21–1.48)
Mother's health good			1.46 (1.30–1.64)	1.45 (1.28–1.63)
[illegible] health fair			1.75 (1.48–2.08)	1.70 (1.42–2.02)
Mother's health poor			2.29 (1.78–2.95)	2.28 (1.75–2.95)

[a]Model 1 was unadjusted.
[b]Model 2 was adjusted for sex, age, race/ethnicity, poverty, education and child weight status.
[c]Model 3 was adjusted for sex, age, race/ethnicity, poverty, education and child weight status, mother's health status and household smoking.
[d]Model 4 was adjusted for sex, age, race/ethnicity, poverty, education and child weight status, mother's health status and household smoking, children's TV viewing and physical activity.

safe had substantially higher levels of reported asthma (OR 1.36, 95% CI 1.21, 1.53) as compared to children living in neighborhoods that were perceived to be always safe. The odds ratios for reporting lifetime asthma among children living in neighborhoods that were perceived to be sometimes or never safe decreased to 1.27 (95% CI 1.11, 1.45) after adjustment for demographic and social covariates. Additionally controlling the models for exposure to second hand tobacco smoke in the household as well as mother's self-rated health reduced the odds of reported asthma associated with living in most unsafe neighborhoods to 1.23 (95% CI 1.08, 1.41). Further inclusion of behavioral variables associated with physical activity and television viewing did not attenuate the effect any further (OR 1.25, 95% CI 1.08–1.43).

Other significant associations for the odds of reported lifetime asthma in the fully adjusted model were found with gender, race/ethnicity, child's weight, and mother's health (Table 2, Model 4). Female children had a substantially lower likelihood of being diagnosed with asthma. Black non-Hispanic children were substantially more likely to be reported with lifetime asthma compared to white non-Hispanic children. Children who were overweight were also more likely to be reported with lifetime asthma compared to children who were in the 'normal' weight range. Poor self-reported health status of mother was strongly associated with reported lifetime asthma. Notably, socioeconomic status was not associated with reported lifetime asthma.

Discussion

Using the most recent, large and nationally representative survey we report a robust inverse association between parental perception of neighborhood safety and reported lifetime asthma among children. Children living in neighborhoods that parents felt were only sometimes or never safe had a 25% increased risk of being reported with lifetime asthma as compared to those living in neighborhoods where parents felt always safe. The effect size observed for this perceived exposure was comparable to the effects observed for non-Hispanic Blacks or for being overweight; both of which have been strong correlates of asthma. Before we interpret the study findings, the following limitations need to be considered.

First, notwithstanding the general challenges of measuring asthma in population-based studies [32], the measurement of asthma in the NSCH has clear limitations. The NSCH measure of asthma prevalence was based on a single question, as opposed to a hierarchy of asthma/wheeze outcomes based on responses to standardized respiratory questionnaires [33], [34]. No effort was made to clinically test for asthma. However, the prevalence estimates observed in NSCH are not very different from other national estimates [1]. Additionally, there were no data on level of allergens in the home. At the same time, the NSCH, by way of collecting extensive social and demographic data (including perception of neighborhood safety), and being nationally representative, provides a unique, if not the only, opportunity to draw descriptive inference on the patterning of asthma risk by exposure to neighborhood stressors in the U.S. The second limitation relates to neighborhood safety data being based on one question about parental perceptions as opposed to objective systematic observation. Arguably, parents' perception could be as important as actual safety data, since they reflect parental attitudes toward the neighborhood and provide an indication of how parent (and child) might interact with the environment [35]. However, their perceptions could also serve as a proxy for other parental characteristics, such as depression or anxiety, which could independently influence child health. However a convenience sample study examined this idea and found no evidence that parental perception of dangerous neighborhoods was linked to maternal depression [36]. Third, given the cross-sectional and observational study design we cannot rule out both reverse causation (i.e., parents with asthmatic/sick kids moving to neighborhoods that are generally more socially stressful and violent) or confounding due to unobserved confounders (i.e., factors that might be a common prior cause of both child's asthma as well as neighborhood safety condition). Notably, an important confounder is parental health that can influence both the child's risk of asthma and can influence selection into certain type of neighborhoods. While mother's subjective health status is not an ideal measure to control for this potential confounding, it certainly improves the faith in the observed finding that we report between perception of neighborhood safety and childhood asthma. Fourth, the low reported response rate in the NSCH does raise concerns related to the representativeness of the survey and the potential for differential bias. In general, telephone survey response rates have been lower than in past decades. Both the NSCH interview completion rate (68.8%), indicating the completion of interviews among households that were reached, and the overall NSCH response rate were similar to other telephone surveys such as the BRFSS, suggesting that the NSCH achieved a response rate typical for current telephone surveys (see http://www.cdc.gov/BRFSS/technical_infodata/2003QualityReport.htm, accessed May 26, 2009). Additionally, the NSCH post-survey weighting procedures attempted to adjust for non-responses to minimize potential bias. Further, in our analysis, we also found that the prevalence estimates of asthma observed in NSCH were not very different from other national estimates [1]. Finally, there is some evidence to suggest that reducing the non-response rate of a survey does not dramatically change the results [37].

Psychosocial factors that can be linked to asthma mortality, morbidity and medication compliance have been well articulated [38]. Exposure to violence (measured here through parental perception of safety) can be a major psychosocial stressor that may impact the pathogenesis of asthma and/or contribute to asthma morbidity by triggering exacerbations through neuroimmunologic mechanisms [13]. Psychological stress has also been posited to intensify asthma symptoms by increasing the body's inflammatory response to asthma triggers, with the hypothalamic-pituitary-adrenal (HPA) axis, the sympathetic-adrenal-medullary (SAM) axis, and the sympathetic (SNS) and parasympathetic (PNS) arms of the autonomic nervous system serving as the biological pathways [39], between stress and bronchoconstriction in asthma [40]. Sustained exposure to violence, or more importantly the chronic fear of violence, may provide the trigger that links psychosocial stress and asthma.

Psychosocial stressors also possibly moderate both humoral and cellular immune function, with such alterations predisposing the individual to respiratory tract infections [41], [42], which may in turn trigger acute asthma episodes. Thus, stress hormones through their influence on immune expression may increase a genetically predisposed individual's risk of developing asthma or perpetuate an existing condition. Violence as a psychosocial stressor, therefore, may be an "adjuvant" to the asthmatic inflammatory response [23]. Current knowledge supports the notion that environmental factors that include viral infection, air pollutants, maternal smoking, breast-feeding, and allergen exposure modulate the expression of the asthmatic phenotype, as related to the immune response. Stress may also accentuate the response to allergens by increasing the release of inflammatory mediators and the subsequent cascade of inflammatory events characteristic of chronic asthma.

Other indirect mechanisms through which exposures to violence (and perhaps other characteristics of stressful social circumstances) may operate is by adopting coping behaviors such as smoking, thus increasing the exposure to a known environmental asthma trigger. For instance, children exposed to second-hand tobacco exposure were at an increased risk for lifetime asthma. However, we were not able to further explore the duration or amount of household smoking because of limitations of the survey data. Nonetheless, we were able to include information from basic questions about household smoking and physical activity, thus controlling for two potential mediators, and a significant effect remained. This suggests that these are not operating as strong mediators in this case. There is some evidence to suggest that even pre-natally, the mother's stress level can influence the programming of the infant's HPA axis and may lead to dysregulation of the infant's immune system [43]. In the case when the stress is caused by fear of violence, such psychosocial exposures may be considered a social determinant of health that actually results in long-term biological changes that contribute to asthma morbidity.

Wolf and colleagues explored biological reasons to explain the association between asthma and parental stress and depression by examining how well these variables could predict children's inflammatory profiles. They found that in both healthy children and children with asthma, parental stress and to a lesser degree, parental depression was associated with changes in the children's inflammatory markers. Specifically, stimulated interlukin 4 (IL-4) production and eosinophil cationic protein (ECP) release significantly increased. These specific markers are relevant to asthma because they are both important parts of the process that leads to an inflammatory event which results in typical symptoms of asthma, such as the constriction of the airway and edema [44].

In addition to this biological response, living in an area plagued by violence can reduce the availability of and access to resources such as doctors, pharmacies and other health care providers. This could make it difficult to successfully manage a child's asthma. It may also lead to negative behavioral changes that are used as coping tools such as parental or child smoking, or an increased indoor and sedentary lifestyle. An increase in sedentary lifestyle could put children at risk for obesity, which is associated with asthma. Indoor living also leads to higher exposure to indoor allergens and exposures that exist in poor quality housing. Additionally, the stress of living in a violent environment may lead to impaired daily functioning, social isolation, feelings of lack of control, and dysfunctional family environment. This may leave families less able to care for asthmatic children [17], [26]. Restriction in socialization could result in increased isolation and decreased social networks and social support which might otherwise serve to ameliorate the effects of community violence. At the same time, because of the cross-sectional design and inability to demonstrate how neighborhood perceptions of safety and asthma are related, the findings from this study need replication in other settings. Further, the causal mechanism posited in the preceding discussion also remains to be tested.

The public health relevance of asthma at the national level is well recognized. While physical environmental factors, supplemented with evidence from gene-environment interaction studies, have advanced our mechanistic understanding of this complex disease, they do not fully account for the substantial variation in asthma. To our knowledge, this study is the first to show in a nationally representative sample an inverse association between perceptions of neighborhood safety and childhood lifetime asthma. Stress-induced mechanisms, partially captured through exposure to stressful psychosocial circumstances, may be a critical explanatory link in furthering our understanding of disparities in asthma.

Authors' Contributions

Conceived and designed the experiments: SVS. Analyzed the data: MK. Contributed reagents/materials/analysis tools: SVS MK. Wrote the paper: SVS MK.

References

1. Moorman JE, Rudd RA, Johnson CA, King M, Minor P, et al. (2007) National surveillance for asthma–United States, 1980–2004. MMWR Surveill Summ 56: 1–54.
2. Maddox L, Schwartz DA (2002) The pathophysiology of asthma. Annu Rev Med 53: 477–498.
3. Walker BJ, Stokes LD, Warren R (2003) Environmental factors associated with asthma. J Natl Med Assoc 95: 152–166.
4. Adler NE, Boyce T, Chesney MA, Cohen S, Folkman S, et al. (1994) Socioeconomic status and health. The challenge of the gradient. Am Psychol 49: 15–24.
5. Busse WJ, Kiecolt-Glaser J, Coe C, Martin R, Weiss S, et al. (1994) Stress and asthma: NHLBI Workshop Summary. American Journal of Respiratory Critical Care Medicine 151: 249–252.
6. Evans G (2001) Environmental stress and health. In: Baum A, Revenson T, Singer J, editors. Handbook of health psychology. Mahwah, NJ: Lawrence Erlbaum Associates Inc. pp. 365–385.
7. Wright RJ, Fisher EB (2003) Putting asthma into context: community influences on risk, behavior, and intervention. In: Kawachi I, Berkman LF, editors. Neighborhoods and health. New York: Oxford University Press.
8. Wright RJ, Rodriguez M, Cohen S (1998) Review of psychosocial stress and asthma: an integrated biopsychosocial approach. Thorax 53: 1066–1074.
9. Sandberg S, Paton JY, McCann DC, McGuiness D, Hillary CR, et al. (2000) The role of acute and chronic stress in asthma attacks in children. Lancet 356: 982–987.
10. Wright RJ, Cohen S, Carey S, Weiss ST, Gold DR (2002) Parental Stress as a Predictor of Wheezing in Infancy. A Prospective Birth-Cohort Study. American Journal of Respiratory and Critical Care Medicine 165: 358–365.
11. Chen E, Fisher EB, Bacharier LB, Strunk RC (2003) Socioeconomic status, stress, and immune markers in adolescents with asthma. Psychosom Med 65: 984–992.

12. Wright RJ, Finn P, Contreras JP, Cohen S, Wright RO, et al. (2004) Chronic caregiver stress and IgE expression, allergen-induced proliferation, and cytokine profiles in a birth cohort predisposed to atopy. J Allergy Clin Immunol 113: 1051–1057.
13. Wright RJ, Cohen RT, Cohen S (2005) The impact of stress on the development and expression of atopy. Curr Opin Allergy Clin Immunol 5: 23–29.
14. Wright RJ, Subramanian SV (2007) Advancing a multilevel framework for epidemiologic research on asthma disparities. Chest 132: 757S–769S.
15. Sandel M, Wright RJ (2006) When home is where the stress is: expanding the dimensions of housing that influence asthma morbidity. Arch Dis Child 91: 942–948.
16. Gupta RS, Zhang X, Sharp LK, Shannon JJ, Weiss KB (2008) Geographic variability in childhood asthma prevalence in Chicago. J Allergy Clin Immunol 121: 639–645 e631.
17. Wright RJ (2006) Health effects of socially toxic neighborhoods: the violence and urban asthma paradigm. Clin Chest Med 27: 413–421, v.
18. Kawachi I, Berkman LF, editors. (2003) Neighborhoods and health. New York: Oxford University Press.
19. Kawachi I, Subramanian SV (2007) Neighbourhood influences on health. J Epidemiol Community Health 61: 3–4.
20. Kawachi I, Subramanian SV (2006) Measuring and modeling the social and geographic context of trauma: a multilevel modeling approach. J Trauma Stress 19: 195–203.
21. O'Campo P (2003) Invited commentary: Advancing theory and methods for multilevel models of residential neighborhoods and health. Am J Epidemiol 157: 9–13.
22. Sampson RJ (2003) The neighborhood context of well-being. Perspect Biol Med 46: S53–64.
23. Wright RJ, Steinbach SF (2001) Violence: An unrecognized environmental exposure that may contribute to greater asthma morbidity in high risk inner-city populations. Environmental Health Perspectives 109:
24. Jeffrey J, Sternfeld I, Tager I (2006) The association between childhood asthma and community violence, Los Angeles County, 2000. Public Health Rep 121: 720–728.
25. Clougherty JE, Levy JI, Kubzansky LD, Ryan PB, Suglia SF, et al. (2007) Synergistic effects of traffic-related air pollution and exposure to violence on urban asthma etiology. Environ Health Perspect 115: 1140–1146.

26. Wright RJ, Mitchell H, Visness CM, Cohen S, Stout J, et al. (2004) Community violence and asthma morbidity: the Inner-City Asthma Study. Am J Public Health 94: 625–632.

27. Ezzati-Rice TM, Blumberg SJ, Madans JH (1999) Use of an Existing Sampling Frame to Collect Broad-based Health and Health-related Data at the State and Local Level. Federal Committee on Statistical Methodology Research Conference.

28. Blumberg SJ, Frankel MR, Osborn L, Srinath KP, Giambo P (2005) Design and operation of the National Survey of Children's Health, 2003. Vital Health Stat 1.

29. Pedlow S, Luke JV, Blumberg SJ (2007) Multiple Imputation of Missing Household Poverty Level Values from the National Survey of Children with Special Health Care Needs, 2001, and the National Survey of Children's Health, 2003. Department of Health and Human Services, Centers for Disease Control and Prevention, National Center for Health Statistics, Division of Health Interview Statistics, Survey Planning and Special Surveys Branch.

30. Mitchell M (2005) Strategically using General Purpose Statistical Packages: A Look at Stata, SAS, and SPSS.

31. UCLA: Academic Technology Services SCG (2008) Applied Survey Data Analysis in Stata 9. [cited January 20, 2008].

32. Pearce N, Beasley R, Burgess C, Crane J (1998) Asthma epidemiology: principles and methods. New York: Oxford University Press.

33. Beasley RE, Asher I (2003) International patterns of prevalence in pediatric asthma: the ISAAC program. Pediatric Clinics of North America 50: 539–553.

34. Ferris BJ (1978) Epidemiology standardization project. American Review of Respiratory Disease 118: 1–88.

35. Carver A, Timperio A, Crawford D (2008) Playing it safe: The influence of neighbourhood safety on children's physical activity-A review. Health Place 14: 217–227.

36. O'Neil R, Parke R, McDowell D (2001) Objective and subjective features of children's neighborhoods: relations to parental regulatory strategies and children's social competence. J Appl Dev Psychol 22: 135–155.

37. Keeter S, Miller C, Kohut A, Groves RM, Presser S (2000) Consequences of Reducing Nonresponse in a National Telephone Survey. The Public Opinion Quarterly 64: 125–148.

38. Harrison BD (1998) Psychosocial aspects of asthma in adults. Thorax 53: 519–525.
39. Chen E, Miller GE (2007) Stress and inflammation in exacerbations of asthma. Brain Behav Immun 21: 993–999.
40. Nadel JA, Barnes PJ (1984) Autonomic regulation of the airways. Ann Rev Med 35: 451–467.
41. Cohen S, Line S, Manuck SB, Rabin BS, Heise ER, et al. (1997) Chronic social stress, social status, and susceptibility to upper respiratory infections in non-human primates. Psychosomatic Medicine 59: 213–221.
42. Graham NMH, Douglas RB, Ryan P (1986) Stress and acute respiratory infection. Am J Epidemiol 124: 389–401.
43. Wright RJ (2007) Prenatal maternal stress and early caregiving experiences: implications for childhood asthma risk. Paediatr Perinat Epidemiol 21: Suppl 38–14.
44. Wolf JM, Miller GE, Chen E (2008) Parent psychological states predict changes in inflammatory markers in children with asthma and healthy children. Brain Behav Immun 22: 433–441.

Environmental Factors Influence Language Development in Children with Autism Spectrum Disorders

Marine Grandgeorge, Martine Hausberger, Sylvie Tordjman, Michel Deleau, Alain Lazartigues and Eric Lemonnier

ABSTRACT

Background

While it is clearly admitted that normal behavioural development is determined by the interplay of genetic and environmental influences, this is much less the case for psychiatric disorders for which more emphasis has been given in the past decades on biological determinism. Thus, previous studies have shown that Autistic Spectrum Disorders (ASD) were not affected by parental style. However, animal research suggests that different behavioural traits can be differentially affected by genetic/environmental factors.

Methodology/ Principal Findings

In the present study we hypothesized that amongst the ASD, language disorders may be more sensitive to social factors as language is a social act that develops under social influences. Using the Autism Diagnostic Interview-Revised, we compared the early characteristics of sensori-motor and language development in a large sample of children with ASD (n = 162) with parents belonging to different levels of education. The results showed that children raised by parents with a high level of education displayed earlier language development. Moreover, they showed earlier first words and phrases if their mother was at a high level of education, which reveals an additional gender effect.

Conclusions/Significance

To our knowledge this study may trigger important new lines of thought and research, help equilibrate social and purely biological perspectives regarding ASD and bring new hopes for environmentally based therapies.

Introduction

Although the nature/nurture debate may seem to belong to past history, the question of how genetic/experiential factors affect behavioural development remains very vivid [1]. Both genetic and environmental factors are involved in the determinism of aspects like temperament, but their relative weights may vary according to the trait being considered [e.g. 2]. As mentioned by Gosling [3], animal studies are very useful as they can reveal the interplay between different factors. Thus, horses with highly sensitive phenotypes [e.g. 4] may develop abnormal behaviour (such as stereotypies) as a consequence of unfavourable environmental conditions [e.g. 5] (See [6], [7] for reviews).

These animal studies provide useful framework to study normal and pathological behaviours of humans as a result of such interplay. Thus, twin studies show that parenting influences children's prosocial behaviours and acts as a "modulation" of genetic influences. This is especially true in the case of psychiatric disorders: despite a strong genetic basis [8], schizophrenia can been shown to be influenced by parenting profiles [9] as well as by factors such as an infectious disease during mid-pregnancy [10]. The weights attributed to genetic/environmental factors by authors are also often subject to variations along with "science history," especially where psychiatric disorders are concerned [11].

Thus, Autistic Spectrum Disorders (ASD) characterized by social and communication deficits and repetitive or stereotypic behaviour [12] have been for a long while attributed to environmental factors such as mothering (i.e. "refrigerator mother" [13]) or diseases (e.g. congenital rubella [14]). After reacting against the theory of lack of maternal affection during the '50s and '60s, research radically turned towards a neural and cognitive hypothesis [e.g. 15]. Since developments of genetic and neurology technologies during the '90s, more emphasis has been clearly given to biological (i.e. genetic) bases for these disorders (see [16] for review). The well-known social withdrawal of children with ASD has been attributed lately to deficits in the superior temporal sulcus voice selective regions: hearing and processing impairments based on developmental biological deficits could lead to social withdrawal [17].

Here again, animal studies suggest a much more complex situation. Thus, social experience is crucial for the development of the central auditory area in young songbirds [18], [19]. More interestingly, social segregation may induce the same deficits in a central auditory area as physical isolation and/or auditory deprivation [20]. Direct social contact with adults and the quality of interactions may strongly influence both vocal and perceptual development both in birds and humans [21], [22].

Researchers generally acknowledge that ASD are not affected by parental style but one can wonder whether as in animals [2], different behavioural traits are differently affected by genetic/environmental factors. The above mentioned results suggest that language development may be strongly affected by social factors and language abnormalities are the first observed deficit observed in more than half the families of children with ASD [23], [24].

Normally language development of children raised by parents with a high level of education is faster than that of children raised by parents with a low level of education (e.g. lexical richness [25]). In addition, parents' monitoring of language interactions with children differs according to their socioeconomic status [e.g. 26]. Moreover, mothers and fathers appear to influence children in different ways [27].

In the present study, we hypothesized that parental characteristics influenced language development in children with ASD. We compared early characteristics of language development (using Autism Diagnostic Interview-Revised, ADI-R; [28]) for a large sample of children with ASD of parents with different levels of education. These data were compared on similarly acquired items on other non language variables. Our results demonstrate for the first time that parental characteristics (i.e. level of education and gender) can influence language development of children with ASD. This finding may trigger important new lines of thought and research (on the mechanisms underlying this influence, stimulate investigations on exact links between parents' level of education and their language inputs to their children), help equilibrate social and purely biological perspectives regarding ASD and bring new hopes for environmentally based therapies.

Methods

Children

All children were recruited from the "Centre de Ressource Autisme," Brest, France (n = 162, 135 males and 27 females, mean age at assessment, in months ±SD (min–max): 98±54 (37–373); other demographic data in Table 1). They all met the criteria of the Diagnostic and Statistical Manual of Mental Disorders 4th edition [12] and International Classification of Diseases [29] for ASD. All the recruited children were French natives, lived in intact families, were physically healthy and were at least 33 months old.

Table 1. Characteristics of the participants and both their mothers and their fathers.

N	162
Gender	
Male	135 (83.3%)
Female	27 (16.7%)
Variables (Mean±SD ; min–max)	
Age at assessment (months)	98±54 (37–373)
Age at birth (weeks)	38.7±2.8 (28–42)
Height at birth (cm)	49.5±3.1 (36–57)
Weight at birth (g)	3290±675 (1220–4770)
Level of education of parents	
Low level of education	
Mother	53 (32.7%)
Father	64 (39.5%)
Mid level of education	
Mother	26 (16.1%)
Father	26 (16.1%)
High level of education	
Mother	83 (51.2%)
Father	72 (44.4%)

Parents

The level of education of each parent was scored independently (Table 1). According to the French INSEE 2003 classification, three categories were considered:

(1) low level of education (low education status or LES mother and LES father; a professional schooling or no education), (2) mid level of education (mid education status or MES mother and MES father; high school and first years at college) and (3) high level of education (high education status or HES mother and HES father; completed college and graduate school). Mothers and fathers could have different or similar levels of education.

Measures

Behavioural assessments were performed using the Autism Diagnostic Interview–Revised (ADI-R) for the children with ASD [28]. The ADI-R, an extensive, semi-structured parental interview, was conducted by trained psychiatrists and administered to the parents together. As both parents responded together, their answers were not independent and the child's score correspond to their common joined response. The ADI-R scale assessed the three major domains of autistic impairments: reciprocal social interactions, verbal and non-verbal communication, stereotyped behaviours and restricted interests. Based on direct clinical observation of each child by independent child psychiatrists, a diagnosis of ASD was made according to the DSM-IV [12] and ICD-10 [28] criteria and was confirmed by the ADI-R ratings.

Parents Were Asked Questions About their Children's Language and Sensori-Motor Development

Language Criteria Used

(a) Age of first single words (in months, first single words refer to words used repeatedly and consistently for the purpose of communication with reference to a particular concept, object or event and keep out "dad" and "mum"; children were considered as delayed when they used their first single words after 24 months old and as normal or non delayed when they used their first single words before 24 months old). (b) Age of first phrases (in months, first phrases must be consist of two words, one of which must be a verb and keep out attribute-noun combinations nor echolalic speech nor phrases that might have been learned as a single word to convey a single meaning; children were considered as delayed when they used their first phrases after 33 months old and as normal or non delayed when they used their first phrases before 33 months old). (c) Overall level of language used by the children was coded in two categories: they either possessed sufficient verbal skills (daily, functional use of three-word phrases that sometimes included a verb) or they did not (no functional use, mostly single words phrases or fewer than five words used on a daily basis). Finally (d) abnormality of development evident at or before 36 months ; each child was given a score that added (1) the age when parents first noticed something was not quite right in their child's language, relationships or behavior (if observed <36 months, score 1), (2) the age when abnormalities first became evident (if observed <36 months, score 1), (3) the interviewer's judgement on the age when developmental abnormalities probably first became manifest (if observed <36 months, score 1), (4) the age of the first single words uttered (if observed >24 months, score 1), and (5) the age of the first phrases uttered (if observed >33 months, score 1). The higher is the score, the higher is the abnormality of development evident at or before 36 months.

Sensori-Motor Criteria Used

(a) Age of sitting unaided on flat surface (in months; the age when the child first sat, without support, on a flat surface. Children were considered as delayed when they first sat after 8 months old and as normal or non delayed when they first sat 8 months old). (b) Age of walking unaided (in months; the age when the child walked without holding on. Children were considered as delayed when they walked unaided after 18 months old and as normal or non delayed when they walked unaided before 18 months old). (c) Age of bladder control acquisition during daytime (in months; the age when the child was first dry for 12 months without accidents), (d) Age of bladder control acquisition during the night (in months; the age when the child was first dry for 12 months without accidents). Finally, (e) age of bowel control acquisition (in months; the age when the child was first continent for 12 months without accidents).

All the data are confirmed by the health card of each child, a medical document filled out at each stage of the life (e.g. weight, height, age of the first walk, diseases). Verbal informed consent was given by parents and the protocol was approved by the ethics committee of Bicêtre Hospital.

Statistical Analyses

The analyses were conducted in four steps, using Minitab© software and an accepted p level of 0.05. Kruskal-Wallis tests compared ages of sitting unaided on flat surface, walking unaided, bladder control acquisition during daytime, bladder control acquisition during the night, age of bowel control acquisition, first single words, first phrases according to the three levels of education of both mothers and fathers. Post hoc pair-wise comparisons were then applied using Mann–Whitney U-tests. Chi-square tests assessed the relationships between the three levels of education of both mothers and fathers and the following qualitative variables: first single words and first phrases (non-delayed and delayed children). ANOVA test and post hoc Tukey's test assessed the relationships between the three levels of education of both mothers and fathers and the quantitative date of abnormality of development evident at or before 36 months (scale with six levels coded 0 to 5). Binary logistic regression assessed the relationships between the overall level of language and the three levels of education of both mothers and fathers, taking into account the age of children at assessment. Factors were used both in independence and in interaction (age×level of education).

Results

A clear influence of the educational levels of parents appeared on language development while no such effect was observed on sensori-motor development.

Language Development

Age of First Single Words

One hundred and forty-eight children (91.4%) had used their first single words and this had occurred on average at 26.4±15.5 months (min: 6; max: 84). Seventy-five children (46.3%) uttered their first single words before 24 months (i.e. non delayed) and 73 children (41.1%) uttered their first single words after 24 months (i.e. delayed). Fourteen children (8.6%) of the cohort had not yet pronounced their first single words when they were assessed even though they were 82.0±68.3 months old (min: 37; max: 309) (Table 2).

Table 2. Range, mean age ±SD of first single words and first phrases pronounced by children, the number and percentage of associated categories (delayed, non delayed and not achieved) according to the level of education of mothers and fathers.

	Mothers' level of education			Fathers' level of education			TOTAL
	Low level of education	Mid level of education	High level of education	Low level of education	Mid level of education	High level of education	
Age of first single words (months)							
Mean±SD	32.2±18.5	23.0±9.8	24.0±14.0	26.4±15.5	28.9±16.3	24.1±10.8	26.4±15.5
Min–Max	8–84	12–48	6–72	6–84	8–84	9–42	6–84
Non delayed (before 24 months old)	18 (11.1%)	14 (8.6%)	43 (26.6%)	75 (46.3%)	23 (14.2%)	11 (6.8%)	75 (46.3%)
Delayed (after 24 months old)	30 (18.5%)	11 (6.8%)	32 (19.8%)	73 (45.1%)	34 (21.0%)	12 (7.4%)	73 (45.1%)
Not achieved	5 (3.1%)	1 (0.6%)	8 (4.9%)	14 (8.6%)	7 (4.3%)	3 (1.8%)	14 (8.6%)
Age of first phrases (months)							
Mean±SD	45.3±14.9	44.6±27.5	34.7±14.5	39.8±18.0	41.2±15.2	44.8±23.4	39.8±18.0
Min–Max	18–72	18–120	11–77	11–120	18–72	16–120	11–120
Non delayed (before 33 months old)	8 (4.9%)	9 (5.6%)	31 (19.1%)	48 (29.6%)	15 (9.5%)	4 (2.5%)	48 (29.6%)
Delayed (after 33 months old)	33 (20.4%)	11 (6.8%)	31 (19.1%)	75 (46.3%)	34 (21.0%)	14 (8.6%)	75 (46.3%)
Not achieved	12 (7.4%)	6 (3.7%)	21 (13.0%)	39 (24.1%)	15 (9.5%)	8 (4.9%)	39 (24.1%)

Fathers' levels of education did not influence significantly age of first single words (Kruskal-Wallis test: n = 147, H = 3.09, p = 0.21; Figure 1A) but mothers' levels of education did (Kruskal-Wallis test: n = 147, H = 7.12, p = 0.03; Figure 1A). Thus, LES mothers' children pronounced their first single words later than HES mothers' children and MES mothers' children ($\bar{X}$ = 32.1±18.5 months, $\bar{X}$ = 24.0±14.0 months, $\bar{X}$ = 23.0±9.8 months respectively, nL = 48 nH = 74 U = 3428 p = 0.01, nL = 48 nM = 25 U = 1942 p = 0.05; Figure 1A) MES mothers' children and HES mothers' children did not differ significantly (nM = 25, nH = 74, U = 1289, p = 0.76; Figure 1A).

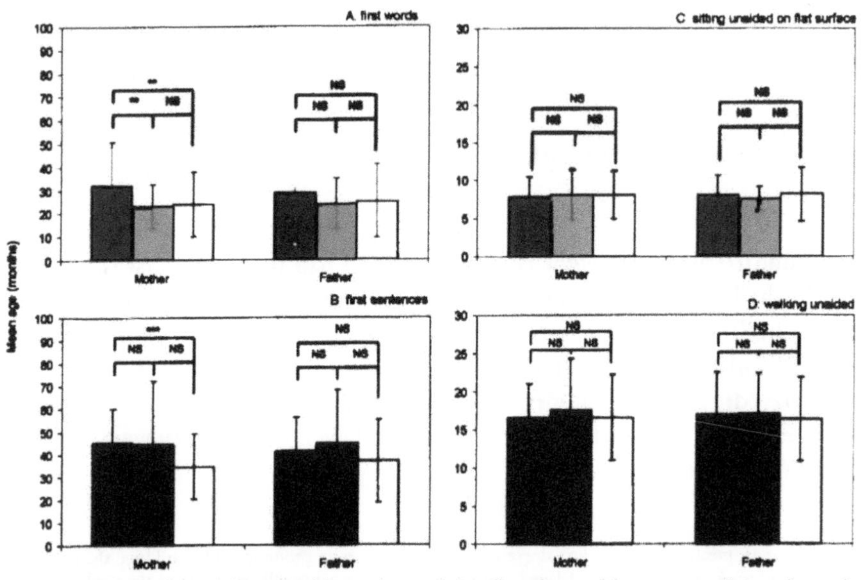

Figure 1. Mean age of the children for first single words (A), first phrases (B) sitting unaided on flat surface (C) and walking unaided (D) according to level of education of mothers and fathers. Error bars show standard deviation. Black bars represent the group of low level of education. Grey bars represent the group of mid level of education. White bars represent the group of high level of education. Level of significance: * p<0.05, ** p<0.01, *** p<0.001 (Mann Whitney U-test).

Eighty-seven children (53.7%) of our cohort appeared to be delayed. The non delayed group and the delayed group differed according to levels of education of both mothers and fathers under random distribution (all $\chi 2$ tests p<0.001). Children of the non delayed group were mostly raised by HES mothers and HES fathers, whereas LES fathers' children, MES mothers' and MES fathers' children were less represented under random distribution (all $\chi 2$ tests p<0.001; Figure 2A). Children of the delayed group were mostly raised by HES mothers,' LES mothers,' and LES fathers' whereas MES mothers' and MES fathers' children were less represented under random distribution (all $\chi 2$ tests p<0.05).

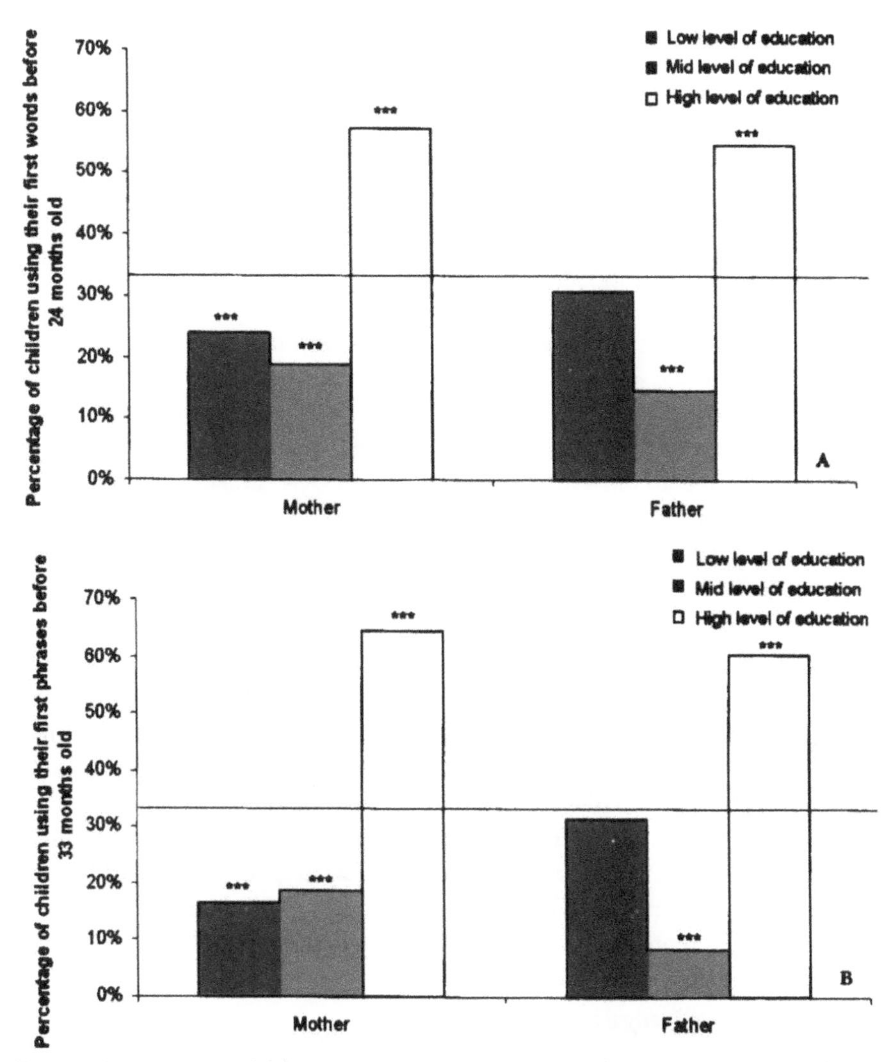

Figure 2. Mean percentages of children A: using their first single words before 24 months (non delayed group), B: using their first phrases before 33 months (non delayed group).

Black bars represent the group of low level of education. Grey bars represent the group of mid level of education. White bars represent the group of high level of education. Black line indicates the mean percentage of children in each category according to level of education of mothers or fathers under random distribution. Values below the line indicate group the less represented in the category, values above the lines indicate group the more represented in the category. Level of significance: * $p<0.05$, ** $p<0.01$, *** $p<0.001$ (Chi square tests were made on real numbers).

Age of First Phrases

One hundred and twenty-three children (75.9%) had uttered their first phrases and pronounced them on average at 39.8±18.0 months (min: 11; max: 120). Forty-eight children (29.6%) uttered their first phrases before 33 months (i.e. non delayed) and 75 children (46.3%) uttered their first phrases after 33 months (i.e. delayed). Thus, 39 children (24.1%) of the cohort had not pronounced their first phrases when they were assessed even though they were 78.2±48.2 months old (min: 37; max: 309) (Table 2).

Ages of first phrases did not differ significantly with fathers' level of education (Kruskal-Wallis test: n = 123, H = 4.01, p = 0.13; Figure 1B), but did differ significantly with mothers' level of education (Kruskal-Wallis test: n = 123, H = 12.38, p = 0.002; Figure 1B). Thus LES mothers' children uttered their first phrases significantly later than did HES mothers' children ($\bar{X}$ = 45.3±14.9 months, $\bar{X}$ = 34.7±14.5 months respectively, nL = 41, nH = 62, U = 2653, p<0.001; Figure 1B), whereas MES mothers' children were intermediate (nM = 20, nL = 41, U = 1355, p = 0.20 and nM = 20, nH = 62, U = 951, p = 0.19, respectively; Figure 1B).

One hundred and fourteen children (70.3%) of our cohort appeared to be delayed. The non delayed group and the delayed group differed significantly according to level of education of both mothers and fathers under random distribution (all $\chi 2$ tests p<0.001). Most children in the non delayed group were raised by HES mothers' and HES fathers,' while LES fathers' children, MES mothers' and MES fathers' children were less represented under random distribution (all $\chi 2$ tests p<0.001; Figure 2B). Most children in the delayed group were raised by HES mothers,' LES mothers,' and LES fathers' while MES mothers' and MES fathers' children were less represented under random distribution (all $\chi 2$ tests p<0.05).

Overall Level of Language

According to ADI-R, children could be divided into two categories of overall level of language. One hundred and four children (64.2%) appeared to have acquired sufficient verbal skills by the time they were assessed and they were then 109.0±56.9 months old (min: 37; max: 373), whereas 58 children (35.8%) used mostly single words or fewer than five words on a daily basis and were assessed when 77.0±42.5 months old (min: 37; max: 309).

A significant effect of the age was observed on the overall level of language (p = 0.01 for both mothers and fathers; Table 3) but both level of education of mothers and fathers and interaction with age did not have influence on the overall level of language (p>0.05; Table 3).

Table 3. Ordinal logistic regression of association between age of children at assessment and both mothers and fathers level of education (factors used both in independence and in interaction; with * in table).

	Global level of language			
	Mothers		Fathers	
	Odds ratio (95% CI)	p	Odds ratio (95% CI)	p
Level of education				
Low level	Reference category		Reference category	
Middle level	3.57 (0.21–61.75)	0.38	0.44 (0.02–9.77)	0.60
High level	2.12 (0.30–15.16)	0.45	3.75 (0.63–22.38)	0.15
Age	1.02 (1.00–1.04)	0.01	1.02 (1.01–1.04)	0.01
Level of education * Age				
Low level * Age	Reference category		Reference category	
Middle level * Age	0.99 (0.96–1.02)	0.44	1.01 (0.98–1.05)	0.48
High level * Age	0.99 (0.97–1.02)	0.57	0.99 (0.97–1.01)	0.22

Level of significance: $p<0.05$.
CI: confidence intervals.

Abnormality of Development Evident at or Before 36 Months

According to ADI-R, the abnormality of development evident at or before 36 months of children could be scored between 0 and 5. 149 children (91.9%), that is most children, were clearly impaired at assessment with scores of 3 and more.

The score of abnormality of development evident at or before 36 months differed according to both mothers ($F(2,159) = 3.36$, $p = 0.037$) and fathers level of education ($F(2,159) = 3.96$, $p = 0.021$). Thus LES mothers children had higher scores than HES mothers children ($\bar{X} = 4.189 \pm 0.942$, $\bar{X} = 3.651 \pm 1.338$ respectively, Tukey's test $p<0.01$) and than MES mothers children ($\bar{X} = 4.189 \pm 0.942$, $\bar{X} = 3.885 \pm 1.071$ respectively, Tukey's test $p<0.01$) whereas HES mothers children and MES mothers children did not differ. Thus LES fathers children had higher scores than HES fathers children ($\bar{X} = 4.231 \pm 1.142$, $\bar{X} = 3.583 \pm 1.254$ respectively, Tukey's test $p<0.01$) while MES fathers children did not differ ($\bar{X} = 4.031 \pm 1.083$, $\bar{X} = 4.231 \pm 1.142$, and $\bar{X} = 4.031 \pm 1.083$, $\bar{X} = 3.583 \pm 1.254$, both Tukey's test $p>0.05$).

Sensori-Motor Development

Mother's and father's levels of education did not influence significantly the age of sitting unaided on flat surface (Kruskal-Wallis test: $n = 159$, $H = 0.23$, $p = 0.89$;

$H = 0.28$, $p = 0.86$ respectively; Figure 1C), the age of walking unaided (Kruskal-Wallis test: $n = 159$, $H = 0.69$, $p = 0.71$; $H = 0.81$, $p = 0.67$ respectively; Figure 1D), the age of bladder control acquisition during daytime (Kruskal-Wallis test: $n = 114$, $H = 1.49$, $p = 0.48$; $H = 0.35$, $p = 0.84$ respectively), the age of bladder control acquisition during the night (Kruskal-Wallis test: $n = 102$, $H = 2.90$, $p = 0.23$; $H = 0.35$, $p = 0.84$ respectively) and the age of bowel control acquisition (Kruskal-Wallis test: $n = 105$, $H = 2.66$, $p = 0.26$; $H = 1.18$, $p = 0.56$ respectively). No significant difference was found between LES, MES and HES mothers' and fathers' children (all Mann Whitney tests $p>0.05$)

Discussion

Our study of early characteristics of language development in a large sample of children with ASD revealed the influence of parents' level of education and a differential influence of mothers and fathers on these characteristics. In addition, general abnormalities appeared to be influenced by parents' level of education. Thus the language of children raised by high level of education parents developed earlier and first single words and first phrases were uttered earlier by children with high level of education mothers. Although some genetic transmission of cognitive abilities cannot be totally excluded at that stage [30], these results strongly suggest the importance of environmental factors, such as parental influence, on behavioural development of children with such disorders. However these influences clearly related to language as sensori-motor stages were not affected. This study constitutes, to our knowledge, the first demonstration of such an influence.

One could argue that this evaluation of dates of first words and phrases may be biased by the retrospective aspect of the survey: parents may not be sure of when these occurred. This is certainly true but was common for all classes of parents and therefore would not explain the differences observed. Also, the general features of language outputs in the ASD children studied here agree with previous reports showing that about half such a population never acquires functional language [31] and confirming that language impairments are one of the first signs of ASD [e.g. 23]. Our large sample shared global deficits with all the other populations studied, which may reveal shared biological sources. However, as in normal children population, inter-individual variation was high and, contrary to expectations from earlier studies [32], strongly associated with parents' socioeconomic status, included level of education. Thus, an earlier review [32] showed only 4 of 12 studies aiming to relate ASD and social class supported the possibility of such a link but concluded that social class was not a risk factor. Our findings suggest indeed that other risk factors are probably important as global deficits are found in children with ASD of parents from all levels of education (deficits are related to the overall level of language but not to class). However our results show that environmental factors such as parent's level of education may influence more refined aspects such as age of first single words or first phrases. Reports show that some behavioural traits in animals may be more open to environmental influences than others and that individual variations result from the interplay between genetic and environmental influences [2].

Because environmental factors may act on very precise aspects, only detailed studies such as our present study could reveal their influence. The current predominance of genetic models for psychiatric disorders may also explain that such aspects have been overlooked [11]. Our results emphasize the importance of remaining focused on this dual influence. As Robert [11] mentioned, "there is no such thing as a genome without a system."

The fact that external factors, especially social environment, has been found to influence language characteristics is not surprising, language being "a social act" [33]. Social influences may help both humans and animals to overcome inhibitions, and to achieve exceptional learning in vocal communication processes [34]. Children need both communicative opportunities and a language model in order to develop language [35], [36]. For example, mother's and father's levels of education are significant predictors of child language [37], [38]. Recent studies suggest that children with ASD share an inherent basis with typical language learners in at least some aspects of language acquisition and that therefore delays might result more from social disinterest than from a core language disability [39]. Tager-Flusberg [40] suggested that language impairments may reflect the lack of attention of these children to their social environment. ASD children can be so unresponsive to voices and speech that they are first believed to be deaf [40].

Perceptual deficits may indeed exist as a consequence of impairments of voice processing in the STS central area [17], but social withdrawal and lack of social attention may well be involved in these central abnormalities [20]. Individual variations in language impairments may therefore reflect variations in social attention/involvement [41].

How could level of education, and more generally socioeconomic status, explain these differences? Socioeconomic status is a compound variable [35] that creates "different basic conditions of life at different levels of the social order" [42]. It involves education level of parents, their income, social network (other people encountered by children) and the individual effects of these components are not well known [43]. However socioeconomic status has a strong impact on typical language learners. High socioeconomic status mothers talk more to their children, use a more varied vocabulary, read books to their children more readily [44], [45]. According to Hoff [46] and Huttenlocker et al. [47], socioeconomic status-related differences in richness of maternal speech explain socioeconomic status differences in the development of young children's vocabulary and syntax (review in [35]). In our study, mothers' level of education appeared to have a major effect on the age of first words and phrases, showing that children with ASD, like normal children, might be sensitive to maternal inputs. Interestingly, fathers' level of education appeared also to have an effect, as being delayed or non delayed in the production of first words and phrases depended on both parents' level of education. Fathers' parenting behaviours have been shown to be predictive of young children's language development [48] and fathers' outputs have been shown to predict language scores of children [37]. Nevertheless, very few studies have investigated the influence of fathers' socioeconomics status on their language inputs to their children [37]. Our data showed that at least the level of education is probably important for children with ASD as well.

However, the processes involved in stimulating language outputs in children with ASD of high level of education parents remain unknown: could these processes include more perceptual stimulations, more triggering of social attention, enriched environments, more language inputs from family members and outside friends [41]? At this stage, the answer is unknown but, our study, which to our knowledge, demonstrates for the first time an impact of parents' level of education on language outputs of children with ASD, should trigger important new lines of thought and research. It suggests an openness of some traits to environmental conditions, and probably social influences that would reveal greater plasticity than expected in these children. The next crucial step involves understanding the processes at stake (social attention, perceptual experience, brain plasticity…?): is perception improved, selective attention developed, what aspects are crucial (book reading? focused language outputs?). The finding that more general "abnormalities of development" are also influenced suggests that environmental conditions, even though they cannot overcome the profound basic biologically-based impairments, may help improve a series of finer behavioural disturbances.

It is the first evidence that language development of children with ASD is at least in part under the influence of social factors. This study may trigger important new lines of thought and research (on the mechanisms underlying this influence; stimulate investigations on the exact links between parents' socioeconomic status and their language inputs to their children), help equilibrate social and purely biological perspectives regarding ASD, and brings new hopes for environmentally based therapies.

Acknowledgements

We are thankful to Dr. Ann Cloarec for improving the English, to families for their participation and to Fondation Adrienne et Pierre Sommer for their support.

Authors' Contributions

Conceived and designed the experiments: MG MH ST MD el. Performed the experiments: MG el. Analyzed the data: MG MH el. Contributed reagents/materials/analysis tools: MG MH al el. Wrote the paper: MG MH ST el.

References

1. Plomin (2001) The genetics of G in human and mouse. Nat Rev Neurosci 2: 136–141.
2. Hausberger M, Bruderer C, Le Scolan N, Pierre JS (2004) Interplay between environmental and genetic factors in temperament/personality traits in horses (Equus caballus). J Comp Psychol 118: 434–446.
3. Gosling SD (2001) From mice to men: what can we learn about personality from animal research? Psychol Bull 127: 45–86.
4. Luescher VA, Mc Keawn DB, Dean H (1998) A cross sectional study on compulsive behaviour in horses. Equine Vet J 27: 14–18.
5. Mc Greevy PD, French NP, Nicol FJ (1995) The prevalence of abnormal behaviours in dressage, eventing and endurance horses in relation to stabling. Vet Record 137: 36–37.
6. Houpt KA, Kusunose R (2001) Genetics of behavior. In: Bowling AT, Ruvinsky A, editors. The genetics of the horse. CABI Publishers. pp. 281–306.
7. Hausberger M, Richard MA (2005) Individual differences in the domestic horse, origins, development and stability. In: Mills D, McDonnell S, editors. The domestic horse. Cambridge: Cambridge University Press. pp. 33–52.
8. Tienari P, Wynne LC, Sorri A, Lahti I, Läksy K, Moring J, et al. (2004) Genotype–environment interaction in schizophrenia-spectrum disorder: long-term follow-up study of Finnish adoptees. Br J Psychiatry 184: 216–222.
9. Tienari P, Wynne LC, Moring J, Lahti I, Naarala M, et al. (1994) The Finnish adoptive family study of schizophrenia: implications for family research. Br J Psychiatry Suppl 16420–26.
10. Franzek E, Stöber G (1995) Maternal infectious diseases during pregnancy and obstetric complications in the etiology of distinct subtypes of schizophrenia: further evidence from maternal hospital records. Eur Psychiatry 10: 326–330.
11. Robert JS (2000) Schizophrenia epigenesis? Theor Med Bioeth 21: 191–215.
12. American Psychiatric Association (2000) Diagnostic and Statistical Manual of Mental Disorders; Volume IV-TR. Washington, DC: American Psychiatric Association.

13. Kanner L (1949) Problems of nosology and psychodynamics of early infantile autism. Am J Ortho 19: 416–426.
14. Chess S (1977) Follow-up report on autism in congenital rubella. J Autism Child Schizophr 7: 69–81.
15. Rimland B (1964) Infantile autism: the syndrome and its implications for a neural theory of behavior. New York: Appleton-Century-Crofts.
16. Muhle R, Trentacoste SV, Rapin I (2004) The genetics of autism. Pediatrics 113: 472–486.
17. Gervais H, Belin P, Boddaert N, Leboyer M, Coez A, et al. (2004) Abnormal cortical voice processing in autism. Nat Neurosci 7: 801–802.
18. Cousillas H, Richard JP, Mathelier M, Henry L, George I, et al. (2004) Experience-dependent neuronal specialization and functional organization in the central auditory area of a songbird. Eur J Neurosci 19: 3343–3352.
19. Cousillas H, George I, Mathelier M, Richard JP, Henry L, et al. (2006) Social experience influences the development of a central auditory area. Naturwissenschaften 93: 588–596.
20. Cousillas H, George I, Henry L, Richard JP, Hausberger M (2008) Linking social and vocal brains: could social segregation prevent a proper development of a central auditory area in a female songbird? Plos One 3(5): e2194. doi:10.1371/journal.pone.0002194.
21. Kuhl (2003) Human speech and birdsong: Communication and the social brain. PNAS 100: 9645–9646.
22. Goldstein MH, King AP, West MJ (2003) Social interaction shapes babbling: testing parallels between birdsong and speech. Proc Nat Acad Sci USA 100: 8030–8035.
23. De Giacomo A, Fombonne E (1998) Parental recognition of developmental abnormalities in autism. Eur Child Adolesc Psychiatr 7: 131–136.
24. De Myer MK (1979) Parents and children in autism. Washington DC: Winston and Sons.
25. Hoff E, Naigles L (2002) How children use input in acquiring a lexicon. Child Dev 73: 418–433.
26. Hoff E, Laursen B, Tardif T (2002) Socioeconomic status and parenting. In: Bornstein MH, editor. Handbook of parenting 2nd ed. Mahwah, NJ: Lawrence Erlbaum Association. pp. 231–252.
27. Hart B, Risley T (1995) Meaningful differences in the everyday experience of young. American children. Baltimore: Brookes.

28. Lord C, Rutter M, Le Couteur A (1994) Autism Diagnostic Interview-Revised—A revised version of a diagnostic interview for caregivers of individuals with possible pervasive developmental disorders. J Autism Dev Disord 24: 659–685.
29. World Health Organization (1994) The composite international diagnostic interview, Version 1.1. Geneva: Researcher's manual.
30. Pinker S (2002) The blank slate: the modern denial of human nature. New York, NY: Viking.
31. Bailey A, Phillips W, Rutter M (1996) Autism: towards an integration of clinical, genetic, neuropsychological, and neurobiological perspectives. J Child Psychol Psychiatry 37: 89–126.
32. Fombonne E (2003) Epidemiological surveys of autism and other pervasive developmental disorders. J Aut Dev Disorders 33: 365–382.
33. Locke JL, Snow C (1997) Social influences on vocal learning in human and non human primates. In: Snowdon CT, Hausberger M, editors. Social influences on vocal development. Cambridge: Cambridge University Press. pp. 274–292.
34. Snowdon CT, Hausberger M (1997) Social influences on vocal development. Cambridge: Cambridge University Press.
35. Hoff E (2006) How social contexts support and shape language development. Dev Rev 26: 55–88.
36. Sanders LD, Stevens C, Coch D, Neville HJ (2006) Selective auditory attention in 3-to-5-year-old children: An event-related potential study. Neuropsychologia 44: 2126–2138.
37. Pancsofar N, Vernon-Feagans L (2006) Mother and father language input to young children: contributions to later language development. J App Dev Psychol 27: 571–587.
38. Hoff-Ginsberg E (1991) Mother–child conversation in different social classes and communicative settings. Child Dev 62: 782–796.
39. Swensen LD, Kelley E, Fein D, Naigles LR (2007) Processes of language acquisition in children with autism: evidence from preferential looking. Child Dev 78: 542–557.
40. Tager-Flusberg H (2000) Differences between neurodevelopmental disorders and acquired lesions. Dev Sci 3: 33–34.
41. Stevens C, Fanning J, Cocha D, Sandersa L, Neville H (2008) Neural mechanisms of selective auditory attention are enhanced by computerized training:

Electrophysiological evidence from language-impaired and typically developing children. Brain Res 1250: 55–69.

42. Kohn ML (1963) Social class and parent-child relationships: An interpretation. Am J Sociol 68: 471–480.
43. Ensminger ME, Fothergill K (2003) A decade of measuring SES: what it tells us and where to go from here? In: Bornstein VA, Bradley RH, editors. Socioeconomic status, parenting, and child development. New Jersey: Lawrence Erlbaum Associates, Publishers.
44. Hoff Ginsberg E (2000) Soziale Umwelt und Sprachlernen. In: Grimm H, editor. Sprachentwicklung. Bern: Hogrefe. pp. 463–494.
45. Fletcher KL, Reese E (2005) Picture book reading with young children: A conceptual framework. Dev Rev 25: 64–103.
46. Hoff E (2003) The specificity of environmental influence: socioeconomic status affects early vocabulary development via maternal speech. Child Dev 74: 1368–1378.
47. Huttenlocher J, Vasilyeva M, Cymerman E, Levine S (2002) Language input at home and at school: relation to child syntax. Cogn Psychol 45: 337–374.
48. Tamis-LeMonda CS, Shannon JD, Cabrera NJ, Lamb ME (2004) Fathers and mothers at play with their 2- and 3-year-olds: contributions to language and cognitive development. Child Dev 75: 1806–1820.

Interactions of Socioeconomic Position with Psychosocial and Environmental Correlates of Children's Physical Activity: An Observational Study of South Australian Families

James Dollman and Nicole R. Lewis

ABSTRACT

Background

Evidence for psychosocial and environmental correlates on children's physical activity is scattered and somewhat unconvincing. Further, the moderating influences of socioeconomic position (SEP) on these influences are largely unexplored. The aim of this study was to examine the interactions of SEP, operationalised by mother education, and predictors of children's physical activity based on the Youth Physical Activity Promotion Model.

Methods

In 2005, a sample of South Australians (10–15 y) was surveyed on psychosocial and environmental correlates of physical activity using the Children's Physical Activity Correlates Questionnaire (n = 3300) and a parent survey (n = 1720). The following constructs were derived: 'is it worth it?' (perceived outcomes); 'am I able?' (perceived competency); 'reinforcing' (parental support); and 'enabling' (parent-perceived barriers). Self-reported physical activity was represented by a global score derived from the Physical Activity Questionnaire for Adolescents. Associations among physical activity and hypothesised correlates were tested among children with mothers of high (university educated) and low (left school at or before 15 y) SEP.

Results

Among high SEP children, 'is it worth it?' emerged as a significant predictor of physical activity for boys and girls. Among low SEP children, 'is it worth it?' predicted boys' physical activity, while among girls, 'reinforcing' was the only significant predictor, explaining ~35% of the total explained variance in physical activity.

Conclusion

While perceived outcomes emerged as a consistent predictor of physical activity in this sample, parental support was a powerful limiting factor among low SEP girls. Interventions among this high risk group should focus on supporting parents to provide both emotional and instrumental support for their daughters to engage in physical activity.

Background

Physical inactivity and high sedentariness have been associated with negative health outcomes for both adults [1] and children [2]. There is widespread evidence for poorer health among adults [3] and children [4-6] of low socioeconomic position (SEP), across a range of health indicators. Gradients in physical activity behaviours that parallel SEP gradients in health have been reported among adults [7] and children [8-12]. A recent study [10] identified inverse associations among SEP and screen-based leisure time in South Australian youth, while the same trend has also been reported from various European countries [11] and the United States [8]. Furthermore, there have been more marked declines in active transport between school and home, school sport, and physical education participation among low SEP compared with high SEP children in Victoria, Australia, between 1985 and 2001 [12].

There is a broad range of social, psychological and environmental factors identified as correlates of youth physical activity [13], and several theoretical models proposed to explain variance in physical activity behaviour. The Youth Physical Activity Promotion Model (YPAP) has recently been offered as an ecological framework for understanding the inter-connectedness of these influences on youth physical activity, by merging several theories into one [14]. Components of the model include psychological attributes ('predisposing factors'), social influences ('reinforcing factors'), and environmental influences ('enablers').

Despite the growing determinants literature, our understanding of influences on youth physical activity remains clouded. There are very few factors that consistently predict youth physical activity across the majority of studies, prompting the authors of the classic review of this area [13] to call for predictors to be studied in more clearly defined socio-demographic groups. The development of intervention strategies has been primarily based on cross-sectional observations of correlates of healthy behaviours in heterogeneous samples, and has accordingly followed a 'one size fits all' approach, with largely disappointing outcomes [15]. The capacity to address socioeconomic disparities in youth physical activity and associated health outcomes will depend on understanding how correlates of physical activity function differentially across SEP strata. For instance, among low SEP children it is reasonable to postulate that stronger parental support is required to overcome the barriers associated with economic disadvantage, such as access to facilities and costs of participation [15,16].

The aim of this study was to examine the interactions of SEP, operationalised by mother education, and correlates of youth physical activity based on the YPAP model [14]. Specifically, it is hypothesised that relationships between psychosocial and environmental factors and children's physical activity are moderated by maternal education.

Methods

Participants

In 2005, a sample of South Australians (10–15 y) was assessed through student (n = 3300) and parent (n = 1720) surveys. To ensure representation across the socioeconomic spectrum, a listing was obtained of all South Australian government and independent schools, along with the School Card Register (SCR; the percentage of students enrolled at the school receiving government support for low-income families). A high SCR score denotes a higher percentage of children receiving government assistance, and thereby is an inverse of SEP at the school level. Four quartile bands were identified according to SCR (0–19, 20–39, 40–59, and ≥ 60%), with the selection of schools from each band proportional to the number of schools per band. Within schools, all children in grades 5 to 10 were invited to participate. Remote and special schools were excluded from the sample due to low student enrolments and issues with self-report in these populations, respectively. Of 70 schools invited, 52 (74.3%) agreed to participate. From these participating schools, 5348 children were eligible, 3754 (70.2%) gave consent to participate and 3300 provided complete survey responses. Representation of schools in each quartile band was 26% (0–19 band), 49% (20–39 band), 19% (40–59 band) and 7% (60+ band), compared with the state-wide percentage of schools in each quartile band (19%, 46%, 24% and 10% respectively).

Measurement Instruments

Two questionnaires collected information on family demographics, child physical activity and its hypothesised predictors. The children's questionnaire was completed in the classroom setting, administered by teachers according to a standardised script. Children took home a questionnaire for parents/caregivers to complete and return to the school.

The Children's Questionnaire

Two sections comprised the Children's Questionnaire. A psychosocial questionnaire, the Children's Physical Activity Correlates (CPAC), includes 44 items that assess various psychosocial correlates of physical activity. The instrument includes 15 items from the Children's Attraction to Physical Activity scale, five items from Harter's perceived competence scale, 6 items from Rosenberg's self-esteem scale and 18 items from a parent socialisation scale [17]. In the YPAP model, Welk [14] has postulated domains derived from the CPAC: parental influence ('reinforcing'); and predisposing factors [attitudes to physical activity ('is it worth it?') and perceived competence ('am I able?')]. In the current study, internal consistency for these three scales was acceptable: 'reinforcing,' $\alpha = 0.74$; 'is it worth it?,' $\alpha = 0.84$; and 'am I able?,' $\alpha = 0.77$).

The Physical Activity Questionnaire for Adolescents (PAQ-A) asks respondents to recall the number of times they performed moderate to vigorous physical activity in the previous week, choosing from a checklist. Seven questions assess physical activity in both school- and out-of-school-hours, covering physical education, lunch, after school, evenings, and the weekend. The items of the PAQ-A are scored on a 5-point scale and a composite index (global PA) is formed as the average of these scores. The questionnaire has exhibited acceptable validity and reliability in previous studies [18-20].

The Parent Questionnaire

The parent survey assessed physical environmental factors, focusing on: risks to safety (strangers and traffic); access to facilities and play opportunities (playgrounds and other children in the neighbourhood); and transport availability. These constructs represented the 'enabling' domain of the YPAP model [14]. The questionnaire was adapted from a study by Sallis and colleagues [21], who reported test-retest reliabilities 0.68–0.89 for neighbourhood characteristics and access to facilities among US college students. Additional questions represented parents' leisure-time physical activity, based on the Transtheoretical Model of Change [22]. Individual items from the Parent Questionnaire are presented in Table 1.

Table 1. Items from the parent survey (and response options) for enabling factors and parent physical activity

Variable	Items	Response options (data code in brackets)
neighbourhood risk (mean of 3 items)	It is not safe for my child to play, walk or ride a bike near the house because of traffic/strangers (2 items)	strongly disagree (1); disagree (2); unsure (3); agree (4); strongly agree (5)
	My child is allowed to play outside of my property alone, or with friends, without adult supervision*	strongly disagree (1); disagree (2); unsure (3); agree (4); strongly agree (5)
access (mean of 3 items)	There are parks, recreation facilities and other play areas within my child's walking distance from home*	strongly disagree (1); disagree (2); unsure (3); agree (4); strongly agree (5)
	It is easy to arrange transport for my child to get to sporting activities*	strongly disagree (1); disagree (2); unsure (3); agree (4); strongly agree (5)
	There are not enough children of similar age near where we live, for my/our child to play with	strongly disagree (1); disagree (2); unsure (3); agree (4); strongly agree (5)
mother/father activity (mean of 2 items)	Which one of the following best characterises your activity habits? Physically active implies that you get at least 30 minutes of moderate physical activity on most days of the week or 20 minutes of vigorous activity on at least 3 days a week.	I am not physically active and I don't plan to start (1); I am not physically active, but I intend to start (2); I am occasionally active, but not regularly (3); I have been physically active recently, but for less than 6 months (4); I have been physically active regularly for the past 6 months (5)

Note:
* Denotes survey item was reverse coded

Parent Education

The parent questionnaire included items on mother's and father's education level, according to the following classifications: still at school (coded 1); left school at 15 years or less (2); left school after age 15 (3); left school after age 15 but still studying (4); trade/Apprenticeship (5); post-secondary certificate/diploma (6); university degree or higher (7). In this analysis, mother education was used to represent SEP, as this attribute is consistently associated with youth overweight and obesity [4], and with adolescent physical activity [23].

Statistical Analysis

Scores for each YPAP construct were derived by averaging individual item responses (coded 1 to 4). For each construct, higher scores represented higher predisposition ('is it worth it?' and 'am I able?'), parental support ('reinforcing') and environmental support ('enabling'). Descriptive statistics (sex-specific means +/- SD) for global PA and all predictors were calculated, and comparisons between boys and girls performed using ANCOVA, adjusting for age.

Stepwise multiple regression models of global PA and predictors based on the YPAP model were established, controlling for age. Because children were sampled in schools, regression modelling was conducted with robust standard errors to account for clustering of measured attributes among children in the same school. These models were tested separately in boys and girls, and included interaction terms of mother education and each of the YPAP variables. As some interaction terms were significant, regression models of global PA and predictors, controlling for age and accounting for design effects, were repeated in separate subsamples based on mother education: low, mother education ≤ 2 (did not complete secondary education); and high, mother education = 7 (completed a University degree). This stratification was confined to high and low mother education categories to maintain a clear distinction between strata; mother education levels 3 to 6 is quite a 'heterogeneous' group, including those who left school before completing the final secondary year, those still pursuing secondary qualifications, and those with post-secondary diplomas and certificates.

Statistical significance was inferred at $p \leq 0.05$. All statistical procedures were conducted using STATA [version 9] (Stata Corporation, College Station, USA, 2003).

Results

Among subjects who completed the student survey, those who did, and did not, return a completed parent survey were compared on parent education, age and sex. Not surprisingly, those who did provide parent data were younger (11.7 vs 13.3 y). There were no differences by sex or parent education.

Compared with girls, boys reported higher levels of global PA and more positive influences on physical activity, with the exceptions of mother's and father's physical activity (see Table 2). From regression models of global PA and YPAP constructs in the whole sample, 'reinforcing' and 'is it worth it?' emerged as significant predictors, among boys and girls (see Table 3). Among girls, interactions of constructs with mother education were also significant. The correlation of 'reinforcing' with global PA was weaker among girls with university-educated mothers, while the correlation of 'is it worth it?' was stronger among girls with university-educated mothers. For boys, 'am I able' and its interaction with mother education were significant predictors of global PA. In all regression models, age emerged as a significant predictor of global PA, with higher levels among younger children.

Table 2. Comparisons between males and females on global physical activity and its hypothesised predictors.

	boys	girls	*p* for comparison
Age	12.03 (1.45)	11.97 (1.41)	0.43
Global PA	3.00 (0.71)	2.72 (0.65)	<0.0001
Is it worth it?	3.29 (0.49)	3.13 (0.49)	<0.0001
Am I able?	3.04 (0.44)	2.86 (0.49)	<0.0001
Reinforcing	3.09 (0.46)	3.05 (0.46)	0.05
Enabling	3.53 (0.68)	3.40 (0.71)	<0.0001
Father PA	3.80 (1.11)	3.75 (1.19)	0.30
Mother PA	3.75 (1.24)	3.70 (1.26)	0.38

Note:
PA = physical activity

Table 3. Predictors of global physical activity in the whole sample.

Predictor	Coefficient	SE	Beta	*p*	Partial R^2
			Boys (n = 691)		
Age	-0.11	0.01	0.22	<0.0001	0.05
Is it worth it?	0.62	0.06	0.43	<0.0001	0.26
Reinforcing	0.22	0.06	0.14	<0.0001	0.02
Enabling	0.07	0.03	0.07	0.004	0.01
Mother education* enabling	-0.007	0.002	-0.07	0.01	0.01
			Model: $F(5,47) = 87.45$, $R^2 = 0.35$, $p < 0.0001$		
			Girls (n = 827)		
Age	-0.11	0.02	-0.24	<0.0001	0.05
Is it worth it?	0.26	0.10	0.19	0.01	0.01
Reinforcing	0.48	0.10	0.32	<0.0001	0.19
Mother education* is it worth it?	0.05	0.02	0.49	0.007	0.02
Mother education* reinforcing	-0.06	0.02	-0.51	0.005	0.02
			Model: $F(5,48) = 37.04$, $R^2 = 0.29$, $p < 0.0001$		

Different patterns of predictors emerged in sub-samples formed on the basis of mother education, particularly for girls (see Tables 4, 5). For children with University educated mothers, 'is it worth it?' emerged as a significant predictor, while for boys in this group, father activity was also a significant but relatively weak predictor of global PA. Among those with poorly educated mothers, 'is it worth it?' predicted boys' global PA, but among girls 'reinforcing' was the only significant predictor, explaining 35% of the total explained variance in global PA (see Table 5).

Table 4. Predictors of global physical activity among children with University educated mothers

Predictor	Coefficient	SE	z	p	Partial R^2
			Boys (n = 130)		
Age	-0.10	0.02	-0.21	<0.0001	0.04
Is it worth it?	0.80	0.14	0.57	<0.0001	0.38
Father PA	0.07	0.03	0.11	0.04	0.01
	Model: $F(3,32) = 16.76$, $R^2 = 0.43$ $p < 0.0001$				
			Girls (n = 130)		
Age	-0.15	0.03	-0.30	<0.0001	0.08
Is it worth it?	0.74	0.09	0.53	<0.0001	0.27
	Model: $F(2,30) = 34.20$, $R^2 = 0.36$, $p < 0.0001$				

Table 5. Predictors of global physical activity among children with mothers who did not complete secondary education

Predictor	Coefficient	SE	Beta	p	Partial R^2
			Boys (n = 40)		
Age	-0.17	0.06	-0.34	0.007	0.09
Is it worth it?	0.72	0.27	0.40	0.02	0.16
	Model: $F(2,22) = 8.39$, $R^2 = 0.26$, $p = 0.002$				
			Girls (n = 59)		
Age	-0.14	0.04	-0.28	0.001	0.08
Reinforcing	0.65	0.15	0.51	<0.0001	0.35
	Model: $F(2,26) = 18.40$, $R^2 = 0.44$, $p < 0.0001$				

Similar patterns of predictors emerged in models for sub-groups formed on the basis of father education; only those based on mother education are reported here.

Discussion

This study examined correlates of physical activity among children from contrasting SEP, as defined by mother education. Consistent with the published literature, the results identified higher levels of physical activity among boys compared with girls, and younger compared with older children [13]. Among low SEP boys and high SEP boys and girls, physical activity was associated with perceptions of outcomes ('is it worth it?'). These findings resonate with a national survey of USA pre-adolescents (9–13 y) that identified outcomes expectations as the only psychosocial predictor of both non-organised and organised physical [24].

The findings of the current study extend the literature by identifying a moderating effect of mother's education on the association of parental influences and girls' physical activity. Specifically, parental support ('reinforcing') was a limiting factor for physical activity among low, but not high, SEP girls. This suggests that parental support plays an important role in assisting girls to overcome barriers to physical activity opportunities that are more potent in low SEP communities [16]. Further, low SEP girls with unsupportive parents appear to be at greatest risk of low habitual physical activity. On the other hand, high SEP girls might be less dependent on supportive parents if surrounded by more abundant opportunities and fewer restrictions to be physically active. It is unclear why parental support was a limiting factor for physical activity among low SEP girls, but not low SEP boys. Higher physical activity levels among boys, consistently reported in the literature [13], may reflect a higher internal drive among boys that predisposes them to greater participation regardless of influences from their social and physical environments.

Family support is a widely reported predictor of physical activity among young females [25,26]. Notably, declines in perceived social support have been reported in girls followed between the 5th and 7th grades [27]. This mirrors the widely reported decline in girls' physical activity through adolescence [28]. Kimm and colleagues [29] reported declines in North American females' physical activity of 64% in White girls and 100% in Black girls over a ten year period. In the same study, girls reporting higher perceived family support in the 8th grade were more physically active in 12th grade, independent of their self-efficacy or perceived behavioural control, suggesting that higher family support may attenuate the age-related decline in adolescent girls' physical activity. Interventions focusing on factors that are most proximal to the targeted behaviour have a greater likelihood of success [16]. The current study underscores the critical role played by parents in helping low SEP girls maintain active lifestyles, and directs attention to family-based interventions among this high-risk group as an urgent priority. The effectiveness of this approach is yet to be confirmed [30], although some studies have been successful in increasing youth physical activity by increasing family support [31-33].

Parental support can manifest in a variety of ways, through instrumental and affective mediation [34]. Cost, safety and access to facilities have been shown to limit physical activity among low SEP youth [35]. Instrumental support of low SEP families might occur through government-funded subsidies for associated costs, such as registration fees and sports uniforms, and the provision of more comprehensive public transport options to improve access to venues. Enhancement of emotional support mechanisms for physical activity participation, especially among young females in low SEP circumstances, is uniquely challenging. In particular, strategies to elevate girls' physical activity as a priority can be seen as incongruous with Australian sporting culture, which continues to marginalise female sport across all levels of competition [36].

Given the difficulties associated with changing family attitudes and practices [30], schools have the potential to address disparities in access to physical activity opportunities [37]. Nevertheless, Australian schools are challenged to meet the demands for instruction time from all learning areas. Regular physical activity as part of the curriculum is also threatened by ageing staff profiles and lack of expertise to conduct effective physical education (PE) and sport programs [38]. In light of a recent report of fewer physical activity opportunities in low SEP Australian schools [12], support for low SEP schools to appoint PE specialists would seem warranted, while more flexible timetabling could be adopted to free time for intra- and inter-school sport.

The YPAP construct 'is it worth it?' is comprised of benefits that relate to health and socialisation outcomes as well as enjoyment. Humbert and colleagues found that Canadian youth from both high and low SEP schools expressed the overwhelming importance of enjoyment as an attractant to physical activity [35]. Strategies that maximise enjoyment among participants are likely to be characterised by self-initiated behaviours and high levels of perceived choice [39]. According to the results of the current study, it is likely that these are important elements of any successful intervention, regardless of SEP.

Strengths and limitations of this study must be acknowledged. The sampling method ensured that schools in neighbourhoods towards the extremes of SEP were included. While response rates for the parent survey substantially reduced the final sample, there were no differences in mother education between those who did, and did not, return parent surveys. A parent survey was used to collect data on aspects of parent education and environmental barriers that are likely to be difficult for children to report. Self-reported physical activity has widely accepted limitations [40]. However, the use of the previous week as the sampling period and the absence of duration as a measured variable are positive features. Confining the sampling period to the previous week, and asking respondents to report frequency of activities during specified time periods, potentially reduce the effect of associative recall bias of self-reports [40] when compared to longer monitoring frames or when specific duration of physical activities are reported.

Finally, as with all cross-sectional studies, causal relationships cannot be inferred from the observed correlations.

Conclusion

Adolescence is a period of life consistently associated with declining physical activity levels, particularly among females [29,41,42]. Low SEP children often do not have the same access to convenient facilities for physical activity, compared to those from higher income families [16], and may be less likely to receive parental support [43]. Interventions to address inequities in physical activity opportunities need to assist low SEP parents to provide both instrumental and emotional support for their children. While acknowledging the limitation of the cross-sectional study design, the current study suggests that low SEP girls may be particularly dependent on social support to be physically active, underscoring the need to more specifically understand how parents stimulate girls' attraction to physical activity.

Competing Interests

The authors declare that they have no competing interests.

Authors' Contributions

JD acquired the funding, supervised the research, participated in the design of the study, performed the statistical analysis and drafted the manuscript. NL performed the research, collected the data, participated in the research design and coordination, and helped to draft and edit the manuscript.

Acknowledgements

This study was supported by a grant from the Department of Recreation and Sport, Government of South Australia. The authors would like to thank the principals, parents and children of the schools involved in the project.

References

1. Bauman AE: Updating the evidence that physical activity is good for health: An epidemiological review 2000–2003. J Sci Med Sport 2004, 7(Suppl 1):6–19.

2. Department of Health, Physical Activity, Health Improvement and Prevention: At Least Five a Week: Evidence on the Impact of Physical Activity and its Relationship to Health. In A Report from the Chief Medical Officer. London: Department of Health; 2004.
3. Australian Institute of Health and Welfare 2004: Australia's health 2004. Canberra: AIHW.
4. Shrewsbury V, Wardle J: Socioeconomic status and adiposity in childhood: A systematic review of cross-sectional studies 1990 to 2005. Obesity 2008, 16(2):275–284.
5. Pulkki L, Keltikangas-Jarvinen L, Ravaja N, Viikari J: Child-rearing attitudes and cardiovascular risk among children: Moderating influence of parental socioeconomic status. Prev Med 2003, 36(1):55–63.
6. Hegewald MJ, Crapo RO: Socioeconomic Status and Lung Function. Chest 2007, 132(5):1608–1614.
7. Gidlow C, Johnston LH, Crone D, Ellis N, James D: A systematic review of the relationship between socio-economic position and physical activity. Health Educ Journal 2006, 65(4):338–67.
8. Gordon-Larsen P, McMurray RG, Popkin B: Determinants of adolescent physical activity and inactivity patterns. Pediatrics 2000, 105(6):83.
9. Kantomaa MT, Tammelin TH, Näyhä S, Taanila AM: Adolescents' physical activity in relation to family income and parents' education. Prev Med 2007, 44(5):410–415.
10. Dollman J, Ridley K, Magarey A, Martin M, Hemphill E: Dietary intake, physical activity and TV viewing as mediators of the association of socioeconomic status with body composition: A cross-sectional analysis of Australian youth. International Journal of Obesity 2007, 31(1):45–52.
11. Veerecken C, Todd C, Roberts C, Mulvihill C, Maes L: TV viewing behaviour and associations with food habits in different countries. Public Health Nutrition 2006, 9(2):244–250.
12. Salmon J, Timperio A, Cleland V, Venn A: Trends in children's physical activity and weight status in high and low socio-economic status areas of Melbourne, Victoria, 1985–2001. Australian and New Zealand Journal of Public Health 2005, 29(4):337–342.
13. Sallis JF, Prochaska JJ, Taylor WC: A review of correlates of physical activity of children and adolescents. Med Sci Sports Exerc 2000, 32(5):963–975.
14. Welk GJ: The youth physical activity promotion model: A conceptual bridge between theory and practice. Quest 1999, 51(1):5–23.

15. Ball K, Crawford D: Socioeconomic factors in obesity: A case of slim chance in a fat world? Asia Pac J Clin Nutr 2006, 15(Suppl):15–20.
16. Sallis JF, Zakarian JM, Hovell MF, Hofstetter CR: Ethnic, socioeconomic, and sex differences in physical activity among adolescents. J Clin Epidemiol 1996, 49(2):125–134.
17. Schaben J, Welk GJ, Joens-Matre R, Hensley L: The predictive utility of the children's physical activity correlates (CPAC) scale across multiple grade levels. Journal of Physical Activity and Health 2006, 3(1):59–69.
18. Crocker PR, Bailey DA, Faulkner RA, Kowalski KC, McGrath R: Measuring general levels of physical activity: Preliminary evidence for the Physical Activity Questionnaire for Older Children. Med Sci Sports Exerc 1997, 29(10):1344–1349.
19. Kowalski KC, Crocker PRE, Faulkner RA: Validation of the Physical Activity Questionnaire for Older Children. Pediatr Exerc Sci 1997, 9(2):174–186.
20. Kowalski KC, Crocker PRE, Kowalski NP: Convergent validity of the physical activity questionnaire for adolescents. Pediatr Exerc Sci 1997, 9(4):342–352.
21. Sallis JF, Johnson MF, Calfas KJ, Caporosa S, Nicols JF: Assessing perceived physical environmental variables that may influence physical activity. Res Q Exerc Sport 1997, 68(4):345–351.
22. Prochaska JO, Velicer WF: The Transtheoretical Model of health behavior change. American Journal of Health Promotion 1997, 12(1):38–48.
23. Ferreira I, Horst K, Wendel-Vos W, Kremers S, van Lenthe FJ, Brug J: Environmental correlates of physical activity in youth—a review and update. Obesity reviews 2006, 8(2):129–154.
24. Heitzler C, Levin Martin S, Duke J, Huhman M: Correlates of physical activity in a national sample of children aged 9—13 years. Prev Med 2006, 42(4):254–260.
25. Saunders RP, Motl RW, Dowda M, Dishman RK, Pate RR: Comparison of social variables for understanding physical activity in adolescent girls. Am J Health Behav 2004, 28(5):426–436.
26. Zakarian JM, Hovell MF, Hofstetter CR, Sallis JF, Keating KJ: Correlates of vigorous exercise in a predominantly low SES and minority high school population. Prev Med 1994, 23(3):314–321.
27. Garcia AW, Broda MA, Frenn M, Coviak C, Pender NJ, Ronis DL: Gender and developmental differences in exercise beliefs among youth and prediction of their exercise behavior. J Sch Health 1995, 65(6):213–219.

28. Dowda M, Dishman RK, Pfeiffer KA, Pate RR: Family support for physical activity in girls from 8th to 12th grade in South Carolina. Prev Med 2007, 44(2):153–159.

29. Kimm SY, Glynn NW, Kriska AM, Barton BA, Kronsberg SS, Daniels SR, Crawford PB, Sabry ZI, Liu K: Decline in physical activity in black girls and white girls during adolescence. N Engl J Med 2002, 347(10):709–715.

30. Kahn EB, Ramsey LT, Brownson RC, Heath GW, Howze EH, Powell KE, Stone EJ, Rajab MW, Corso P: The effectiveness of interventions to increase physical activity: A systematic review. Am J Prev Med 2002, 22:73–107.

31. Neumark-Sztainer D, Story M, Hannan PJ, Tharp T, Rex J: Factors associated with changes in physical activity: A cohort study of inactive adolescent girls. Arch Pediatr Adolesc Med 2003, 157(8):803–810.

32. Sääkslahti A, Numminen P, Salo P, Tuominen J, Helenius H, Välimäki I: Effects of a three-year intervention on children's physical activity from age 4 to 7. Ped Exerc Sci 2004, 16(2):167–180.

33. Jamner MS, Spruijt-Metz D, Bassin S, Cooper DM: A controlled evaluation of school-based intervention to promote physical activity among sedentary adolescent females: Project FAB. J Adolesc Health 2004, 34(4):279–289.

34. Weiss MR, Hayashi CT: All in the family: Parent-child influences in competitive youth gymnastics. Pediatric Exercise Science 1995, 7(1):36–48.

35. Humbert ML, Chad KE, Spink KS, Muhajarine N: Factors that influence physical activity participation among high- and low-SES youth. Qual Health Res 2006, 16(4):467–483.

36. Wright J: Analysing sports media texts: Developing resistant reading positions. In Critical inquiry and problem-solving in physical education. Edited by: Wright J, McDonald D, Burrows L. London: Routledge; 2004:183–196.

37. Sallis JF, McKenzie TL, Alcaraz JE, Kolody B, Faucette N, Hovell MF: The effects of a two year physical education program (SPARK) on physical activity and fitness in elementary school students. American Journal of Public Health 1997, 87(8):1328–1334.

38. Dollman J, Dodd G, Boshoff K: The relationship between curriculum time for physical education and literacy and numeracy standards in South Australian primary schools. European Physical Education Review 2006, 12(2):151–163.

39. Wilson DK, Kitzman-Ulrich H, Williams JE, Saunders R, Griffin S, Pate R, Lee Van Horn M, Evans A, Hutto B, Addy CL, Mixon G, Sisson SB: An overview of "The Active by Choice Today" (ACT) trial for increasing physical activity. Contemporary Clinical Trials 2008, 29(1):21–31.

40. Sirard JR, Pate RR: Physical activity assessment in children and adolescents. Sports Med 2001, 31(6):439–454.
41. Klasson-Heggebo L, Anderssen SA: Gender and age differences in relation to the recommendations of physical activity among Norwegian children and youth. Scand J Med Sci Sports 2003, 13(5):293–298.
42. Anderssen N, Wold B, Torsheim T: Tracking of physical activity in adolescence. Res Q Exerc Sport 2005, 76(2):119–129.
43. Vilhjalmsson R, Thorlindsson T: Factors related to physical activity: A study of adolescents. Social Science & Medicine 1998, 47(5):665–675.

Associations among Parental Feeding Styles and Children's Food Intake in Families with Limited Incomes

Sharon L. Hoerr, Sheryl O. Hughes, Jennifer O. Fisher, Theresa A. Nicklas, Yan Liu and Richard M. Shewchuk

ABSTRACT

Background

Although general parenting styles and restrictive parental feeding practices have been associated with children's weight status, few studies have examined the association between feeding styles and proximal outcomes such as children's food intake, especially in multi-ethnic families with limited incomes. The purpose of this study was to evaluate the association of parental feeding styles and young children's evening food intake in a multiethnic sample of families in Head Start.

Methods

Participants were 715 Head Start children and their parents from Texas and Alabama representing three ethnic groups: African-American (43%), Hispanic (29%), and White (28%). The Caregivers Feeding Styles Questionnaire (Hughes) was used to characterize authoritative, authoritarian (referent), indulgent or uninvolved feeding styles. Food intake in several food groups was calculated from 3 days of dietary recalls for the child for evening food intakes from 3 PM until bedtime.

Results

Compared to children of authoritarian parents, intakes of fruits, juice and vegetables were lowest among children of indulgent or uninvolved parents (1.77 ± 0.09 vs 1.45 ± 0.09 and 1.42 ± 0.11 cups) as were intakes of dairy foods (0.84 ± 0.05 vs 0.67 ± 0.05 and 0.63+0.06 cups), respectively.

Conclusion

Findings suggest that permissive parent feeding styles like indulgent or uninvolved relate negatively to children's intake of nutrient-rich foods fruit, 100% fruit juice, vegetables and dairy foods from 3 PM until bedtime.

Background

The interactive behavioral processes occurring between parents and children surrounding eating have become a recognized influence on children's eating behaviors and weight status [1-3]. Unfortunately, the research literature linking parenting behaviors to child eating and weight status has followed two separate and distinct paths resulting in some confusion in the field. One path involves a series of studies (laboratory, cross-sectional, and longitudinal) on a circumscribed set of parental feeding practices (restriction, monitoring, and pressure to eat). Parenting practices, by definition, are considered behaviors that parents use to get children to do something specific, in this case to control children's eating. In general, laboratory studies have demonstrated negative effects of high levels of restriction and pressure to eat on aspects of children's self-regulation of energy intake and satiety [4-6]. Moreover, restriction, in particular, a highly controlling feeding practice, has also been consistently associated with overweight and weight gain in children across multiple studies [7-9].

A separate research path has emerged recently in the literature associating general parenting styles with children's overweight status. General parenting style is a global and stable characteristic of parenting reflecting both the degree of demands/control on the child as well as the parental responsiveness to the child

[10]. Parenting styles are considered distinct from parenting practices, because parenting practices and the meaning of those practices for children's development are embedded in the larger parenting style [9]. In a large survey sample in which general parenting styles were measured, Rhee [11]found child overweight most prevalent in those with authoritarian parents (highly demanding, but not very responsive), a finding consistent with earlier work on high parental control of children's food intake and child self-regulation [12]. In Rhee and other studies, permissive parenting styles, involving high parental responsiveness to the child, but few demands, were also associated with increased risk for child overweight [11,13,14]. These findings on permissive parenting styles, associating low parental demandingness/control with child overweight, contrast with earlier work showing high parental control related to poor eating self-regulation and overweight in children. As such, it is unclear what specific mechanisms in general parenting styles lead to overweight status in children. Unfortunately, this set of literature on general parenting styles lacks the specificity of context (see Costanzo and Woody, 1985 for an overview of context specific parenting)[15] and leaves many possibilities for intervening influences.

More recently, the concept of feeding styles has been introduced into the literature which embeds how parents interact with children around eating within a general parenting style framework. Studies using this new conceptualization in which general parenting styles are characterized within the context of child feeding show positive associations between permissive feeding styles and children's weight status [16,17]. These findings are similar to those of Rhee [11] and others [13,14] suggesting that too little demandingness may not be adaptive in the current dietary climate. In contrast with studies showing a positive association between children's weight status and authoritarian feeding practices (see Clark, 2007 for review)[7] as well as authoritarian parenting styles [11], Hughes and colleagues found authoritarian feeding styles to be negatively related to children's weight status [17,18]. Findings from studies using this new concept of feeding styles suggest that some level of parental demandingness is probably necessary to promote optimal eating and ultimately weight outcomes in young children [3,16,19].

To date, there have been relatively few studies showing associations between parenting/feeding styles and children's food intake [9,20]. Furthermore, a noteworthy qualification of many studies involving parenting styles and feeding practices is that they have been conducted, for the most part, with middle-class White samples [9,21]. Therefore, the aim of this study was to determine if and how feeding styles of parents are related to what children eat, specifically in children from families with limited incomes and of diverse race-ethnicity. The authoritarian feeding style, high in demandingness, was the referent in this study out of interest in integrating the existing literature on authoritarian feeding practices with authoritarian parenting/feeding styles. Based on preliminary work on feeding styles and child weight [17,18], the hypothesis was that children of parents with authoritarian feeding styles would consume higher levels of most fruits, vegetables and dairy foods and lower levels of energy-dense diets as compared to children of parents who were permissive in their feeding styles, such as indulgent and uninvolved.

Methods

Sample

Parent-child dyads selected were from a study designed to investigate the facilitators and barriers to fruit and vegetable intake of parent-child dyads in Head Start families recruited from Head Start centers in Alabama and Texas. Selection criteria included being a non-pregnant primary caregiver, having a child enrolled in Head Start in his or her first year of participation and between 3-5 years of age, having an income at or below 100% of the poverty index, and self-identification of race/ethnicity as African-American (AA); Hispanic American (HA) or White (W). The primary caregiver was the person most often responsible for what the Head Start child consumed outside of preschool. Of these caregivers, 95% were female (93% mothers, 6% grandmothers, 1% other) and 5% were male. Because only a few caregivers were not parents; all are referred to as 'parents' through out the remainder of this study. Of those who agreed to be interviewed, 715 parents completed the feeding styles questionnaire and reported dietary data on their children.

Procedures

Following approval by the Institutional Review Boards from the Baylor College of Medicine and Temple University and in compliance with the Declaration of Helsinki (1996), Head Start personnel sent recruitment flyers to parents about their interest in participation. Parents and children who fit the selection criteria signed consent forms and were interviewed at the Head Start centers during fall 2004 to fall 2005. Bilingual data collectors were matched by race/ethnicity to the parents they interviewed. Demographic data were elicited such as marital status, level of education, and race-ethnicity. The interviewers measured heights and weights and collected three dietary recalls on the parent and their preschool child. A packet of questionnaires including feedings styles was sent home with the parents who returned the completed forms to the Head Start centers in sealed envelopes. Upon data completion, parents received incentives of cash and food coupons.

Anthropometric Measures

Data collectors collected weight and height measurements twice on each parent and child without shoes and dressed in light clothing using standardized protocols [22]. Weight was measured to the closest 0.1 kg on a digital platform scale accurate to 500 kg within ± 0.05 kg. (Befour Model PS-6600, Saukville, WI). Height was measured to the closest 0.1 cm using an adult height measuring board (Shorr Productions Growth Unlimited, Olney, MD). Body Mass Index was calculated (BMI = wt in kg/ht m2). Height and weight scores for the children were averaged and converted to age- and gender-specific BMI Z scores using the 2000 growth charts from the Centers of Disease Control and Prevention [23].

Feeding Styles

The self-administered Caregivers Feeding Styles Questionnaire (CFSQ) was used to assess the parental feeding style of parents during the dinner meal [17]. In this typological approach, two scores were derived for demandingness/control and responsiveness/warmth from 12 parent-centered feeding items and 7 child-centered feeding items (response scores were 1 = never to 5 = always). Median splits calculated on these two dimensions into high and low permitted categorization into one of four parenting styles-authoritative, authoritarian, indulgent, and uninvolved-as follows. Because all feeding items assessed the degree to which parents reported doing something to encourage or discourage children's eating behavior, the mean of all 19 items formed the demandingness/control score (a measure of the degree to which parents tried to get their child to eat, regardless of the type of feeding strategy used). To derive the score for responsiveness (a measure of the type of strategy that controlled for the level of demandingness), the mean of the seven child-centered items was divided by the mean of the 19 items for each parent, resulting in a measure of the degree to which the parent used child-centered versus parent-centered techniques for child-eating behaviors. Evidence of test-retest reliability, internal consistency, convergent validity, and predictive validity has been shown with a low-income sample [17].

Dietary Variables

Dietary intakes from three days-one weekend day and two non-sequential weekdays—from 3 PM until bedtime were averaged for the children. Only the dietary intakes of the children from 3 PM in the afternoon until the child's bedtime were used (referred to hereafter as evening foods), because the study focused on the dinner time feeding styles of the parents. The children attended Head Start from 7 AM until 2 PM, so on weekdays their dietary intakes were unavailable for breakfast, lunch and morning snacks. For 42 children only two days of dietary recalls were available and these were averaged. The USDA multiple pass protocol was used to collect the dietary recall data because it is considered a standardized method [24]; two-dimensional food models assisted parents with accuracy in portion size recall [25]. Each parent provided information about the foods and beverages consumed by their children within the previous 24 hours when the children were not at school. Although dietary supplement information was collected, such data were not included because the study focused on the children's food intake. For each food or beverage the mother provided time of consumption, amount ingested, the location of purchase or preparation, and identified each food and beverage occasion as a specific meal. Dietary data were collected and analyzed using Nutrient Data System for Research (version 5.0_35, 2004 developed by the Nutrition Coordinating Center, University of Minnesota, Minneapolis, MN). The NDS-R database contains 18,000 foods including ethnic foods and quantifies nutrient intakes and food group servings.

Mean intakes of foods and beverages of interest were reported as the five main food groups of fruit including 100% fruit juice, vegetables, dairy foods, grains, meats. The food group serving sizes were those in MyPyramid [26], that is 1 cup equivalents for fruit and vegetables; 300 mg calcium equivalents for dairy; 1 oz flour equivalents for grains; 1 oz meat equivalents. Energy density (kcal/g of food/beverage) was calculated three ways dividing the average daily calories consumed after 3 pm by the food weight in grams. The energy density of foods consumed was calculated for: (1) all foods and beverages; (2) all foods and energy containing beverages including milk, juices, sodas, sweet ice tea, juice drinks, etc; and (3) all foods, but not beverages [27]. To date there is no agreement on the best energy density measure to use so all are reported in this study.

Data Analysis

All statistical analyses conducted with the data were run using Statistical Analysis Software (SAS) version 9.1.3 (SAS Institute Inc, Cary, NC, 2006). The significance level was set to 0.05 for overall analyses. Means and standard deviations as well as frequency distributions of participants' characteristics were generated. Missing data were handled on a case by case basis to maximize the information derived for analysis. Dietary intake variables included 24-hr energy intake (kcal), energy densities (kcal/g), and amount of foods from selected food groups (in servings/

per day). Analysis of variance (ANOVA), with a priori specific contrasts (using parents with authoritarian feeding style as the referent) adjusting for the number of contrasts so that family-wise error rate would equal 0.05, was performed to examine differences in dietary intake data between the feeding styles. Thus the three comparisons had to exceed a P value of < 0.0167 to attain significance. To examine differences in feeding style classifications for categorical variables (e.g., race) Chi square analyses were conducted. Least square means were obtained by using the SAS procedure GLM, adjusted for the child's BMI Z-score.

Results

Most parents had feeding styles categorized as authoritarian (30.6%) or indulgent (33.3%). For AA parents the indulgent feeding style predominated and for the HA, the authoritarian feeding style did. There were small, significant differences in the age of the child and the household size by parental feeding style, but these were not clinically meaningful. The average BMI of the parents was within the obese range, ≥ 30 kg/m2. The BMI Z scores of these Head Start children were high with children of indulgent parents tending to have the highest BMI Z scores and children of authoritarian parents, the lowest (a priori comparison $P = 0.056$). The standard deviations in the BMI Z scores were large.

Compared to authoritarian feeders, those with uninvolved ($P = 0.023$) or with indulgent feeding styles ($P = 0.030$) had children who consumed evening foods with a higher energy density.

Preschool children of indulgent or uninvolved parents had lower evening intakes of fruits, 100% fruit juices and vegetables compared to those whose parents had authoritarian feeding styles (1.77 ± 0.09 vs. 1.45 ± 0.09 and 1.42 ± 0.11 cups,) as well as the lowest intakes of dairy foods (0.84 ± 0.05 vs. 0.67 ± 0.05 and 0.63 ± 0.06 cups), respectively. The average evening intakes of children of authoritative parents were between those of children with authoritarian and permissive (indulgent or uninvolved) feeding styles. Only for grains, did the children of authoritative parents have evening intakes lower than that of authoritarian parents ($p < 0.0167$).

Discussion

This study integrated and moved beyond the existing literature on restrictive feeding practices and general parenting styles, by using parental feeding styles with limited income families and by setting authoritarian feeding style as the referent. Results demonstrated that young children of parents with permissive feeding styles (indulgent or uninvolved) had the most energy-dense diets and this study was one of the few to report energy density of young children [28]. Results also showed that children of parents with permissive feeding styles had the lowest evening intakes of fruits, fruit juices, vegetables and dairy foods, as compared to the authoritarian feeding style. Such findings extend the literature on parent-child interactions and children's intake/weight status beyond studies focusing solely on restrictive feeding practices or general parenting styles among predominantly White, middle-class parents [11,12]. It was also important to extend this research to families with limited incomes and diverse race-ethnicities, due to differences in cultural practices and because their children are at a higher risk of becoming overweight compared to those of middle income white families [29]. The present study was unique in examination of the influence of feeding styles (context specific parenting styles) on children's dietary intake to determine if and how feeding styles of low-income parents were related to what children actually ate. A major advantage of the present study was that the selected measures were linked to constructs of interest thus highlighting the specific mechanism (i.e., feeding styles instead of general parenting styles) to the children's food intake that can indirectly affect weight status.

In contrast to studies showing authoritative parenting associated with positive child outcomes[30], our findings demonstrated that an authoritarian feeding style associated with better eating behaviors in low-income children. While authoritarian parenting may not support children's internalization of parental values as described in the general parenting literature, it may be more effective in the eating domain by facilitating moderation of intake in children and consumption of nutrient-rich foods. For example, while the child is young, the parent has greater control of the food environment compared to when the child is older and often eating outside the home [31].

Spurrier and colleagues recently found that parental restriction of access by children to less healthy foods was associated with better dietary patterns while coercive behaviors were associated with poorer ones [32]. Furthermore, among samples of families with limited incomes and those with different cultural backgrounds, authoritarian parenting has been shown to be effective and beneficial. For example, Baldwin, Baldwin, and Cole [33] found that, for African-American youth living in disadvantaged circumstances, authoritarian parenting behaviors were associated with better cognitive, socio-emotional, and health outcomes in children. Similarly, Lamborn, Dornbush, and Steinberg [34] found better adjustment in African-American adolescents compared to Whites when their parents used authoritarian parenting in decision making processes.

Therefore, it might be that within the eating context, parental demandingess or direction is positive especially with young children and those from some cultural and socioeconomic groups. Demandingness was defined in this study as the degree to which parents reported doing something to encourage or discourage children's eating. Therefore, high demandingness in the eating context has to do with the amount of directives used by the parenting during eating. The fact that afternoon and evening food intakes were improved when parents exhibited higher demandingness gives support to the idea that more encouragement (or discouragement) in the feeding context may facilitate better quality intake. Given the current dietary environment with fast food restaurants plentiful and energy dense foods commonly available in shopping malls and movie theatres [35], feeding styles that place demands on children to guide their eating behaviors might be adaptive. Furthermore, feeding styles may be a proxy for other aspects of how parents approach feeding such as organization and structure of the eating environment and availability of foods in the home [31,32]. Having scheduled mealtimes and serving foods that are appetizing and nutrient-rich likely permits the introduction and development of better eating habits in young children [32].

There were several limitations to the present study. Use of a convenience sample, however large, limits the generalizability of findings, and Asian families were not included in the sample. All data were parent-reported, making social desirability a potential issue, even though the questionnaires were completed voluntarily and responses were confidentially coded. Cross-sectional studies cannot address causality, so it remains unclear the degree to which the parental feeding behaviors are an influence or result of children's intake. Furthermore, longitudinal data are needed to address how this type of parental demandingness relates to the development of children's preferences and intake of specific foods over time. Whether this type of demandingness continues to be effective as children age is yet to be determined. While children's intake in the late afternoon and evening may not necessarily represent the full picture of their daily intake, this is the time of day when parents are with their children and, during the dinner meal, interactions are focused specifically on eating. Therefore, later afternoon and evening meals afford a unique opportunity for a closer examination of the influence of parent-child behavioral processes on the specific intake of the child. Intuitively it makes sense to eliminate non-caloric beverages from the calculation of energy density. It might be, however, that parents who served water and/or non-caloric beverages for evening meals and snacks were able to keep energy intakes balanced with energy needs. The significantly higher intakes of beverages by children whose parents exhibited an authoritarian feeding style might support this perspective. Finally, the relationships between feeding styles and the child's food consumption might be confounded by the parent's BMI. However, Wardle found that obese mothers were no more likely than normal weight mothers to practice restrictive feeding with their children, suggesting that "the stereotype of the obese mother who uses food in nonnutritive ways is likely a myth" [36].

Conclusion and Implications

Findings suggest that permissive feeding styles like indulgent or uninvolved were associated with lower children's intakes of nutrient-rich foods like fruit, 100% juice, vegetables and dairy in the afternoon and evening. It is time to move research in this area beyond general parenting style to context specific feeding style and beyond food restriction to the broader feeding context to understand how mealtime feeding behaviors relate to the child's food intake, nutritional and weight status. It seems that restrictive feeding studies may be too limited and the research on general parenting styles and overweight are too broad. We suggest the examination of context specific parenting food practices to know how these influence what children eat and understand how best to intervene.

Competing Interests

The authors declare that they have no competing interests.

Authors' Contributions

SLH, SOH, JOF, TAN, YL, RMS made substantial contributions to the conception, design, acquisition, analysis and interpretation of the data.

SLH, SOH, JOF have been involved in drafting the manuscript or revising it critically for important intellectual content.

SLH, SOH, JOF, TAN, YL, RMS have given final approval of the version to be published.

Acknowledgements

The authors wish to thank Michelle Feese, Project Coordinators in Birmingham, Sandra Lopez, head interviewer and recruiter for the Hispanic participants, Frank Franklin the PI in Birmingham, AL. All were instrumental in the collection of data. We also extend a special thanks to the children and parents of Head Start who participated in the study. This work is a publication of the United States department of Agriculture (USDA\ARS) Children's Nutrition Research Center, Department of Pediatrics, Baylor College of Medicine in Houston, Texas. The contents do not necessarily reflect the views or policies of the USDA.

Funding Disclosure

This research was supported by funds from the National Cancer Institute (Grant R01 CA102671.) The first author was supported in part by the Michigan Agricultural Experiment Station.

References

1. Larson NI, Neumark-Sztainer D, Hannan PJ, Story M: Family meals during adolescence are associated with higher diet quality and healthful meal patterns during young adulthood. J Am Diet Assoc 2007, 107:1502–1510.
2. Orrell-Valente JK, Hill LG, Brechwald WA, Dodge KA, Pettit GS, Bates JE: "Just three more bites": an observational analysis of parents' socialization of children's eating at mealtime. Appetite 2007, 48:37–45.
3. Wardle J, Carnell S, Cooke L: Parental control over feeding and children's fruit and vegetable intake: how are they related? J Am Diet Assoc 2005, 105:227–232.
4. Fisher JO, Birch LL: Restricting access to palatable foods affects children's behavioral response, food selection, and intake. Am J Clin Nutr 1999, 69:1264–1272.
5. Fisher JO, Birch LL: Restricting access to foods and children's eating. Appetite 1999, 32:405–419.
6. Fisher JO, Birch LL: Eating in the absence of hunger and overweight in girls from 5 to 7 y of age. Am J Clin Nutr 2002, 76:226–231.
7. Clark HR, Goyder E, Bissell P, Blank L, Peters J: How do parents' child-feeding behaviours influence child weight? Implications for childhood obesity policy. J Public Health (Oxf) 2007, 29:132–141.
8. Faith MS, Berkowitz RI, Stallings VA, Kerns J, Storey M, Stunkard AJ: Parental feeding attitudes and styles and child body mass index: prospective analysis of a gene-environment interaction. Pediatrics 2004, 114:e429–436.
9. Ventura AK, Birch LL: Does parenting affect children's eating and weight status? Int J Behav Nutr Phys Act 2008, 5:15.
10. Baumrind D: Rearing Competent Children. In Child Development Today and Tomorrow. Edited by: Damon W. San Francisco, CA: Jossey-Bass; 1989.
11. Rhee KE, Lumeng JC, Appugliese DP, Kaciroti N, Bradley RH: Parenting styles and overweight status in first grade. Pediatrics 2006, 117:2047–2054.

12. Birch LL, Fisher JO, Davison KK: Learning to overeat: maternal use of restrictive feeding practices promotes girls' eating in the absence of hunger. Am J Clin Nutr 2003, 78:215–220.

13. Chen JL, Kennedy C: Factors associated with obesity in Chinese-American children. Pediatr Nurs 2005, 31:110–115.

14. Wake M, Nicholson JM, Hardy P, Smith K: Preschooler obesity and parenting styles of mothers and fathers: Australian national population study. Pediatrics 2007, 120:e1520–1527.

15. Costanzo PR, Woody EZ: Domain-specific parenting styles and their impact on the child's development of particular deviance: the example of obesity proneness. J Soc and Clin Psychol 1985, 4:425–445.

16. Hughes SO, O'Conner TM, Power TG: Parenting and children's eating patterns: Examining control in a broader context. Int J Child Adoles Health 2008, 1:323–330.

17. Hughes SO, Power TG, Orlet Fisher J, Mueller S, Nicklas TA: Revisiting a neglected construct: parenting styles in a child-feeding context. Appetite 2005, 44:83–92.

18. Hughes SO, Shewchuk RM, Baskin ML, Nicklas TA, Qu H: Indulgent feeding style and children's weight status in preschool. J Dev Behav Pediatr 2008, 29:403–410.

19. Vereecken CA, Keukelier E, Maes L: Influence of mother's educational level on food parenting practices and food habits of young children. Appetite 2004, 43:93–103.

20. Kremers SP, Brug J, de Vries H, Engels RC: Parenting style and adolescent fruit consumption. Appetite 2003, 41:43–50.

21. Faith MS, Scanlon KS, Birch LL, Francis LA, Sherry B: Parent-child feeding strategies and their relationships to child eating and weight status. Obes Res 2004, 12:1711–1722.

22. Lohman TG, Roche AF, Martorell M: Anthropometric standardization reference manual. Champayne, IL: Human Kinetics; 1988.

23. Kuczmarski RJ, Ogden CL, Guo SS, Grummer-Strawn LM, Flegal KM, Mei Z, Wei R, Curtin LR, Roche AF, Johnson CL: 2000 CDC Growth Charts for the United States: methods and development. Vital Health Stat 11 2002, (246):1–190.

24. Conway JM, Ingwersen LA, Vinyard BT, Moshfegh AJ: Effectiveness of the US Department of Agriculture 5-step multiple-pass method in assessing food intake in obese and nonobese women. Am J Clin Nutr 2003, 77:1171–1178.

25. Posner BM, Smigelski C, Duggal A, Morgan JL, Cobb J, Cupples LA: Validation of two-dimensional models for estimation of portion size in nutrition research. J Am Diet Assoc 1992, 92:738–741.
26. MyPyramid for professionals [http://www.mypyramid.gov/professionals/]
27. Kant AK, Graubard BI: Energy density of diets reported by American adults: association with food group intake, nutrient intake, and body weight. Int J Obes (Lond) 2005, 29:950–956.
28. Mendoza JA, Drewnowski A, Cheadle A, Christakis DA: Dietary energy density is associated with selected predictors of obesity in U.S. Children. J Nutr 2006, 136:1318–1322.
29. Ogden CL, Carroll MD, Curtin LR, McDowell MA, Tabak CJ, Flegal KM: Prevalence of overweight and obesity in the United States, 1999-2004. JAMA 2006, 295:1549–1555.
30. Darling N, Steinberg L: Parenting style as context: An integrative model. Psychological Bulletin 1993, 113:487–496.
31. Utter J, Scragg R, Schaaf D, Mhurchu CN: Relationships between frequency of family meals, BMI and nutritional aspects of the home food environment among New Zealand adolescents. Int J Behav Nutr Phys Act 2008, 5:50.
32. Spurrier NJ, Magarey AA, Golley R, Curnow F, Sawyer MG: Relationships between the home environment and physical activity and dietary patterns of preschool children: a cross-sectional study. Int J Behav Nutr Phys Act 2008, 5:31.
33. Baldwin AL, Baldwin C, Cole RE: Stress resistant families and stress resistant children. In Risk and protective factors in the development of psychopathology. Edited by: Rolf JE, Masten AS, Ciccheti D. New York: Cambridge University Press; 1990:257–280.
34. Lamborn SD, Dornbosch SM, Steinberg L: Ethnicity and community context as moderators of the relations between family decision making and adolescent adjustment. Child Dev 1996, 67:283–301.
35. Harnack LJ, French SA: Effect of point-of-purchase calorie labeling on restaurant and cafeteria food choices: A review of the literature. Int J Behav Nutr Phys Act 2008, 5:51.
36. Wardle J, Sanderson S, Guthrie CA, Rapoport L, Plomin R: Parental feeding style and the inter-generational transmission of obesity risk. Obes Res 2002, 10:453–462.

A Pilot Study Evaluating a Support Programme for Parents of Young People with Suicidal Behaviour

Lorna Power, Sophia Morgan, Sinead Byrne, Carole Boylan, Andreé Carthy, Sinead Crowley, Carol Fitzpatrick and Suzanne Guerin

ABSTRACT

Background

Deliberate self harm (DSH) is a major public health concern and has increased among young people in Ireland. While DSH is undoubtedly the result of interacting factors, studies have identified an association between DSH and family dysfunction as well as the protective role of positive family relationships. Following a focus group meeting held to identify the needs of parents and carers of young people with DSH, a support programme (SPACE) was developed. The aims of the current study are to evaluate the effectiveness of the SPACE programme in decreasing parental psychological distress, reducing parental report of young peoples' difficulties, increasing parental satisfaction and increasing parents' ratings of their own defined challenges and goals.

Methods

Participants were recruited from a Mental Health Service within a paediatric hospital, Community Child and Adolescent Mental Health Teams and family support services. All services were located within the greater Dublin area in Ireland. Forty-six parents of children who had engaged in or expressed thoughts of self harm attended the programme and participated in the evaluation study. The programme ran once a week over an 8-week period and included topics such as information on self harm in young people, parenting adolescents, communication and parental self-care. Seventy percent (N = 32) of the original sample at Time 1 completed measures at Time 2 (directly following the programme) and 37% (N = 17) of the original sample at Time 1 completed them at Time 3 (6 months following the programme).

A repeated measures design was used to identify changes in parental wellbeing after attendance at the programme as well as changes in parental reports of their children's difficulties.

Results

Participants had lower levels of psychological distress, increased parental satisfaction, lower ratings of their own defined challenges and higher ratings of their goals directly after the programme. These changes were maintained at 6-month follow up in the 37% of participants who could be followed up. Furthermore the young people who had engaged in or expressed thoughts of self harm had lower levels of difficulties, as reported by their parents, following the programme.

Conclusion

These findings suggest that the SPACE programme is a promising development in supporting the parents of young people with suicidal behaviour. The programme may also reduce parental reports of their children's difficulties. Further evaluation using a randomized controlled trial is indicated.

Background

Deliberate Self Harm (DSH) is a major public health concern [1] which has become increasingly more common among young people. The term DSH describes "an act with a nonfatal outcome in which an individual deliberately did one or more of the following: initiated behaviour (for example, self-cutting, jumping from a height), which they intended to cause self-harm; ingested a substance in excess of the prescribed or generally recognised therapeutic dose; ingested a recreational or illicit drug that was an act that the person regarded as self-harm; ingested a non-ingestible substance or object" [2].

In a recent study of 4,583 adolescents in Ireland, a lifetime history of DSH was reported by 9.1% of respondents [1]. This is in contrast to a large self report survey of schools in England where a lifetime history of DSH was reported by 13.2% of respondents [2]. The Irish study was consistent with previous findings in that DSH was more common among females (13.9%) than males (4.3%). The most common methods used were cutting (66%) and taking an overdose (35.2%). Of those who had harmed themselves, only a minority (11.3%) had attended hospital afterwards. This is in line with the self-report survey in England where it was found that only 12.6% of episodes of self harm resulted in hospital presentation [2].

DSH is a significant risk factor for suicide in that individuals who take overdoses or deliberately inflict injury on themselves and survive are at a particularly high risk of eventually dying by suicide [3]. In addition there is an association between DSH in young people and a range of poor psychosocial outcomes as adults. A recent study [4] details the early adult outcomes of 132 adolescents who had deliberately self-poisoned. Participants were compared with a matched control group who had never harmed themselves and were randomly selected from the waiting lists of primary care physicians. Results indicated that rates of psychopathology, in particular depression, were higher among those who had self-poisoned—rates of current mental disorders were 16% in the control group and 39% in the self-poisoning group. Furthermore the self-poisoning group also differed to controls on a number of other measures of social functioning and adversity. The self-poisoning group were more likely than controls to have experienced sexual abuse, disrupted education, left school early without qualifications and to have left home, cohabited and become parents at an earlier age. Considering these associations, it is imperative to develop effective means of identifying and managing self harm in young people.

An episode of DSH is likely to have a negative effect on the families of the individual involved. In a qualitative study aimed at investigating parents' experiences after an episode of DSH in young people, parents reported feeling very distressed, helpless and anxious about the possibility of future episodes [5]. These concerns were perpetuated by a perceived lack of support and information from some health professionals. This suggests that parental support and education should be an integral part of aftercare.

In a similar study, aimed at investigating parents' emotional and behavioural responses to adolescents' suicide attempts [6], 22 mothers and 12 fathers were assessed soon after the event using both open-ended and structured interviews. Mothers' reactions included an increase in sad, caring and anxious feelings with approximately half feeling hostile after the suicide attempt. However, few verbalized this hostility and many reported being careful about what they said following the suicide attempt. The authors suggest that intervention with parents should focus on normalising their feelings and responses as well as developing the family's communication skills with a focus on increasing positive feedback and reducing hostile or critical statements.

While DSH is undoubtedly the result of multiple interacting factors with no one causal factor [7], there has been much research to support an association between DSH and poor family functioning. In a study of 20 individuals who had engaged in DSH but who had no further episodes in the two years prior to the interview, participants recalled unpredictability in family life at the time of self harming [8]. They also reported having felt unsupported, not heard and that their story was of no importance to their family. Poor communication in particular has been found to be associated with DSH. In a quantitative study comparing 52 adolescents who had presented to Accident and Emergency departments following DSH with 52 hospital-based controls whom had been admitted to the hospital and had no psychiatric history or self harm, there was a strong association between the absence of a family confidant and adolescent self harm [9]. The authors suggested that poor communication within the family may lead the young person to feel socially isolated, their problems to appear insurmountable with DSH being perceived as their only option. Likewise, positive parental behaviours can serve to protect young people from DSH. In a study of 451 families, who were participating in a longitudinal research project examining rural families in the United States, the family processes that may lead to adolescent suicidality were investigated [10]. Structural equation modelling was used to examine the hypothesis that parents' behaviour would predict their adolescents' emotional distress and subsequent suicidal behaviour. Findings indicated that warm and communicative behaviours conveyed by mothers had a direct negative association with adolescents' reporting of suicidality. These findings again emphasize the importance of parental involvement in treating DSH and suggest that parents may benefit from support which would include learning how to develop and foster effective communication skills with their adolescent.

Design of SPACE Programme

Temple Street Children's University Hospital is a tertiary referral teaching paediatric hospital in the centre of Dublin, Ireland. Due to a marked increase in the number of children and adolescents presenting following DSH, a consultant-led DSH team was established in 2002. Between 2002 and October 2008, the team assessed 458 young people aged 16 years and under, following an episode of DSH. During this time, the team identified a strong need for a programme to support parents and carers of young people who have engaged in DSH.

The SPACE (Supporting Parents and Carers) programme was designed as a support programme for parents and carers of children who have engaged in DSH. According to the Report of the Expert Group on Mental Health Policy in Ireland—A Vision for Change [11], service users should become involved in every aspect of service development and delivery. Considering this, as well as international endorsement of service user involvement, a Focus Group Meeting was held in order to directly establish the needs of parents and carers of young people who have engaged in DSH [12]. Twenty-five participants attended the meeting of whom 15 were parents and ten were carers. Participants were divided into subgroups and presented with two open ended questions a) What areas do you think a support group should address? b) What would you like to gain and/or learn from participating in such a support group? Using conceptual analysis, one central theme which emerged was a strong need for support. Parents felt that there was a lack of available support and that the opportunity to avail of peer support would be extremely beneficial. Another theme which emerged was information and education—parents were interested in learning more about young people's mental health as well as DSH statistics, aetiology and treatments. Other themes which surfaced included re-establishing family communication and boundaries, dealing with adolescent discipline issues, and handling further threats or incidents of self harm.

Using information from the Focus Group Meeting, the SPACE (Supporting Parents and Carers) programme was developed. Its content is guided by the needs of the parents and carers and reflects the themes which emerged. It is a group programme which runs over an eight-week time period. Each session involves 10–12 parents meeting with two facilitators, who are experienced mental health professional from the DSH team, for 90 minutes per week. The approach is psycho-educational. Psycho-educational programs are defined as time-limited, closed groups, conducted by health professionals, for the purpose of educating and providing support to its lay membership [13]. Using presentations, video footage modelling effective communication with adolescents, group discussion and exercises the parents are provided with information regarding DSH, support, and the opportunity for communication skills development. The topics for each session are drawn from themes which emerged at the Focus Group Meeting and include the following:

- Information on self harm in young people
- Depression in young people
- Medication—information about medication used in treating depressive disorders in young people
- Parenting adolescents including positive communication, setting boundaries and dealing with emotional and behavioural difficulties that can arise.
- Help with re-establishing family relationships and boundaries after an incident of self harm.
- Advice on how to handle threats or further incidents of self harm
- Self-care—looking after one-self as a parent by achieving balance in life. The importance of parents taking time off to rest and renew themselves.
- Information on resources within the community such as help-lines, counselling services, family support services.

A typical session involves an introduction, feedback from the previous week, introduction of a new topic, small group discussion, feedback to the larger group and a 'thought for the week,' where parents are encouraged to consider how the content of the week's programme might be helpful in their interaction with their child.

Aims of the SPACE Programme

The aims of the current study are to evaluate the effectiveness of the SPACE programme in decreasing parental psychological distress as measured by the General Health Questionnaire-12; reducing young peoples' difficulties as measured by the Parental Version of the Strengths and Difficulties Questionnaire; increasing parental satisfaction as measured by the Kansas Parenting Satisfaction Scale and increasing parents' ratings of their own defined challenges and goals as measured by Challenges and Goals Scales.

Methods

Study Design

This study used a repeated-measures design to identify significant changes in well-being after treatment. The main independent variable was time, with assessment occurring before (Time 1) and after (Time 2) the SPACE programme. In addition, participants were assessed at a 6-month follow-up session (Time 3). The dependent variables include measures of parent psychological distress, parental satisfaction with their role as parents, parents' ratings of child's difficulties, ratings of challenges and goal achievement.

The study was approved by the Ethics Committee of the Children's University Hospital, Temple Street.

Participants

Participants were 46 parents of children and adolescents (aged 16 years and under) who were attending mental health services having engaged in or expressed thoughts of DSH. Fourteen child care staff looking after children in residential centres also participated in the study, but as they differed significantly from the parents on several baseline measures, it was concluded that they were two distinct groups, and they have been omitted from the remainder of the study. The resulting sample therefore consisted of 46 parents, of whom 31 (67%) were mothers and 15 (33%) were fathers. They were attending services in relation to 32 young people.

With regard to the number of participants at each time-point—46 participants completed measures at Time 1. 70% (N = 32) of that original sample completed the measures at Time 2. These numbers decreased again at Time 3—37% (N = 17) of the original sample at Time 1 completed measures at Time 3.

Parents were recruited from Temple Street Children's University Hospital, 24 Child and Adolescent Mental Health Teams (CAMHS) and 10 family support services through-out Dublin. With regard to Temple Street Hospital, letters were sent to all parents/carers of young people who had attended the Accident & Emergency department over a four year period (2004–2007) with self harm or suicidal behaviour informing them about the programme and inviting them to attend. Residential homes, CAMHS and family support services were also informed by letter of the programme and invited to refer parents and carers. Once a parent or carer was referred, they were contacted by a researcher and invited to attend the SPACE programme. Over the course of the study 64 parents were referred, of whom 46 (72%) subsequently attended the programme.

Young Person Characteristics

Of the 32 young people whose parents attended SPACE, 8 (25%) were male and 24 (75%) were female, with a mean age of 13.71 years. When referrals were made to the SPACE programme, referrers were asked to detail the type of self harm that the young person engaged in as well as whether the young person had engaged in previous episodes of self harm. When participants were recruited from the hospital, this information was obtained from the DSH database. With regard to method of DSH, overdose was the most common (50%, N = 16) followed by cutting (26.7%, N = 8). 13.3% (N = 4) presented with other types such as attempted hanging or self biting and 10% (N = 3) presented with suicidal ideation only. In one instance the type of self harm was not specified. 20 (62.5%) of the young people had a history of repeated self harm.

The SPACE programme ran over four cycles. Ten participants were from the first cycle, 13 from the second, 11 from the third, and 12 from the fourth.

Measures

Parents completed the following measures about themselves:

- The General Health Questionnaire (GHQ 12) is a widely used self report screening tool used for the assessment of mental well-being [14]. It is a measure of common mental health problems/domains of depression, anxiety, somatic symptoms, and social withdrawal. It has well established reliability and validity and has been shown to have internal consistency reliability coefficients of 0.82 to 0.86 in most studies [14,15].
- The Kansas Parenting Satisfaction Scale (KPS) is a 3 item self report measure designed to assess parent-satisfaction with themselves as a parent, satisfaction with the behaviour of their children and satisfaction with their relationship with their children. The scale is reported to have good concurrent validity—significant correlations have been found with the Kansas Marital Satisfaction Scale and the Rosenberg Self Esteem Scale (0.23 to 0.55) [16].
- In addition, parents also completed Challenges and Goals Scales [17] which required them to identify and rate their challenges and goals at the present time.
- Parents completed the following measure about their children.
- The Strengths and Difficulties Questionnaire (SDQ) [18] is a brief behavioural screening questionnaire for 3 to 16 year olds. It consists of 5 subscales—emotional symptoms, conduct problems, hyperactivity, peer relationship problems and pro-social behaviour. All subscales except the pro-social behaviour subscale are added together to generate a total difficulties score. The SDQ subscales have a mean internal consistency reliability co-efficient of 0.71, mean test retest reliability co-efficient over 6 months of 0.62 and demonstrates good criterion validity for predicting psychological disorders [19].

Procedures

At the first session of the programme, parents provided informed consent and completed the survey instruments. At the last session the same measures were administered again. Six months later parents were contacted and asked to attend a booster session, at the start of which they completed the measures for the final time.

Planned Analyses

In order to identify any significant change in the dependent variables over time, a series of one-way repeated Analysis of Variance tests (ANOVA) were used and alpha was set at .05. Where significant differences were found paired sample t-tests were used to further examine where the differences were. Effect sizes were calculated using Cohen's d. The sample size varied slightly across the analyses due to instances of missing data. In order to account for this, intention to treat analysis was also conducted. Using this analysis it was assumed that no change occurred across time for participants who did not complete measures at one or all data collection points.

Results

As is common in longitudinal research, this study was the subject to the effects of attrition. While there were 46 participants at Time 1, this reduced to 32 at Time 2 and 17 at Time 3. In order to evaluate the effects of attrition, a series of independent samples t-tests were conducted to compare scores on each measure between those who completed measures at Time 1 only with those who completed measures at Time 1 and Time 2 as well as those who completed measures at all three data collection points. There was no significant differences between these groups on any measures apart from Rating of Challenges where there was a significant difference ($t\ (58) = 2.58$, $p = .012$) between those who completed Time 1 and Time 2 ($M = 12.97$) and those who completed Time 1 only ($M = 9.94$). Parents who completed Time 1 and Time 2 had higher ratings of their challenges than those who completed Time 1 only.

General Health Questionnaire

Before the SPACE programme, mean parental scores on the General Health Questionnaire 12 (GHQ-12) ($M = 6.33$) were in the caseness range for psychological distress (a score of 3 or above). This had fallen significantly by the end of the programme (M = 3.16) with a further fall to within the normal range (M = 0.88) by Time 3 (Table 1).

Table 1. Parental Scores on the General Health Questionnaire 12 before SPACE, after SPACE and at 6-month follow up

Time	Mean	SD	N
1	6.33	4.1	45
2	3.16	3.9	32
3	0.88	2.47	17

A one-way repeated measures ANOVA was used to identify any significant difference over time. A significant difference was found (F(2,12) = 34.8, p = .000). Intention to treat analysis was also conducted whereby missing data was replaced with the means for Time1 and a significant difference was found (F(2, 44) = 13.92, p = .000).

A series of paired sample t-tests were then conducted in order to identify where the differences were. Given that multiple t-tests were used, Bonferroni correction for multiple comparisons was made whereby the alpha level was divided by the number of comparisons. Using this more conservative alpha level of 0.017, the means for Time 2 were significantly lower than the means for Time 1 with Cohen's d of 1.39 indicating a large effect size [20]. The 95% confidence interval on the difference between means was 1.89—7.68. Furthermore the means for Time 3 were significantly lower than the means for Time 1(95% CI: 4.99—8.87) with Cohen's d = 2.79 indicating a very large effect.

The relationship between child and parent factors was investigated using a series of Mann Whitney U Tests. When referrals were made to the SPACE programme, referrers were asked to document whether the child had a previous history of repeated self harm or whether this was a first instance. Cases which had presented to Temple Street hospital only were checked on the DSH database which also documents whether there have been previous episodes of DSH or not. There was a significant association between parental psychological distress at baseline (as measured by the GHQ-12) and previous history of repeated self harm in the child (Z = -2.23, n = 32, p = .026) with high levels of psychological distress associated with previous episodes of DSH.

Strengths and Difficulties Questionnaire

The Strengths and Difficulties Questionnaire (SDQ) was completed by each parent about their child. With regard to child's Total Difficulties, the mean score was in the abnormal range (17–40) at the beginning of the programme and had decreased into the borderline range (14–16) at the 6 month follow-up session. For each subscale of the SDQ (pro-social behaviour, hyperactivity, conduct problems, emotional difficulties, peer problems), as well as the Total Difficulties score, means and standard deviations for the overall group were calculated and are given in Table 2.

Table 2. Parental scores on the Strengths and Difficulties Questionnaire before SPACE, after SPACE and at 6 month follow up.

	Time 1			Time 2			Time 3		
	Mean	*SD*	*N*	*Mean*	*SD*	*N*	*Mean*	*SD*	*N*
Total Difficulties	20.15	5.19	46	17.41	6.31	32	15.06	8.36	17
Pro-Social Behaviour	6.09	2.3	46	6.38	2.67	32	6.2	3.22	17
Hyperactivity	5.8	2.19	46	5.44	2.42	32	4.71	2.73	17
Conduct Problems	4	2.07	46	3.69	2.15	32	2.76	2.44	17
Emotional Difficulties	6.93	1.89	46	5.63	2.86	32	4.88	2.96	17
Peer Problems	3.41	2.15	46	2.66	2.35	32	2.53	2.24	17

The initial analysis focused on change over time and a series of one-way repeated measures ANOVAs were used to identify any significant differences. A significant difference was found for the Total Difficulties Scale ($F(2,12) = 11.827$, $p = 0.001$). Using paired samples t-tests, it was found that the means for Time 2 and Time 3 were significantly lower than the means for Time 1. This difference remained significant using Intention to Treat analysis.

A significant difference was also found for the Hyperactivity subscale ($F(2,12) = 4.289$ $p = 0.039$) and the Emotional Difficulties subscale ($F(2, 12) = 10.264$, $p = 0.003$.

Kansas Parenting Satisfaction Scale

Parental satisfaction increased across the three time periods. Means and standard deviations for the overall group on the Kansas Parenting Scale (KPS) were calculated and are presented in Table 3.

Table 3. Parental scores on the Kansas Parenting Satisfaction Scale before SPACE, after SPACE and at 6 month follow up.

Time	Mean	SD	N
1	12.13	3.21	45
2	13.48	3.48	31
3	15	3.48	17

A one way repeated ANOVA was used to identify any significant difference over time. A significant difference was found ($F(2,12) = 30.01$, $p = .000$). The means for Time 3 were significantly higher than the means for Time 1 and Time 2. Using Intention to Treat analysis, these statistically significant results remained.

Challenges & Goals Scale

Parents' ratings of their challenges decreased across the three time periods. Means and standard deviations for the overall group on the Challenges and Goals scale were calculated and are displayed in Table 4.

Table 4. Parental ratings of Challenges & Goals before SPACE, after SPACE and 6 months following SPACE.

	Time 1			Time 2			Time 3		
	Mean	*SD*	*N*	*Mean*	*SD*	*N*	*Mean*	*SD*	*N*
Challenges	12.76	3.84	46	8.88	3.95	24	6.65	3.82	13
Goals	6.14	2.93	46	10.66	4.18	22	11.35	4.65	13

With regard to parents ratings of their challenges, a one way repeated ANOVA revealed that there was a significant difference over time ($F(2,5) = 13.68$, $p = 0.009$). The means for Time 1 were significantly higher than the means for Time 2 and Time 3. Parents' ratings of their goals also increased across the three time periods. A one way repeated ANOVA revealed that there was a significant difference over time ($F(2,5) = 6.003$, $p = .047$). The means for Time 2 and Time 3 were significantly higher than the means for Time 1. These results remained significant using Intention to Treat analysis.

Discussion

This article describes a pilot study evaluating the SPACE programme, a group programme designed to support parents and carers of children and adolescents with Deliberate Self Harm. The study indicates positive results for the parents who completed the programme. Parents had lower levels of psychological distress, higher levels of parental satisfaction, lower ratings on their own defined challenges, and higher ratings of their goals after the programme, and these gains were maintained 6 months after the programme. Parents also reported that their young people had lower levels of total difficulties, hyperactivity and emotional problems following their parent's attendance of the programme. When the principle of Intention to Treat was applied, results remained statistically significant. The fact that this more conservative approach yielded significance only serves to heighten the impact of these findings.

Of particular note is the high number of parents who met the criteria for psychological distress (76%) at Time 1. While this level of psychological distress may reflect parents' reaction to their child's self-harm, it is also possible that it reflects underlying parental psychiatric disorder and psychological distress, which have been shown to be common in families of suicidal young people [21,22]. Either way this psychological distress is likely to be perpetuated by the lack of support available to parents and carers of young people with DSH. It also emphasizes the importance of developing a programme such as SPACE to provide this much needed support to parents and carers and help to alleviate such feelings. Furthermore, parents whose child had a previous history of self harm were found to have higher levels of psychological distress so perhaps the SPACE programme would be particularly beneficial to these individuals.

With regard to the profile of DSH in the 32 young people involved in the study, the male to female ratio reflects that of a large survey of Irish adolescents [1], whereby females were three times more likely to harm themselves than males. In the present study, overdose was the most common method of DSH. This is in contrast to the Irish survey where cutting was the most frequently used method. However this discrepancy is consistent with previous studies whereby cutting was found to be more prevalent amongst a community sample of adolescents [2], while rates of overdosing have been found to be higher in a clinical sample [23].

The findings of the current study suggest that the SPACE programme may be beneficial to parents. However the study was subject to limitations and further evaluation is required in order to conclude this. The study has three significant limitations; the lack of a control group, the lack of information about other interventions which the families may have been receiving, and the high attrition rate. In order to conclude that the SPACE programme is effective it would be necessary to compare it with a group of parents receiving no treatment and/or a group who receive a different treatment. Participants would need to be randomly assigned to conditions in order to ensure that any group biases are evenly distributed. Future research should attempt to identify and quantify any other treatment or support that parents receive during the course of the SPACE programme and subsequently control for this. Such a study is currently planned.

As with most longitudinal research carried out in 'real world' clinical settings, the current study was subject to the effects of attrition. Analyses indicated that those who only completed measures at Time 1 did not differ significantly from other participants on any measures other than Rating of Challenges. Those who completed measures at Time 1 and Time 2 had higher ratings of their own defined challenges compared with those who completed measures at Time 1 only.

It is possible that that those who perceived their challenges as greater felt that the group was an important support to them and so continued attending in order to help them to overcome their difficulties. By applying an 'intention to treat' analysis which also showed significant results, we have attempted to address the difficulty presented by the high attrition rate. It may be beneficial for future evaluation studies to adopt a preventive approach to attrition. One possibility would be to obtain a facilitator's rating of parent participation in the group. This data would be used to form a profile of parents who drop out and could be used to alert facilitators to potential 'drop outs' through observation of the group [24].

The participants in this study were parents whose child had been referred to a specialist service or who had presented to an Accident and Emergency department. This makes it difficult to generalize the findings to other parents whose children engage in DSH. Considering that only a minority of adolescents (11.3%) attend medical services following DSH, our sample may represent a small proportion of parents of young people with DSH, and may not relate to parents of young people with DSH which does not come to medical attention.

It is anticipated that these limitations will be addressed in a future study which is being developed. This will involve a randomized controlled trial in which the SPACE programme will be extended to the wider community and will be available to parents who are not in contact with services as well as those who are. Such a study will be able to ensure the generalisability of findings and will address questions regarding comparisons between families who attend health services following DSH with those who do not. Considering that a need identified at the Focus Group Meeting was that of re-establishing communication, the randomized controlled trial will include the communication subscale of the McMaster Family Assessment which incorporates items pertaining to communication within the family [25].

Conclusion

In conclusion, these findings suggest that the SPACE programme may be an effective means of support for parents of young people with Deliberate Self Harm. Parents who completed the programme experienced positive gains afterwards which were maintained 6 months later. Future research will aim to address some of the limitations associated with the present study such as the ability to generalise from this sample, the lack of a control group, the small sample size and high attrition rate. However these preliminary results suggest that SPACE is a promising development in providing support for parents of young people with Deliberate Self Harm.

Competing Interests

The authors declare that they have no competing interests.

Authors' Contributions

LP was involved in the data collection, statistical analysis and drafted the manuscript. SM and SB participated in the design of the study, data collection and statistical analysis. CB, SC, AC and CF ran the SPACE groups. CF was involved in the conception of the study, participated in its supervision, design and co-ordination as well as critically revising the final draft of the manuscript. SG participated in the design of the study, statistical analysis and interpretation of data. All authors have read and approved the final manuscript.

Acknowledgements

The authors would like to acknowledge the Fundraising Department of the Children's University Hospital, Temple Street and the ESB in funding this project.

References

1. Morey C, Corcoran P, Arensman A, Perry IJ: The prevalence of self reported deliberate self harm in Irish adolescents. BMC Public Health 2008, 8:79.
2. Hawton K, Rodham K, Evans E, Weatherall R: Deliberate self harm in adolescents: Self report survey in schools in England. BMJ 2002, 325:1207–11.
3. Cooper J, Kapur N, Webb R, Lawlor M, Guthrie E, Mackway-Jones K, Appleby L: Suicide after Deliberate Self Harm: A 4 year cohort study. Am J Psychiatry 2005, 162:297–303.
4. Harrington R, Pickles A, Aglan A, Harrington V, Burroughs H, Kerfoot M: Early adult outcomes of adolescents who deliberately poisoned themselves. J Am Acad Child Adolesc Psychiatry. 2006, 45(3):337–345.
5. Raphael H, Clarke G, Kumar S: Exploring parents' responses to their child's deliberate self harm. Health Educ J 2006, 106:9–20.
6. Wagner BM, Aiken C, Mullaley PM, Tobin JJ: Parent's reactions to adolescents' suicide attempts. J Am Acad Child Adoles Psychiatry 2000, 39:429–436.
7. Fergusson DM, Woodward LJ, Horwood LJ: Risk factors and life processes associated with the onset of suicidal behaviour during adolescence and early adulthood. Psychol Med 2000, 30:23–39.

8. Sinclair J, Green J: Understanding resolution of deliberate self harm: Qualitative interview study of patients' experiences. BMJ 2005, 330:1112.
9. Tulloch AL, Blizzard L, Pinkus Z: Adolescent-parent communication in self harm. J Adolesc Health 1997, 21:267–275.
10. Connor JJ, Rueter MA: Parent-child relationships as systems of support or risk for adolescent suicidality. J Fam Psychol 2006, 20:143–155.
11. Department of Health and Children: A Vision for Change: Report of the Expert Group on Mental Health Policy. Dublin: Health Service Executive; 2006.
12. Byrne S, Morgan S, Fitzpatrick C, Boylan C, Crowley S, Gahan H, Howley J, Staunton D, Guerin S: Deliberate self harm in children and adolescents: A qualitative study exploring the needs of parents and carers. Clin Child Psychol Psychiatry 2008, 13(4):493–504.
13. Walsh J: Methods of psycho-educational program evaluation in mental health settings. Patient Educ Couns 1992, 19:205–218.
14. Goldberg D, Williams H: General Health Questionnaire (GHQ-12). Windsor, Nfer-Nelson; 1988.
15. Goldberg D, Gater R, Sartorious N, Uston T, Piccinelli M, Gureje O, Rutter C: The validity of two versions of the GHQ in the WHO study of mental illness in general health care. Psychol Med 1997, 27:191–197.
16. James DE, Schumm WR, Kennedy CE, Grigsby CC, Shectman KL, Nichols CW: Characteristics of the Kansas parental satisfaction scale among two samples of married parents. Psychol Rep 1985, 57:163–69.
17. Sharry J, Guerin S, Griffin C, Drumm M: An evaluation of the Parents Plus early years programme: A video based early intervention for parents of preschool children with behavioural and developmental difficulties. Clin Child Psychol & Psychiatr 2005, 10:319–336.
18. Goodman R: The Strengths and Difficulties Questionnaire: A Research Note. J Child Psychol Psychiatry. 1997, 38(5):581–586.
19. Goodman R: Psychometric properties of the Strengths and Difficulties Questionnaire (SDQ). J Am Acad Child Psy 2001, 40:1337–1345.
20. Cohen J: A power primer. Psychol Bull 1992, 112:155–159.
21. Houston K, Hawton K, Shepperd R: Suicide in young people aged 15–24: A psychological autopsy study. J Affect Disorders 2001, 63:159–170.
22. Brent DA: Risk factors for adolescent suicide and suicidal behaviour: Mental and substance abuse disorders, family environmental factors, and life stress. Suicide & Life-Threatening Behavior 1995, 25:52–63.

23. Hawton K, Fagg J, Simkin S, Bale E, Bond A: Deliberate self-harm in adolescents in Oxford, 1985—1995. J Adolesc 2000, 23:47–55.
24. Oei TP, Kazmierczal T: Factors associated with drop-out in a group cognitive behaviour therapy for mood disorders. Behav Res Ther 1997, 35:1025–1030.
25. Epstein NB, Baldwin LM, Bishop DS: The McMaster Family Assessment Device. J Marital Fam Ther 1983, 9:171–180.

Assessment of Intensity, Prevalence and Duration of Everyday Activities in Swiss School Children: A Cross-Sectional Analysis of Accelerometer and Diary Data

Bettina Bringolf-Isler, Leticia Grize, Urs Mäder, Nicole Ruch, Felix H. Sennhauser and Charlotte Braun-Fahrländer

ABSTRACT

Background

Appropriately measuring habitual physical activity (PA) in children is a major challenge. Questionnaires and accelerometers are the most widely used instruments but both have well-known limitations. The aims of this study were to determine activity type/mode and to quantify intensity and duration of children's everyday PA by combining information of a time activity diary with accelerometer measurements and to assess differences by gender and age.

Methods

School children (n = 189) aged 6/7 years, 9/10 years and 13/14 years wore accelerometers during one week in winter 2004 and one in summer 2005. Simultaneously, they completed a newly developed time-activity diary during 4 days per week recording different activities performed during each 15 min interval. For each specific activity, the mean intensity (accelerometer counts/min), mean duration per day (min/d) and proportion of involved children were calculated using linear regression models.

Results

For the full range of activities, boys accumulated more mean counts/min than girls. Adolescents spent more time in high intensity sports activities than younger children ($p < 0.001$) but this increase was compensated by a reduction in time spent playing vigorously ($p = 0.04$). In addition, adolescents spent significantly more time in sedentary activities ($p < 0.001$) and accumulated less counts/min during these activities than younger children ($p = 0.007$). Among moderate to vigorous activities, children spent most time with vigorous play (43 min/day) and active transportation (56 min/day).

Conclusion

The combination of accelerometers and time activity diaries provides insight into age and gender related differences in PA. This information is warranted to efficiently guide and evaluate PA promotion.

Background

Childhood overweight and obesity are increasing in many countries including Switzerland [1] and there is growing concern that decreasing levels of physical activity (PA) may contribute to this development. Still, appropriately measuring PA in children is a major challenge. Questionnaires and accelerometer measurements are the most widely used instruments [2,3]. Self- or proxy reports provide information about mode/type and duration of PA but show limited validity in assessing PA levels and are susceptible to reporting bias by social desirability [4]. On the other hand, accelerometer measurements provide valid overall estimates of intensity of PA [5,6]. Nevertheless, they neither determine which activities contribute most or least to PA in children nor the variation in the type and duration of habitual activities over time. Yet, this information is of great importance for public health authorities in order to efficiently guide PA promotion and to evaluate changes in PA levels over time, by gender or with age. So far, only few studies have combined accelerometer measurements with self-report data, taking advantage of the unique pieces of information that each instrument provides [7-9]. However, the self-report part in the above cited studies is based on questionnaires or activity logs but not on continuous physical activity records which, based on the exact time specification, can be compared minute by minute to accelerometer outputs. This allows to calculate the mean intensity for each activity and the summed time spent in it and thus to evaluate which activity contributes most or least to physical activity by gender and age.

In the framework of a pilot study for a monitoring programme of PA levels in Swiss school children, we developed a time-activity diary including a list of 21 typical everyday activities and asked parents (or adolescents) to allocate to each of 15-minute intervals of the child's (or their) day a specified activity. Concomitant to completion of the diary, the children wore an accelerometer device. The aims of this study were to determine activity type/mode and to quantify intensity and duration of children's everyday PA by combining information of a time-activity diary with accelerometer measurements and to assess differences by gender and age.

Methods

Sample

Participants were part of a larger cross-sectional study [10] which included three age groups of children (kindergarten/1st grade, 4th grade and 8th grade) living in three communities (Bern, Biel-Bienne and Payerne). For the present study, children in a random sample of 19 school-classes (at least two classes per grade and per community) were invited to wear an accelerometer device and to complete a time-activity diary. A participation flow chart is shown in figure 1. Personal and social characteristics of the children in invited classes did not differ significantly from those in non-invited classes. The study protocol was approved by the ethics committee of the University of Bern, and parents gave written consent.

Accelerometer

Objective assessment of PA was obtained over two seven day periods, one in winter 2004 and one in spring/summer 2005, using Actigraph accelerometers (Model AM7164, formerly Computer Science and Applications (CSA), now Manufacturing Technology Inc. (MTI), Fort Walton Beach, FL). The device measures the change in body position taking 40 measurements per second and integrates acceleration signals continuously (epoch time 1 minute). The summed values (activity counts) were stored in the device memory and downloaded to a computer. Participants were instructed to wear the accelerometer rigidly fixed at the waist with a belt. It was not worn during sleeping hours, bathing or other water activities.

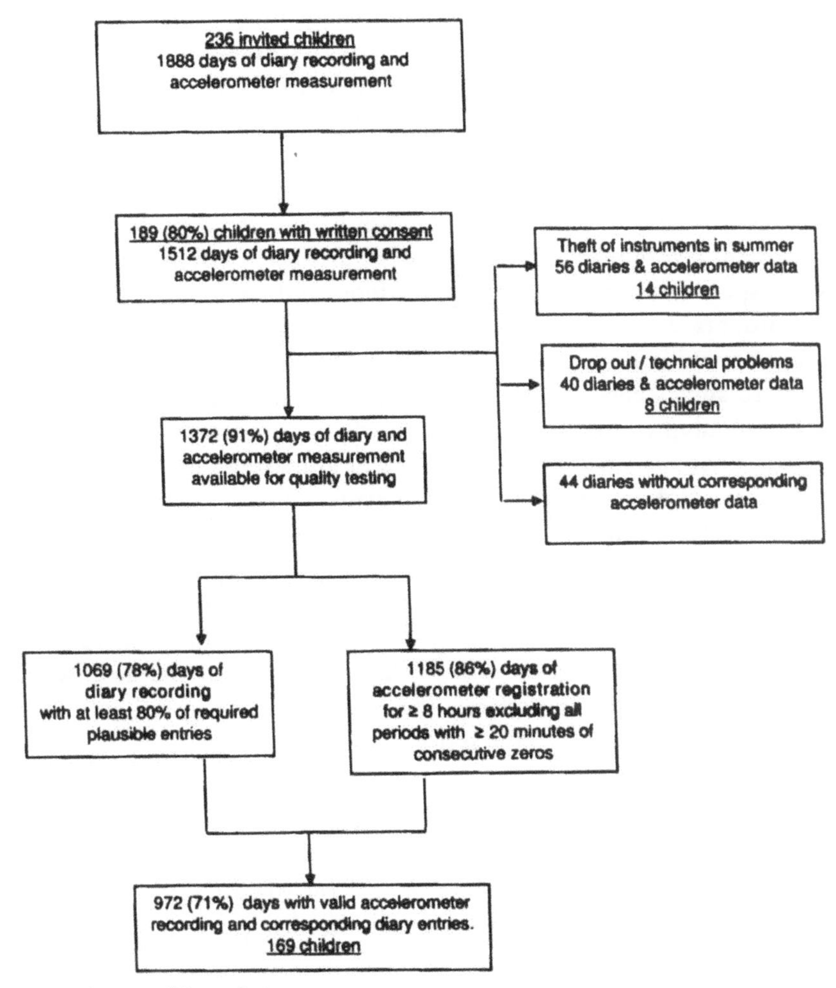

Figure 1. Schematic of the study design.

Diary (Physical Activity Record)

The diary, highly structured, required the continuous record of activities performed during each 15-minutes intervals between 6:00 and 22:00 h and each hour between 23:00 and 6:00 h. Its format has been adapted from an existing diary used for the sleeping and feeding time assessment in babies [11]. An English version of the time-activity diary is presented in Additional file 1. It has been suggested that physical activity records provide more detail about the type, intensity and patterns of activity completed during the day than physical activity logs [12]. In addition, the given continuous timeline facilitates the recording of time spent in a specific activity, thus reducing potential recall bias. Activities, which were expected to represent low intensity levels [13] included: watching TV, sitting at a computer, playing a music instrument, reading, attending class, performing homework, playing quietly, eating, travelling by car, travelling by public transport and going out. Activities expected to be of moderate intensity [13] included: cycling, walking, attending recess, moderately intense playing indoors and outdoors. Finally, activities assumed to represent vigorous intensities [13] were: attending physical education (PE) classes, sports training indoors and outdoors and vigorously playing indoors and outdoors. A first selection of activities based on existing diaries [13]has been pre-tested in two primary school classes with different social background and in 10 adolescents. After an interview with these children, the list of activities was adopted to capture activities, which differ among age groups. Less specific activities such as attending recess or attending PE-class describe contexts where children are expected to be active. "Attending recess" means the time (15 to 30 minutes) between lessons spent in the school recreation area. As in the Swiss school system the number of lessons per week increases with age, also time attending recess increases. PE classes are integrated in the class schedule. By law, all school children (but not those in the kindergarten) receive three 45 min PE classes per week. In contrast to training (which in Switzerland is organized completely independent of the school activities) PE classes do not focus on a specific sport.

Students (8th grade) or parents (kindergarten, 1st grade and 4th grade) were instructed to mark at least one activity for any given 15 minutes interval. The diary also allowed marking sleeping time, an activity difficult to classify and the removal of the accelerometer. The diary was completed during two week days and a weekend, concomitant to the accelerometer recording periods.

Questionnaire

Information on children's age, sex, nationality, parental education, weight and height and leisure time habits were extracted from the main survey parents' completed questionnaire. Parents filled in the questionnaire two months before the accelerometer measurements and completitions of diaries.

Meteorological Data

For each community and time periods of accelerometer and diary recording, daily meteorological data were provided by MeteoSwiss [14]. Measurements included mean, maximum and minimum temperatures (range -9.5 to 17.1°C), atmospheric pressure (945–971 hPa), relative humidity (57–97%), sun radiation (10–318 watt/km2), sum of sunshine duration (0–798 min), and sum of precipitation (0–18.2 mm).

Procedure

The study was conducted during the school year 2004/2005 and was organized within the framework of the School Health Services. Children and their families received written instructions on how to use the accelerometer and how to fill in the diary. In addition, trained staff instructed the children at school. The 13–14 years olds completed the diary on their own, the younger ones with their parents.

Data Processing

The completeness and plausibility of all diary entries and MTI outputs were checked. If two or more activities indicated during the same 15 min interval were logically not compatible (e.g. sleeping and eating), they were set to missing. If data were missing between 23:00 and 6:00 h, the activity was considered to be 'sleeping.' Diaries having 80% of the 15 minutes blocks filled in with logically plausible entries were considered valid for analyses. If two or more activities were indicated the more intense was considered.

The length of an accelerometer-recording day was individually defined. A day started if 10 consecutive minutes had at least one value of 800 counts/min and not more than one value with zero counts/min. The individual day stopped if after the measurement, which was considered to be the last of the day, there was an inactive period of four consecutive hours. MTI outputs equal to zero for more than 20 continuous minutes were excluded, assuming that the device was not worn during this period. Special attention was given to artefacts in accelerometer measures, which can occur if accelerometers are hit resulting in extremely high counts/min. Therefore, all values above 30,000 counts/min were substituted by the mean counts/min of the respective age group. Only days with at least 8 hours of registration were considered valid for the analyses.

For the analysis of the intensity of single activities valid diary and accelerometer days were matched by exact point in time. The information about the duration of specific activities is based on the diary.

Validation of the Diary

Total time spent in moderate to vigorous physical activity (MVPA) assessed by the time-activity diary was validated using accelerometer measurements. Metabolic equivalents (MET) based on accelerometer counts were calculated using the cut off levels of Freedson/Trost [15]. MET levels equal or above 3 were defined as MVPA. Spearman correlation between total time spent in MVPA per day based on an a priori classification of specific activities in the diary and total time with MET levels ≥ 3 was moderate ($r = 0.52$) and statistically significant ($p \leq 0.001$).

Statistical Analysis

All analyses were conducted with STATA 9.0 [16]. Univariate logistic regression models were used to examine differences in personal and social factors between participants and non-participants.

Mixed linear regression analysis was used to determine the association between accelerometer counts/min and socio-demographic and environmental characteristics. The models included age, sex, maternal education, nationality, the day of the week (weekday/weekend), season (winter/summer), mean daily temperature, the sum of precipitation and a random effect for subject.

For each child, the mean intensity for every given activity (counts/min) over all 15 minutes were calculated. Differences in intensity of a specific activity between age groups and between gender as well as interactions between age and gender were assessed by linear regression analysis.

Mixed linear regression models with a random effect for subject were generated to evaluate age and gender differences in mean duration spent in a given activity while simultaneously taking into account the effect of the day of the week, the season and meteorological factors. As the distribution of the residuals was skewed, standard errors of regression estimates were determined using a bootstrap (with 1000 replications) [17]. The final multivariate model to estimate activity duration (min/day) included the following variables: sex, age, maternal education, sum of precipitation above a threshold of 4 mm, and maximum temperature.

Results

Study Population

Of the189 participants, 164 completed diaries in both seasons, 3 only in summer and 22 only in winter. If parents were non-Swiss (53.9% vs. 85.7% valid diaries) or less educated (62% vs. 82%), more diaries had to be excluded because of low quality.

Sociodemographic Characteristics

Sociodemographic and environmental characteristics and their association with mean counts/min are shown in table 1. Mean counts/min decreased significantly with age and were significantly lower in girls. The small number of overweight children in this sample (n = 12) did not allow the evaluation of overweight related differences in activity levels. In addition, PA increased with temperature (7.8 counts/min per 1°C increase) and decreased with the sum of precipitation (-6.2 counts/min per 1 mm of precipitation over a threshold of 4 mm/day).

Table 1. Socioeconomic and environmental characteristics and their association with accelerometer counts/min.

Social and environmental factors	Total n (%)	Accelerometer counts/min: adjusted§ mean (95%CI)
Grade (Age)		
1st (6/7 years) (reference group = ref.)	47 (28)	751 (712–790)
4th (9/10 years)	60 (36)	662 (626–698)**
8th (13/14 years)	62 (38)	546 (508–583)***
Sex		
Boys (ref.)	81 (48)	737 (706–767)
Girls	88 (52)	569 (539–599)***
Education Mother		
Low (ref.)	22 (14)	636 (584–717)
Middle	80 (49)	672 (638–698)
High	61 (37)	633 (599–668)
Nationality		
Swiss (ref.)	129 (78)	649 (624–675)
Non Swiss	37 (22)	661 (602–720)
Day of the week		
Weekend (ref.)	414 (44)	631 (603–659)
Weekday	532 (56)	670 (645–696)*
Season		
Winter (ref.)	502 (53)	638 (604–671)
Summer	444 (47)	670 (634–706)

§ Mutually adjusted estimates based on multivariate regression models including all variables presented in the table as well as mean daily temperature, sum of precipitation, and a random effect for subject

* $p \leq 0.05$, ** $p \leq 0.01$, *** $p \leq 0.001$; compared to reference group

Intensity

Intensities varied significantly among activities. Highest raw mean counts/min were achieved when children were involved in sports training outdoors (1513 counts/min) and lowest when playing a music instrument (392 counts/min). Watching television (407 counts/min) yielded similarly low counts/min as sedentary activities such as reading (467 counts/min).

Figure 2 gives the adjusted mean counts/min for each specific activity by gender. For all activities, boys' mean counts/min clearly exceeded those of girls. These differences were statistically significant for 10 out of 21 activities. The variation of intensity was smaller in sedentary activities and higher in more active ones as well as such, that are mainly a description of the context (as recess). There was no interaction between age and gender with respect to intensity. The three self-reported intensity levels of playing in the diary corresponded to accumulated accelerometer counts/min in both sexes. Outdoors activities were associated with higher counts/min (1017 (935–1080) and 1259 (971–1586) for playing moderately and vigorously outdoors) than the corresponding indoor activity (734 (582–867) and 1097 (718–1629) for playing moderately and vigorously indoors).

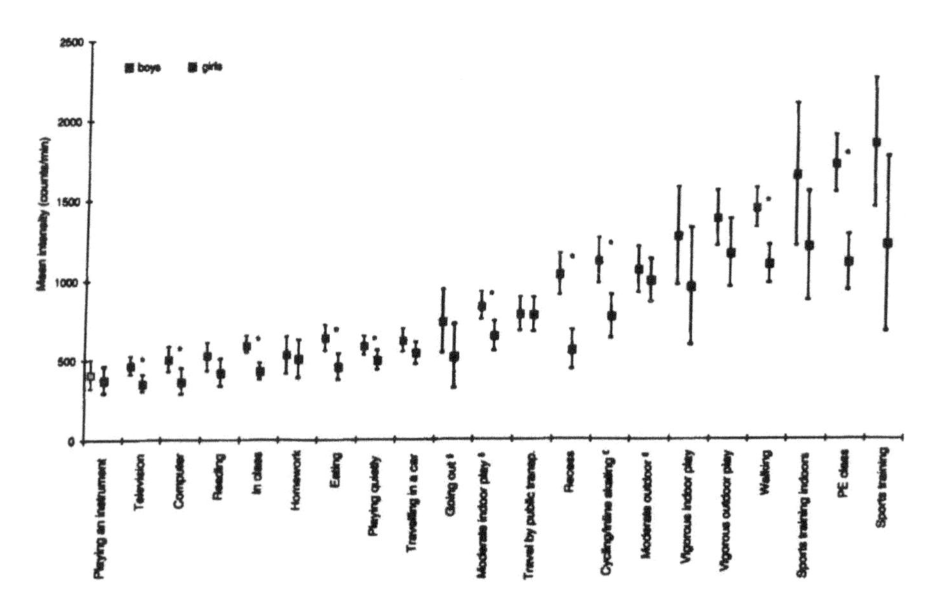

Figure 2. Age adjusted intensities for specific activities by gender. (*$p < 0.05$; [a]only assessed in 8th graders; [b]only assessed in the two younger age groups; [c]cycling not appropriately measured by accelerometer). Some inaccuracy estimating the intensity of each activity may have occurred because the start and end of the activities were based on the activity log.

Figure 3 displays adjusted mean intensities by age group. Individual activities were summarized into broader categories (see legend figure 3). For screen recreation and the other quiet activities, 8th graders accumulated significantly ($p < 0.001$) less mean counts/min when compared to the youngest age group. However, during sports training, they achieved significantly higher counts/min ($p = 0.002$). In the two younger age groups, mean counts/min accumulated during walking and vigorously playing indoors/outdoors were similar to those achieved during physical education classes at school. In the youngest age group walking (1296 counts/min) and vigorously playing (1358.2 counts/min) yielded higher mean counts/min than sports training indoors (894 counts/min) or outdoors (1031 counts/min).

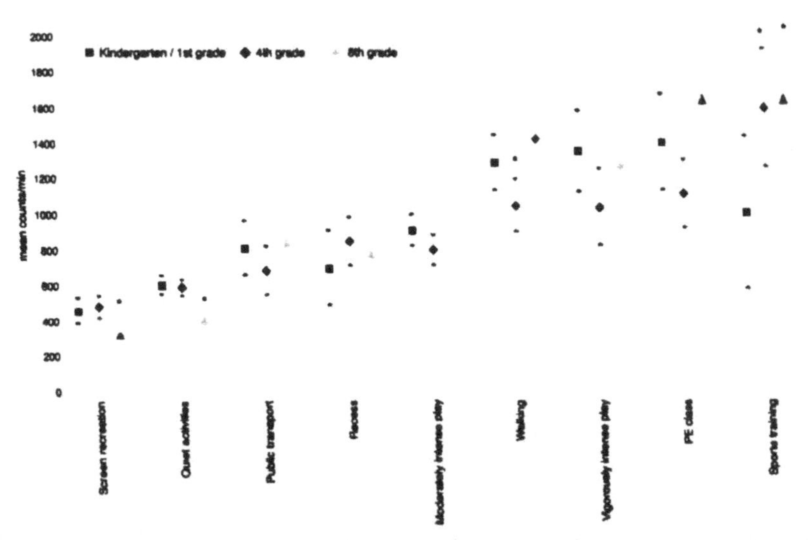

Figure 3. Gender adjusted intensities for specific activities by age group. Screen recreation: watching television andsitting at a computer. Quiet activities: playing a music instrument, reading, in class, homework, eating, playing quietly and travelling by car. Moderately intense play: moderately intense play indoors and moderately intense play outdoors. Vigorously intense play: vigorously intense play indoors and vigorously intense play outdoors. Sport training: indoors sports training andoutdoors sports training. Cycling was excluded, as it can't be measured by accelerometer in an appropriate mode.

Duration and Prevalence

Next, the mean time spent in a given activity (table 2) and the prevalence of children engaged in the respective activity were evaluated. Most time was spent with sedentary activities. Among moderate to vigorous activities, children spent most time in active transportation or playing outdoors. Only 37% of the children attended sports training during the four diary reporting days, hence, mean duration of sports training over all children was very short.

Table 2. Adjusted mean duration of specific activities by age group. Boys spent significantly more time sitting at a computer than girls (adj. mean duration (95% CI): 38.3 (33.7–42.8) min/day and 18.0 (14.5–21.4) min/day, $p < 0.001$, respectively). Girls, on the other hand reported a longer duration of playing quietly than boys (102.8 (95.4–110.3) min/day and 82.7 (75.3–90.1) min/day, $p < 0.001$, respectively). Significantly more boys (85.2%) reported vigorously playing outdoors compared to girls (59.1%) and they spent more time with this activity than girls (47.9 (40.1–55.8) min/d and 18.5 (14.1–23.0) min/day $p < 0.001$, respectively).

Activity		Activity duration in min/day All Children	Kindergarten/1st grade (6/7 years old)	4th grade (9/10 years old)	8th grade (9/10 years old)	p[§]
		Mean[ǂ] (95% CI)	Mean[ǂ] (95% CI)	Mean[ǂ] (95% CI)	Mean[ǂ] (95% CI)	
Screen recreation		104.4 (98.8–110.1)	71.0 (62.3–79.7)	105.9 (96.9–115.0)	135.8 (123.2–148.5)	< 0.001
	Television	76.2 (71.2–81.0)	56.8 (49.1–64.5)	77.0 (68.5–85.5)	94.3 (83.1–105.5)	< 0.001
	Computer	28.2 (25.3–31.1)	14.2 (10.7–17.6)	28.8 (24.2–33.4)	41.4 (35.0–47.7)	< 0.001
Any other quiet activity		391.7 (384.0–399.3)	364.1 (350.3–378.0)	391.6 (378.9–404.2)	418.9 (403.9–434.0)	0.002
	Leisure time activities[¢]	238.4 (231.2–245.7)	249.9 (237.1–262.7)	247.3 (235.6–259.1)	216.9 (203.9–230.0)	< 0.001
	Attending school/ homework[#]	261.2 (254.0–268.5)	200.2 (189.3–211.1)	256.1 (244.9–267.2)	327.5 (313.7–341.3)	< 0.001
Active Transportation		56.0 (52.3–59.7)	54.9 (48.7–61.1)	51.1 (45.4–56.7)	62.52 (54.8–70.3)	0.029[&]
	Walking	40.6 (37.3–42.8)	41.9 (36.5–47.4)	33.5 (28.6–38.4)	47.2 (40.1–54.4)	0.198
	Cycling/inline skating	15.4 (13.3–17.6)	13.0 (9.6–16.4)	17.6 (13.6–21.5)	15.4 (11.4–19.3)	0.394
Recess[#]		25.1 (22.6–27.7)	9.9 (7.2–12.7)	27.5 (23.6–31.4)	37.6 (31.0–44.3)	< 0.001
Vigorously intense unstructured play		42.9 (38.0–47.8)	42.7 (34.7–50.7)	49.6 (41.4–57.7)	35.7 (27.4–44.0)	0.035[§]
	Indoors	9.7 (7.5–11.8)	14.1 (9.7–18.6)	7.1 (3.7–10.4)	8.1 (4.5–11.8	0.037
	Outdoors	33.3 (28.9–37.7)	28.5 (21.6–35.5)	42.5 (34.4–50.6)	27.6 (19.5–35.7)	0.853
Attending PE[*] classes[#]		27.4 (23.1–31.7)	21.2 (13.2–29.2)	24.8 (19.3–30.2)	36.3 (28.5–44.2)	0.008
Sports training		10.7 (8.3–13.0)	4.7 (2.3–7.1)	10.8 (6.7–15.0)	16.5 (11.3–21.7)	< 0.001

[ǂ] Estimates adjusted for sex, maternal education, season, day of the week, mean daily temperature, sum of precipitation and subject (as random effect)
[§] Test for trend
[¢] Leisure time activities: playing a music instrument, reading, quietly playing and riding a car
[#] Duration calculated excluding weekends
[&] 8th graders compared to kindergarten/1st graders and 4th graders
[*] Physical education

With increasing age, the duration of screen recreation and school related quiet activities increased significantly and there was a shift from unorganized PA (vigorously playing) to organized PA (PE classes or attending sports) (table 2). Furthermore, eighth graders spent significantly more time in active modes of transportation and in attending recess at school.

With decreasing level of maternal education, children's time watching TV increased (adj. mean duration (95% CI): 66 (60–72) min/day, 79 (71 -88) min/day, and 98 (80–115) min/day, respectively for low, middle and high levels of maternal education) whereas time spent reading books decreased (31 (26–35) min/day, 23 (19–26) min/day and, 17 (11–24) min/day, respectively). Further, children of mothers with low educational levels spent less time playing vigorously intense compared to children of mothers with high educational levels (adj. mean duration (95% CI): 45 (36–53) min/day, 45(38–52) min/day, and 27 (16–37) min/day, respectively, for high, middle and low level of maternal education). Yet, mean counts/min did not differ significantly by maternal education (table 1).

The duration of many activities was influenced by season, the day of the week and meteorological parameters. There was a significant decrease in time spent vigorously playing outdoors (-1.7 (-2.7 to -0.7) min/day per 1 mm rainfall, $p < 0.001$), biking (-0.88 (-1.3 to -0.5) min/day per mm rainfall, $p < 0.001$) and going out (-2.4(-4.6 to -0.2) min/day per mm rainfall (activity assessed in adolescents only), $p < 0.03$) with the sum of precipitation whereas activities typically performed indoors such as sitting at a computer increased (1.0 min/day per mm rainfall, $p = 0.037$). In summer, children spent significantly more time playing vigorously outdoors (42.3 (32.9–51.7) min/day) than in winter (25.2 (17.0–33.4) min/day; $p = 0.02$). On weekends, significantly more time was spent in quiet leisure activities (284.4 (271.0–297.9) min/day and 191 (182.5–200.4) min/day for weekends and weekdays respectively, $p = 0.001$), and in vigorous play (54.0 (44.8–63.3) min/day and 34.2 (28.4–39.4) min/day for weekends and weekdays respectively, $p \leq 0.001$) than on weekdays.

Discussion

The present study combined objectively measured accelerometer data with a detailed assessment of the exact time point and the duration of single activities in a time-activity diary (physical activity record), allowing precisely estimated intensity and duration of specific activities. This combination enabled the unique pieces of information that each instrument provides to be integrated and showed significant differences in intensity and duration of specific activities by gender and age. Compared with physical activity logs [13] the registration of each activity along a time-line provided more in-depth insight into physical activity patterns of school-aged children.

In line with other research, the present study found girls to be less active than boys [7,18-20]. It has previously been reported that during PE classes [9,21] and during recess [22], boys accumulate more counts/min than girls, yet, the results of the present study indicated that girls collected systematically less counts/min for the full range of everyday activities. However, the observed gender difference in PA was also due to differences in activity patterns such as girls spending significantly less time playing vigorously. In contrast to a previous study [7], the present study did not observe significant gender differences in the prevalence or duration of structured sports or PE classes.

The decrease in PA with age followed a more complex pattern. On one hand, more adolescents were engaged in high intensity activities such as structured sports, spending significantly more time in these activities than younger children. On the other hand, younger children spent more time playing vigorously than adolescents. In addition, adolescents spent significantly more time in sedentary activities but accumulated less accelerometer counts/min during sedentary activities than younger children. The combination of these factors resulted in significantly less mean counts/min for adolescents.

Younger children accumulated high count values predominantly during unstructured moderate and vigorous intense play whereas structured sports activities were less important. This is consistent with previous studies [18,23], reporting that time spent playing outdoors was an important contributor to PA in children

Walking was another important activity contributing to PA in all age groups. The present study underlines the importance of active commuting for PA levels of children [10,24,25]. In the Swiss context and most likely in other European countries, it would thus not be sufficient to only focus on playing outdoors to assess PA in younger children as has been suggested by Burdette et al. [23].

The detailed assessment of different activities with the time-activity diary also provided insight into subtle activity differences related to the educational background of the child's mother. Total mean counts/d did not vary significantly with maternal education, but the type of passive activities (TV and reading books) varied. Watching television is not only problematic because of its low intensity but also because of its association with overweight [26,27]. This association might not only be due to the lack of PA, but also result from poorer eating habits [28].

Consistent with previous findings [29,30], the present analyses also illustrate the importance of meteorological conditions for the assessment of different outdoor activities. Moreover, it is noteworthy that self-reported intensities of playing corresponded well with accelerometer measurements and differentiating between playing "indoors" and "outdoors" added useful information to the activity assessment.

Compared to previous studies [7,8] the combination of the time-activity diary with simultaneous accelerometer measurements provided precisely estimated intensity and duration of a large range of everyday activities and allowed to detect systematic differences in intensity between gender or between age-groups. In contrast to PA questionnaires and to physical activity logs, the fixed 24-hour timeline of our time-activity diary facilitated the assessment of specific activities by the parents and their children. Repeating such accelerometer measurements along with a time-activity diary as part of a monitoring programme will allow to determine changes in physical activity pattern and the relative contribution of specific activities to overall childhood physical activity levels over time.

Although the use of the time-activity diary in the present study provided valuable information about children's activity pattern, several limitations became apparent. First, it has to be acknowledged that the proportion of good quality filled diaries depended on maternal education and nationality, thus leading to an under-representation of lower social classes and non-Swiss populations limiting the generalizability of our findings. Second, for future use of the diary, it is suggested to assess exactly the same activities for all children, as comparisons across age groups are otherwise limited. Third, the sample size was rather small for stratified analyses. Fourth, attending recess or attending PE class are a description of a context which is expected to be active but we did not collect exact data on the type of activity for this period. Last, the documentation of only four days of the week potentially underestimates activities like sports training. In the parents' questionnaire, 59% of the children were reported to attend sports training whereas only 37% indicated this activity during the 4 days of diary recording.

Conclusion

The combination of accelerometer and time activity diaries allowed the precise quantification of both the intensity and the duration of children's everyday PA and provided insight into age and gender related differences. This information is warranted to efficiently guide and evaluate PA promotion. Moreover the results underline the relevance of providing opportunities for unstructured play to promote PA in young children and to promote structured sports activities to increase PA in adolescents. It also emphasizes the important contribution of active forms of commuting.

Competing Interests

The authors declare that they have no competing interests.

Authors' Contributions

BB, UM, FHS and CB contributed to the study conception and design. BB, NR, and UM collected the data and were responsible for the collaboration with the school health services and teachers. UM and NR prepared the accelerometer data for analyses. BB, LG and CB conducted statistical analyses and interpreted the data. BB and CB wrote the paper. All authors critically revised the draft versions of the manuscript, provided critical feedback and approved the final version.

Acknowledgements

The authors are grateful to their many colleagues in the School Health Services of Bern, Biel and Payerne who organized the survey. We should also like to thank the children, parents and teachers for their enthusiastic co-operation, which made this investigation possible. The study was supported by a grant of the Federal Commission of Sport (ESK).

References

1. Zimmermann MB, Gubeli C, Puntener C, Molinari L: Overweight and obesity in 6–12 year old children in Switzerland. Swiss Med Wkly 2004, 134:523–528.
2. Corder K, Ekelund U, Steele RM, Wareham NJ, Brage S: Assessment of physical activity in youth. J Appl Physiol 2008, 105:977–987.
3. Rowlands AV: Accelerometer assessment of physical activity in children: an update. Pediatr Exerc Sci 2007, 19:252–266.
4. Sallis JF, Saelens BE: Assessment of physical activity by self-report: status, limitations, and future directions. Res Q Exerc Sport 2000, 71:S1–14.
5. Janz KF: Validation of the CSA accelerometer for assessing children's physical activity. Med Sci Sports Exerc 1994, 26:369–375.
6. Trost SG, Ward DS, Moorehead SM, Watson PD, Riner W, Burke JR: Validity of the computer science and applications (CSA) activity monitor in children. Med Sci Sports Exerc 1998, 30:629–633.
7. Jago R, Anderson CB, Baranowski T, Watson K: Adolescent patterns of physical activity differences by gender, day, and time of day. Am J Prev Med 2005, 28:447–452.
8. Telford A, Salmon J, Timperio A, Crawford D: Examining physical activity among 5- to 6- and 10- to 12-year-old children: The Children's Leisure Activities study. Pediatric Exercise Science 2005, 17:266–280.
9. McKenzie TL, Marshall SJ, Sallis JF, Conway TL: Student activity levels, lesson context, and teacher behavior during middle school physical education. Res Q Exerc Sport 2000, 71:249–259.
10. Bringolf-Isler B, Grize L, Mader U, Ruch N, Sennhauser FH, Braun-Fahrländer C: Personal and environmental factors associated with active commuting to school in Switzerland. Prev Med 2008, 46:67–73.
11. Largo RH: Babyjahre. Die frühkindliche Entwicklung aus biologischer Sicht. Piper Verlag GmbH; 2001.
12. Welk GJ: Physical Activity Assessments for Health-Related Research. New York: Oxford University Press; 2002.

13. Weston AT, Petosa R, Pate RR: Validation of an instrument for measurement of physical activity in youth. Med Sci Sports Exerc 1997, 29:138–143.
14. MeteoSwiss, [http://www.meteoschweiz.admin.ch/web/en/weather.html].
15. Freedson P, Pober D, Janz KF: Calibration of accelerometer output for children. Med Sci Sports Exerc 2005, 37:S523–530.
16. STATA: Statistical Software. Book Statistical Software (Editor), Release 9 edition. City 2005.
17. Kirkwood B, Stern J: Essential Medical Statistics. 2nd edition. Oxford: Blackwell Science; 2003.
18. Sallis JF, Prochaska JJ, Taylor WC: A review of correlates of physical activity of children and adolescents. Med Sci Sports Exerc 2000, 32:963–975.
19. Trost SG, Pate RR, Sallis JF, Freedson PS, Taylor WC, Dowda M, Sirard J: Age and gender differences in objectively measured physical activity in youth. Med Sci Sports Exerc 2002, 34:350–355.
20. Horst K, Paw MJ, Twisk JW, Van Mechelen W: A brief review on correlates of physical activity and sedentariness in youth. Med Sci Sports Exerc 2007, 39:1241–1250.
21. Nader PR: Frequency and intensity of activity of third-grade children in physical education. Arch Pediatr Adolesc Med 2003, 157:185–190.
22. Ridgers ND, Stratton G, Fairclough SJ: Assessing physical activity during recess using accelerometry. Prev Med 2005, 41:102–107.
23. Burdette HL, Whitaker RC, Daniels SR: Parental report of outdoor playtime as a measure of physical activity in preschool-aged children. Arch Pediatr Adolesc Med 2004, 158:353–357.
24. Merom D, Tudor-Locke C, Bauman A, Rissel C: Active commuting to school among NSW primary school children: implications for public health. Health Place 2006, 12(4):678–87.
25. Timperio A, Ball K, Salmon J, Roberts R, Giles-Corti B, Simmons D, Baur LA, Crawford D: Personal, family, social, and environmental correlates of active commuting to school. Am J Prev Med 2006, 30:45–51.
26. Andersen RE, Crespo CJ, Bartlett SJ, Cheskin LJ, Pratt M: Relationship of physical activity and television watching with body weight and level of fatness among children: results from the Third National Health and Nutrition Examination Survey. Jama 1998, 279:938–942.
27. Marshall SJ, Biddle SJ, Gorely T, Cameron N, Murdey I: Relationships between media use, body fatness and physical activity in children and youth: a meta-analysis. Int J Obes Relat Metab Disord 2004, 28:1238–1246.

28. Barr-Anderson D, Story M, Neumark-Szainer D: Longitudinal trends in television viewing and dietary intake of older adolescents and young adults: Findings from Project EYT (Eating Among Teens). 2nd ICPAPH; Amsterdam 2008.
29. Duncan JS, Hopkins WG, Schofield G, Duncan EK: Effects of weather on pedometer-determined physical activity in children. Med Sci Sports Exerc 2008, 40:1432–1438.
30. Belanger M, Gray-Donald K, O'Loughlin J, Paradis G, Hanley J: Influence of weather conditions and season on physical activity in adolescents. Ann Epidemiol 2009, 19:180–186.

The Specificity and the Development of Social-Emotional Competence in a Multi-Ethnic-Classroom

Katja Petrowski, Ulf Herold, Peter Joraschky, Agnes von Wyl, and Manfred Cierpka

ABSTRACT

Background

Ethnic diversity in schools increases due to globalization. Thus, the children's social-emotional competence development must be considered in the context of a multi-ethnic classroom.

Methods

In this study, the social-emotional competence of 65 Asian-American and Latin-American children was observed at the beginning and the end of their kindergarten year.

Results

Initially, significant differences existed among these ethnic groups in respect to moral reasoning. Furthermore, the male children showed more dysregulated aggression but the female children implemented more moral reasoning than their male counterparts. These ethnic specificities did not disappear over the course of the year. In addition, a significant change in avoidance strategies as well as expressed emotions in the narrative took place over the course of one year.

Conclusion

Ethnic specificity in social-emotional competence does exist independent of gender at the beginning as well as at the end of the kindergarten year in a multi-ethnic kindergarten classroom.

Introduction

Multi-ethnic kindergarten classrooms can be found in many parts of the world, but especially in the USA [1]. Since social development is based on observation and imitation in respect to the social learning theory [2], and the child's peers play a role in social development [3], the influence of the multi-ethnic classroom on social-emotional competence must be taken into serious consideration.

Behaviorism and the social learning theory explained the social-emotional development through processes such as reinforcement, punishment, conditioning, observations and limitations [2,4,5]. This development took place at the preschool/kindergarten age (effect sizes of the increase in social-emotional competence = .24 and .33, [6,7]), not only inside but also outside the family [2]. Therefore, exposure to prosocial or antisocial peers was related to changes in social behavior over the course of the preschool/kindergarten year [3,8]. Furthermore, the development from age two to five was crucial for the manifestation of early childhood aggression [9,10] as well as the development of emotion and self-regulation which prevented early childhood aggression [11,12]. Children of this age primarily tend to use hedonistic reasoning or needs-oriented (primitive empathetic) social reasoning [13]. However, the development of social competence was affected by the varying degree of the normativity of the society and subcultural variations [14].

Individualism and collectivism were possible antecedent values in societies for the explanation of social-emotional development [15]. Individualism was more prevalent in Western societies than in the more traditional societies of developing countries, where collectivism was the dominant value [e.g. [16,17]]. An orientation of a society toward individualism, on the one hand, or collectivism, on the other, implied basic psychological functioning such as the expression of emotions, moral reasoning, the style of conflict resolution and social competence.

The importance of social responsibility and moral reasoning differed across cultures and subcultures. In the USA, interpersonal responsiveness and caring was viewed as less obligatory and more of a personal choice [18]. European-American and Mexican-American children did not differ in regard to the degree of obligation to the family. However, European-American children equated obligation to the family with relationship quality and closeness to family whereas to Mexican-American children obligation to the family and to the collective was a part of being a family or a group member [19].

Collectivism and individualism also influenced collaboration and conflict resolution style. European-American children high in individualism preferred confronting others and immediately taking a turn rather than waiting for a turn in group interaction tasks [20,21]. Furthermore, they communicated and resolved conflicts in an individualistic mode. In contrast, Mexican-American children high in collectivism preferred accommodation as a mode of handling conflicts with family and friends [22]. European-American children were less likely to use equality norms in interactions with in-group members than Chinese children [23].

Focusing on social competence, socialization not only inside but also outside the family and its cultural background must be considered based on the social learning theory [2]. In the social-emotional and moral development empathy played an essential part of social behavior [24,25]. According to the "main effect model" [26] empathy inhibited aggression and anger, which was connected positively to emotional expressions and essential for further development [27]. Mexican-American and European-American children did not differ in sharing candy with a classmate [28]. However, Mexican-American children were generally more inclined to share something with a peer than European-American children were [29,30]. Furthermore, in Chinese kindergarten classrooms the incidences of sharing and comforting were higher than in American kindergarten classrooms [31] since Chinese societies generally emphasized responsibility and prosocial behavior towards others [32]. Based on the literature, ethnically specific social-emotional development can be assumed even for the subcultures of a country.

Concerning multi-ethnic classrooms, the socialization within the family and the peers in the preschool/kindergarten class may augment or counteract cultural influences. However, results on the reciprocal influence of the ethnic background and the peers on child development are not yet available. Only one study evaluated the status quo and found ethnic differences in normative beliefs, expressed emotions and interpersonal conflicts in the multi-ethnic classroom [33]. Latin-American children reported higher levels of normative beliefs about aggression and expressed more aggressive fantasies but reported less fights than

African-American children did [33]. However, we assume that these cultural differences concerning aggression are dependent of gender specificities as some of the most well-supported findings in the research literature showed that boys were more aggressive than girls [34-36]. Also, there are no data available on how the social-emotional competencies develop in respect to ethnic backgrounds over the course of a year.

Based on the literature, the following hypotheses can be stated:

1. It can be presumed that the different ethnic groups in the classroom differ in the degree of social-emotional competence at the beginning of the kindergarten year. The Latin-American children will more likely show more moral reasoning, more expressed emotions and less interpersonal conflicts than the Asian-American children.
2. By the same token, at the end of kindergarten year the two ethnic groups may display the same level of moral reasoning, more expressed emotions and less interpersonal conflicts due to the reciprocal adaptation.
3. At both, the beginning and the end of the kindergarten year, the boys may show more aggressive behavior than the girls.
4. Furthermore, the social-emotional competence in this age group may increase during the course of the year. Hereby, the two ethnic groups may develop differently, depending on their initial level.

Detailed assessment of social-emotional competence in a multi-ethnic classroom would present an interesting scope. The USA as a country that attracts large numbers of immigrants from all over the world, seem predestinated for conducting studies in the multi-ethnic classroom [37]. Since the preschool/kindergarten age is the sensitive age-range for the development of social-emotional competence and kindergarten is a child's first official contact with the American school system, the kindergarten year was chosen for evaluating social-emotional competence and its development over the course of one year.

Method

The study was conducted in four kindergarten classes located in Oakland, California, USA. These kindergarten classes were chosen due to their high ethnic diversity. All of the children in these classes were the offspring of parents who had been brought up in their country of origin, later immigrating to the USA (first generation immigrants). Among them, there was a large group of Asian-American and Latin-American children. The parents of the children of these Oakland kindergarten classes were asked to give written consent to allow their children to participate in the study. In the first test, at the beginning of the kindergarten year, 65 of 90 (84%) children, and in the second test, at the end of the kindergarten year, 65 of the original 80 (95%) children were tested, respectively. The remaining children were Afro-American children or the parents did not agree on the study participation. Children who were not tested at the beginning as well as at the end of the kindergarten year were not considered in the calculations. The first test was conducted in August/September, at which time all of the children had reached the age of five (31 by August). The second test took place in May of the following year. The school year as such proceeded as in any normal public classroom in the USA does.

The sample consisted of 65 children of which 30 were female and 35 were male. In this sample 46 percent were Asian-Americans (N = 27) from China, 53 percent Latin-Americans (N = 38) from Mexico. This was not a representative sample, the small sample sizes being due to the distribution in this kindergarten. The school board did not permit the assessment of detailed information of the sample concerning socio-demographic characteristics such as income, family situation, year of immigration and religion. The distribution of the social-emotional behavior at the beginning and at the end of the kindergarten year is presented for the male and female children of the different ethnic groups in Table 1.

Table 1. Ethnic and gender specificity at the beginning of the kindergarten year

	At the beginning	At the end	At the beginning of the kindergarten year				At the end of the kindergarten year			
Variable	M (SD)	M (SD)	M(SD) Asian-American		M(SD) Latin-American		M(SD) Asian-American		M(SD) Latin-American	
	N = 65	N = 65	male n = 17	female n = 10	male n = 18	female n = 20	male n = 17	female n = 10	male n = 18	female n = 20
Empathetic relation	7.81 (2.39)	6.96 (2.77)	7.95 (2.34)	7.46 (2.78)	6.71 (1.42)	8.44 (2.34)	5.75 (2.63)	7.42 (4.37)	7.47 (1.81)	7.76 (3.18)
Avoidance strategies	12.09 (5.83)	12.82 (6.44)	11.39 (5.85)	10.65 (3.40)	12.56 (5.20)	12.65 (6.47)	14.67 (6.01)	11.97 (6.14)	12.22 (6.63)	11.90 (6.69)
Moral themes	14.58 (5.20)	15.07 (5.09)	10.90 (3.99)	14.75 (4.44)	14.14 (4.04)	17.58 (4.45)	12.21 (5.13)	15.00 (5.29)	14.54 (5.20)	17.79 (4.16)
Interpersonal conflict	1.11 (1.39)	0.87 (1.08)	0.48 (0.52)	0.80 (1.03)	1.52 (1.67)	1.31 (1.50)	0.83 (1.03)	1.05 (1.27)	0.46 (0.64)	1.12 (1.32)
Dissociation codes	1.45 (2.37)	1.26 (2.09)	1.37 (1.91)	0.41 (0.71)	2.60 (3.47)	0.87 (1.46)	1.24 (1.89)	0.87 (0.71)	1.58 (2.68)	0.92 (1.64)
Dysregulated aggression	6.77 (9.41)	5.75 (8.54)	8.92 (9.43)	2.74 (3.55)	10.43 (12.48)	3.38 (5.93)	7.80 (8.49)	2.32 (1.26)	7.09 (10.39)	2.49 (5.07)
Narrative emotions	6.73 (4.63)	9.13 (5.35)	7.68 (5.00)	6.16 (3.60)	7.33 (4.78)	5.41 (3.97)	10.24 (5.08)	10.37 (3.60)	9.11 (5.85)	7.02 (5.16)

Since the aim was to observe social-emotional behavior in the doll play at the beginning and the end of the kindergarten year, the Mac Arthur Story Stem Battery (MSSB) was implemented [38,39]. The Mac Arthur Story Stem Battery is a method [38,39] that reveals the inner world and representations of young children of age three and older (developed by Emde, Bretherton and colleagues; cf. [40-43]). This method bases on a standardized sequence of the beginnings of stories like short, half-structured conflict situations. The interviewer starts by telling and playing the beginning of one story with puppets including a conflict. Then, the child is being asked to show and to tell what happens next in this story. These 30 minute-interviews were video-taped and coded in respect to the different content and behavior addressed in the coding manual. The MSSB is a valid and reliable instrument (for further details see [39]).

In this study, ten stories were chosen according to the targeted social-emotional competence to be tested (one warm-up story and nine conflict stories): Susan/George's Birthday (warm up story), The Hot Soup, Barney's Disappearance, The Departure, The Return of the Parents, The Lost Key, The Exclusion, The Mother's Headache, Three is a Crowd, and The Sand Castle. The warm up story is a beginning of a conflict story as are the other stories. With this warm-up story the child gets accommodated to the setting and the procedure so that the child can follow through the following stories without any assistance. Each child completed the nine narratives from the MacArthur Story-Stem Battery (Table 2 presents a brief description of each story, original stories and examples described by [41]).

Table 2. Survey of the administered MSSB stories described by Warren et al. (1996).

0) The Birthday of Susanne/George (warm-up story): The family celebrates the birthday of Susanne/George.
1) The Hot Soup: Although the mother had forbidden it, the child grasps at the pot with the hot soup, pouring it out and burning her hand.
2) Barney's Disappearance: The child goes to the garden for playing with the dog Barney, but Barney is not there.
3) The Departure: The parents drive on a trip overnight and the children remain with the grandmother.
4) The Return of the Parents: The parents return from their trip.
5) The Lost Key: The child enters the room and hears mother and father arguing over a lost key.
6) The Exclusion: Mother and father want to be alone and send the child to its room to play.
7) Mother's Headache: The mother has a headache and asks Susanne/George to switch off the television. There the friend comes by and absolutely wants to watch television.
8) Three is a Crowd: The child and the friend play with his new ball. There comes the small brother from the house and wants to join in, but the friend doesn't want that at all.
9) The Sand Castle: A small child in a park built a sand castle. The friend says to the child: Come on, we break the little guy's sand castle.

Instruction after playing the respective scene: *"Show and tell, what happens next!"*

The children's play narratives were videotaped in a playroom laboratory and coded with the MacArthur Narrative Coding System [44] (for further details of the coding manuals see [43,45,46]).—The author attended a training which was provided by the original author of the coding manual JoAnn Robinson. JoAnn Robinson developed and validated the coding system and published it [43]. In this coding manual, the social-emotional competence was evaluated by several additional global scales besides those of empathetic relation and dysregulated aggression. Thus, the social problem-solving can be analyzed in more detail in order to specify strategies besides aggression. The following content themes expressed by the children's stories were coded (global scales): Empathic relation, avoidance strategies, moral themes, interpersonal conflict, dysregulated aggression, dissociation codes and narrative emotions (see [46]). The content themes were coded as 'present' or 'absent.' Under interpersonal conflict, the aspects of competition, rivalry/jealousy, exclusion, active refusal of empathy/help and verbal conflict were summarized. For empathetic relation in the stories, sharing, empathy/helping, affiliation and affection had to be present. To show avoidance strategies, the doll play had to include self-exclusion, repetition, denial, passive refusal of empathy, sudden sleep onset and sensomotor/mechanical preoccupation. Moral themes were coded if scenes included non-/compliance, shame, blame, teasing/taunting, dishonesty, punishment/discipline, reparation/guilt and politeness. Aggressive themes were coded if the subject engaged in physically aggressive acts directed toward another character, prop, or object by the subject or the puppet character. Such interactions had a negative quality to them and included hostile or destructive gestures and forms of physical violence such as an object being thrown at another character with the intent to cause pain. Personal injury was coded whenever there was an instance of a character's being physically hurt or injured. The focus needed to be on the injury or pain and not just on the act of aggression itself. Atypical negative responses were coded for any atypical or disorganized response with a negative tone. The narrative emotion code summarizes all expressed emotions in the narrative.

The doll play was conducted by a trained interviewer. The stories and the actual doll play behavior of the children were transcribed from the tape and blind-coded. Coding was carried out by two especially trained individuals. The inter-rater agreement was calculated based on 20 children out of the original 80 children (9 stories with 7 scales per child respectively). Herefore, each rating-scale (7 of them) was added up for the nine stories of a child in order to gain seven general ratings for this child. For the inter-rater agreement Kappas were calculated for the 7 global scales for each of the 20 children at the two measurement points. Hereby, the inter-rater agreement ranged between Kappa = .74 to .82. In the case of disagreement between the two reliable rater a consensus rating with a third reliable rater was implemented.

For the Statistical Analysis, SPSS.15.0 was Used.

Hypothesis 1 +2+3

Ethnic differences in social-emotional competence at the beginning and at the end of the kindergarten year:

A two factorial (gender, ethnic group) multivariate analysis of variance was calculated to compare the ethnic groups at the beginning of the kindergarten year. The same procedure was chosen to compare the ethnic groups at the end of the kindergarten year. Since the gender had an influence on the social-emotional competence, and the gender distribution in the ethnic groups was diverse, the gender needed to be considered in the calculations as a second factor. Regarding the accumulation of error effects the statistical program SPSS automatically corrected the range of significance and the usually used range of values can be applied ($p < 0.05$). For the multivariate overall effect of ethnicity Wilks-Lambda was calculated. For the scale-specific effects the corrected model was used in order to minimize the effect of the sample size. In the Tables 2 and 3 the interaction effect of ethnicity and gender is not displayed since there are no significant effects.

Hypothesis 4

Ethnic differences in social-emotional competence over the development of one year:

To determine the changes over the course of one year, a three-factorial (time, gender, ethnic group) univariate analysis of variance with repeated measurement points was calculated separately for each scale of the MSSB. Greenhouse-Geisser correction was applied to account for violation of the sphericity assumption. In the Table 5 the interaction effect of ethnicity and gender is not displayed since there are no significant effects.

Results

The first consideration concerned the ethnic and gender differences in social-emotional competence at the beginning and the end of the kindergarten year. Later on, the changes in the social-emotional competence were focused on in order to evaluate ethnic specific development.

Beginning of the Kindergarten Year:

In general the two ethnic groups differed significantly in their social-emotional competence at the beginning of the kindergarten year (see Table 3). In Table 3 a two factorial (gender, ethnic group) multivariate analysis of variance is displayed which shows the significant effects of the multiple comparisons with corrections for the error effects. The descriptive statistics of the MSSB codes are presented in Table 1. Since there was no significant interaction of gender and ethnicity these effects were not displayed in the Table. Comparing the two ethnic groups, the Latin-American children most frequently utilized a moral approach while the Asian-American children employed this method the least.

In general the gender differed significantly in their social-emotional competence at the beginning of the kindergarten year (see Table 3). The female children included more moral themes in their play than the male children did. The most dysregulated aggression and dissociation codes were employed by the male children. Therefore, the two ethnic groups and the different genders entered the kindergarten year with small differences in social-emotional competence.

Table 3. Ethnic and gender specificity at the beginning of the kindergarten year (multivariate two factorial analysis of variance).

	F (df = 1;2)[d]		p (F)				Wilks-Lambda[c]		p	
Variable	ethnic	gender	ethnic	gender	η^2_{ethnic}[a]	η^2_{gender}[b]	ethnic	gender	ethnic	gender
							0.78	0.71	.04	<.01
Empathetic relation	0.03	0.79	.38	.47	<.01	.02				
Avoidance strategies	1.17	0.18	.29	.67	.02	<.01				
Moral themes	5.29	6.20	.03	.02	.11	.15				
Interpersonal conflict	2.55	0.13	.12	.72	.08	<.01				
Dissociation codes	1.18	3.80	.28	.05	.04	.09				
Dysregulated aggression	0.01	5.48	.92	.02	<.01	.13				
Narrative emotions	1.06	0.32	.31	.57	<.01	.04				

[a, b] Effect sizes for the ethnic groups and gender specific change; [c] Exact test; [d] corrected model; Ethnic*gender interaction not significant

End of the Kindergarten Year

In general the two ethnic groups differed significantly in their social-emotional competence at the end of the kindergarten year (see Table 4). The Latin-American children most frequently utilized a moral approach while the Asian-American children employed this method the least. The gender generally did not differ in their social-emotional competence at the end of the kindergarten year (see Table 4). However, the most dysregulated aggression was employed by the male children (Table 4).

Table 4. Ethnic and gender specificity at the end of the kindergarten year (multivariate two factorial analysis of variance).

	F (df = 1;2)[d]		P (F)				Wilks-Lambda[c]		p	
Variable	ethnic	gender	ethnic	gender	η^2_{ethnic}[a]	η^2_{gender}[b]	ethnic	gender	ethnic	gender
							2.19	0.22	.04	.11
Empathetic relation	2.15	0.54	.15	.47	.03	.02				
Avoidance strategies	1.33	0.16	.26	.69	.01	.01				
Moral themes	5.09	2.59	.03	.11	.05	.07				
Interpersonal conflict	0.09	0.34	.93	.56	<.01	.03				
Dissociation codes	0.19	1.29	.67	.26	<.01	.01				
Dysregulated aggression	0.00	5.50	.95	.02	<.01	.08				
Narrative emotions	2.79	0.74	.10	.40	.04	.01				

[a, b] Effect sizes for the ethnic groups and gender specific change; [c] Exact test; [d] corrected model; Ethnic*gender interaction not significant

The change in the social-emotional competence and its ethnic specificity: Herefore, a three-factorial (time, gender, ethnic group) univariate analysis of variance with repeated measurement points was calculated separately for each scale of the MSSB for the MSSB-codes at the beginning and at the end of the kindergarten year (Table 5). Greenhouse-Geisser correction was applied to account for violation of the sphericity assumption.

The data showed no significant changes in the MSSB-codes with the exception of the expressed emotions in the narratives. The children expressed more emotions at the end of the kindergarten year than at the beginning (see Table 5).

Second, the ethnic specificities in the change of social-emotional competence were examined (Table 5). For the avoidance strategies, there was an ethnic group-by-time interaction effect present. The avoidance strategies of the Asian-American and Latin-American children developed differently: Those of the Asian-American children increased, and the Latin-American children's decreased.

Third, the gender specificities in the change of social-emotional competence were analyzed (Table 5). There were no developments in a gender-specific way over the course of the kindergarten year.

Table 5. Changes in social-emotional competence.

	F (df = 1; 2)[c]						p(F)								
	time	ethnic	gender	time* ethnic	time* gender	time* ethnic*gender	time	ethnic	gender	time* ethnic	time* gender	time* ethnic*gender	η^2_{time}	η^2_{ethnic}[a]	η^2_{gender}[b]
Empathetic relation	1.70	1.05	1.37	1.40	0.00	1.30	.20	.31	.25	.23	.98	.26	.03	.02	.03
Avoidance strategies	1.03	<.01	0.14	4.85	0.00	0.17	.31	.99	.71	.03	.97	.68	.02	<.01	<.01
Moral themes	0.11	7.82	6.30	0.00	0.50	.30	.74	.01	.02	.99	.48	.59	<.01	.13	.11
Interpersonal conflict	2.52	1.35	.31	2.16	0.00	5.17	.12	.25	.58	.15	.99	.03	.05	.03	.01
Dissociation codes	0.56	0.88	2.87	0.63	1.08	1.03	.46	.35	.10	.43	.30	.32	.01	.02	.05
Dysregulated aggression	1.43	0.04	6.20	0.00	0.06	.67	.24	.85	.02	.96	.81	.42	.03	<.01	.11
Narrative emotions	12.95	1.99	0.47	0.79	0.17	.00	.00	.16	.50	.38	.68	.94	.20	.04	.01

A three factorial (time, gender, ethnicity) univariate analysis of variance with repeated measurements.
[a] Effect sizes for the time, ethnic and gender specific change; [c] Greenhouse-Geisser corrected; (M and SD see Table 1); Ethnic*gender interaction not significant

Fourth, there was a significant time*ethnic*gender interaction effect tor interpersonal conflict. The Asian Americans, girls as well as boys, increased the implementation of interpersonal conflicts in contrast to the decrease in the Latin-American children. The male children do it more intense than the girls do.

Discussion

The core findings of this study were:

1. The two ethnic groups in the classroom differed in moral reasoning at the beginning of the kindergarten year. A moral approach was utilized the most by the Latin-American children and the least by the Asian-American children.
2. The two ethnic groups did also differ in moral reasoning at the end of the kindergarten year.
3. At the beginning of the kindergarten year the gender groups differed in moral themes as well as in dysregulated aggression and dissociation codes. The boys showed more aggressive behavior than the girls, who displayed more moral reasoning. At the end of the kindergarten year the gender groups differed in dysregulated aggression since the boys implemented it more often than the girls.
4. The two ethnic groups developed differently concerning avoidance strategies and displayed at the end more expressed emotions in their narratives than at the beginning of the kindergarten year. The Asian-American children increased and the Latin-American children decreased the avoidance of conflicts over the course of the kindergarten year.

At the beginning of the kindergarten year (Hypotheses 1), compared to the Asian-American children, the Latin-American children more frequently applied moral themes to solving the stories. Using questionnaires and observations, Samples and colleagues were also able to observe more moral reasoning in Latin-American children than in their African-American counterparts [33]. This ethnic specificity might be explained by the strong Catholic background and the emphasis on a moral value system typical of their upbringing as reported in the literature [47]. Collectivism and obligation to the family was found to be prevalent in Mexican-American children and might explain the high usage of moral reasoning in this ethnic group [19].

At the end of the kindergarten year (Hypotheses 2), the ethnic differences in moral reasoning were still present. The results did not support the hypothesis that not only the ethnic background shapes moral reasoning or rather the value system during the kindergarten year. The results clearly showed that during one year the moral value system can not be enriched by experiences in the kindergarten. However, these data did not specify to which extend the moral value system was focused on in the kindergarten classroom. Since moral development is important for the development of social-emotional competence, the moral value system need to be addressed already in kindergarten. Based on the present data it can be further assumed that the "main effect model" might be independent of the ethnicity [26]. The ethnic-specific usage of aggression was not accompanied by ethnic-specific implementation of empathy and expressed emotions. However, the ethnic-specificity of the model has to be evaluated in more detail before conclusions can be drawn.

Concerning aggression, gender specificity was prevalent (Hypotheses 3). Hereby, the male children most frequently utilized dysregulated aggression compared to the female children. Gender differences concerning aggression are already well documented in the literature [48-51]. In addition to the ethnic differences at the beginning of the kindergarten year, female children used more moral reasoning than the male children. Also, the female children of the Wittenberg's study showed more moral reasoning, fairness and empathy than their male counterparts [52].

In a pre-post comparison (Hypotheses 4) the children of the two ethnic groups showed a significant increase in expressed emotions in the doll play narratives. As found in the literature, the exposure to prosocial peers—in this study the high amount of expressed emotions in the Latin-American children—was related to improved social behavior in antisocial peers one year later [8]. To detect changes in the other areas of social competences the observation phase of one year as well as the exposure of six hours per day had to be expanded. Furthermore, ethnically specific development took place exclusively in the avoidance strategies. The use of avoidance strategies increased in the Asian-American children whereas it decreased in the Latin-American children. The same contrasting development took place in tendencies in interpersonal conflicts. This decrease in avoidance strategies and interpersonal conflicts in Latin-American children replicated the process that the level of negative behavior decreased in children with initially higher levels when exposed to their more prosocial peers [3].

The results of this study rely on the observed doll play behavior, which is a good predictor for classroom behavior [41]. Even so, an additional observation of the classroom behavior would be helpful for drawing conclusions on child behavior in general. One limitation to this study was the relatively small sample size which was considered in the calculations. This can be explained by the distribution of ethnic groups in this particular Oakland district. A second limitation was the lack of information concerning the year of immigration and socio-demographic information of these children such as family situation, country of birth and religion. Information on the year of immigration would have been helpful to further specify the degree of integration and to weight the ethnic differences. Presumably, there may have been a varied degree of integration in this sample as the children came from a natural setting. Furthermore, the cultural background and its influence on child-rearing might possibly be linked to the religion practiced by the individual family group. Also, income and family situation could be of help to further understand the ethnic differences better. Respectively, one needs to proceed with caution when trying to transfer these conclusions to children of other ethnic groups or countries.

The strengths of this study are: Until now there have been only a few studies examining the effects of multi-ethnic-classrooms on child development. A strength of this study is the chosen setting. The kindergarten was located in a high risk area and attended exclusively by the children of families with an immigration background. In addition, the children were observed twice: at the beginning and at the end of their kindergarten year. The social-emotional competence was observed by an independent person who did not belong to the kindergarten staff.

As a recommendation for future studies, the design for examining social-emotional competence should definitely include doll play and observations of classroom behavior observation in multi-ethnic and non-multi-ethnic classrooms with large sample sizes. In a longitudinal design, additional development in the social-emotional competence of the different ethnic groups might be found. As an additional focus, school performance as well as language skills should be contained as well. It would be essential to also include an aggression prevention program as well as a program to increase social-emotional competence and their longitudinal outcome. There are more and more kindergartens where for example Spanish or Chinese are the classroom language for one ethnicity. The development of children attending such a classroom needs to be compared to children of multi-ethnic classrooms. This should be followed by a study of the effect of the exposure of multi-ethnicity during the children's later developments.

In summary, after taking the gender effects into account a specific ethnic diversity in social-emotional competence did exist in the particular sample. Considering the different highs and lows in the competencies of the ethnic groups in this study leads to the assumption that a natural model learning and imitation process did occur, e.g. the large amount of expressed emotions of the Latin-American children presented a model for the less emotionally expressive Asian-American children. Therefore, model learning as well as learning social-emotional competence in the early years is essential to prevent problematic behavior in multi-ethnic classrooms. These multi-ethnic classrooms build an environment with until now unknown effects on the child development. The more knowledge of early learning can be assimilated, the easier it will be to spot and prevent problematic behavior. Based on this knowledge programs can be developed to prevent and eliminate problematic behavior. In order to get to the early roots of problematic behavior these programs should especially focus on moral development and empathetic competence [53,54].

Competing Interests

The authors declare that they have no competing interests.

Authors' Contributions

KP did the first and final draft of the manuscript and critically revised it for the intellectual content. AvW has given final the approval of the version to be published. UH substantially contributed to the analysis and the interpretation of data. MC and PJ were responsible for the general supervision of the research group. They substantially contributed to the conception and the design of the study as well as the acquisition of the funding. All authors read and approved the final manuscript.

Acknowledgements

The authors would like to thank the Garfield school and their staff in Oakland, Ca for their patience and cooperation in this study. Furthermore, we would like to thank the Lego group cooperation for providing the necessary Lego figures for the study. Especially Liz Orrison contributed to this article since she reviewed several times the English language.

References

1. Lederer H: Structural Integration. In Effectiveness of National Integration Strategies towards Second Generation Migrant Youth in a Comparative European Perspective. Final Report to the European Commission, Annex Volume II: Results of the Field Survey (in cooperation with the EFFNATIS research team). Bamberg Edited by: Heckmann F, Lederer H, Worbs S. 2000, 28–30.
2. Bandura A: Social foundations of Thought and action: A social cognitive theory. Englewood Cliffs, New Jersey: Prentice-Hall; 1986.
3. Eisenberg N, Fabes RA, Spinrad TL: Prosocial Development. In Handbook of child psychology: Social, emotional, and personality development. Volume 3. Edited by: Eisenberg N, Fabes R, Spinrad T. Hoboken, New Jersey: John Wiley & Sons Inc; 2006:646–718.
4. Hartmann DP, Gelfand DM, Smith CL, Paul SC, Cromer CC, Page BC: Factors affecting the aquisition and elimination of children's donation behavior. J Exp Child Psychol 1976, 21:328–338.
5. Aaronfreed J: The socialisation of altruistic and sympathetic behavior: Some theoretical and experimental analyses. In Altruism and helping behaviour. Edited by: Macaulay J, Berkowitz L. New York, NY: Academic Press; 1970:103–126.

6. Eisenberg N, Fabes RA: Prosocial Development. In Handbook of child psychology: Social, emotional, and personality development. Volume 3. 5th edition. Edited by: Damon W (Editor-in-Chief), Eisenberg N. New York: Wiley; 1998:701–778.
7. Benson JF, Markovitz H, Roy R, Denko P: Behavioural rules underlying learning to share: Effects of development and context. International Journal of Behavioral Development 2003, 27:457–466.
8. Fabes RA, Moss A, Reesing A, Martin CL, Hanisch LD: The effects of peer prosocial exposure on the quality of young children's social interactions. Data presented at the annual conference of the National Council on Family Relations, Phoenix, AZ; 2005.
9. Gauthier Y: Infant Mental Health as enter the third Millennium: can we prevent aggression? Infant Mental Health Journal 2003, 24:296–308.
10. Tremblay RE: The development of aggressive behaviour during childhood: what have we learned in the past century? International Journal of Behavioral Development 2000, 24:129–141.
11. Juen F, Peham D, Juen B, Benecke C: Emotion, aggression, and the meaning of prevention in early childhood. In Emotions and aggressive behavior. Edited by: Steffgen G, Gollwitzer M. Ashland, OH: Hogrefe & Huber Publishers; 2007:201–214.
12. Donnellan MB, Trzesniewski KH, Robins RW, Moffit TE, Caspi A: Low Self Esteem is related to aggression, antisocial behaviour and delinquency. Psychological Science 2005, 15:328–335.
13. Lennon R, Eisenberg N: Emotional Displays Associated with Prescooler's Prosocial Behavior. Child Dev 1987, 58:992–1000.
14. Nadler A, Romek E, Shapira-Friedman A: Giving in the Kibbutz: Pro-Social Behavior of City and Kibbutz Children as Affected by Social Responsibility and Social Pressure. Journal of Cross-Cultural Psychology 1979, 10:57–72.
15. Hofstede G: Cultures consequences: International differences in work-related values. London: Sage; 1980.
16. Inglehart R: Modernization and postmodernizationw. London: Sage; 1997.
17. Sampson EE: Reinterpreting individualism and collectivism: Their religious roots and monologic versus dialogic person-other relationship. Am Psychol 2001, 55:1425–1432.
18. Miller JG, Bersoff DM, Harwood RL: Perceptions of social responsibilities in India and in the United States: moral imperatives or personal decisions? J Pers Soc Psychol 1990, 58:33–47.

19. Freeberg AL, Stein CH: Felt obligations towards parents in Mexican-American and Anglo-American young adults. Journal of Social and Personal Relationships 1996, 13:457–471.
20. Oetzel JG: Explaining Individual Communication Processes in Homogeneous and Heterogeneous Groups Through Individualism-Collectivism and Self-Construal. Human Communication Research 1998, 25:202–224.
21. Oetzel JG: Culturally homogeneous and heterogeneous groups: explaining communication processes through individualism-collectivism and self-construal. International Journal of Intercultural Relations 1998, 22:135–161.
22. Gabrielidis C, Stephan WG, Ybarra O, Dos Santos Pearson VM, Villareal L: Preferred Styles of Conflict Resolution. Journal of Cross-Cultural Psychology 1997, 28:661–677.
23. Leung K, Bond MH: How Chinese and Americans reward task-related contributions: a preliminary study. Psychologia 1982, 25:32–39.
24. Eisenberg N: The Development of Empathy-Related Responding. Nebraska Symposium on Motivation 2005, 51:73–117.
25. Smith A: Cognitive Empathy and Emotional Empathy in Human Behavior and Evolution. The Psychological Record 2006, 59:3–21.
26. Strayer J, Roberts W: Empathy and Observed Anger and Aggression in Five Year Olds. Social Development 2004, 13:11–13.
27. Pepler D, Craig W, Roberts W: Observations of aggressive and nonaggresive children on the school playground. Merill Palmer Quarterly 1998, 44:55–76.
28. Hansen BK, Bryant BK: Peer influence on sharing behavior of Mexican-American and Anglo-American boys. Journal of Social Psychology 1980, 110:135–136.
29. Kagan S, Knight GP: Social Motives among Anglo American and Mexican American Children: Experimental and Projective Measures. Journal of Research in Personality 1981, 15:93–106.
30. Knight GP, Nelson W, Kagan S, Gumbiner J: Cooperative-competitive social orientation and school achievement among Anglo American and Mexican American Children. Contemporary Educational Psychology 1982, 7:97–106.
31. Stevenson HW: The development of prosocial behavior in large-scale collective societies: China and Japan. In Cooperation and prosocial behaviour. Edited by: Hinde RA, Grovel J. Cambridge, England: Cambridge University Press; 1991:89–105.
32. Hieshima JA, Schneider B: Intergenerational effects on the cultural and cognitive socialization of third- and fourth-generation Japanese Americans: Diversity

and development of Asian Americans. Journal of Applied Developmental Psychology 1994, 15:319–327.

33. Samples FL: Cognitions, behaviours and psychological symptomatology: Relationships and pathways among African American and Latino children. Journal of Negro Education 1997, 66:172–188.

34. Coie LD, Dodge KA: Aggression and antisocial behaviour. In Handbook of Child Psychology, Social, Emotional and Personality Development. Volume 3. Edited by: Damon W, Eisenberg N. New York: Wiley; 1998.

35. Crick NR, Casas JF, Ku H: Relational and physical forms of peer victimization in preschool. Dev Psychol 1999, 35:376–385.

36. Eisenberg N: Emotion, regulation, and moral development. Annual Review Psychology 2000, 51:665–697.

37. Einwanderungsland USA [http://www.american-future.de/].

38. Bretherton I, Ridgeway D, Cassidy J: Assessing Internal Working Models of the Attachment Relationship. In Attachment in the preschool years: Theory, Research and intervention. Edited by: Greenberg MT. Chicago: Chicago University Press; 1990:273–299.

39. Bretherton I, Oppenheim D: The McArthur Story Stem Battery: Development, Administration, Reliability, Validity and reflections about meaning. In Revealing the inner world of young children. Edited by: Emde RN, Wolf DP, Oppenheim D. Oxford, New York: Oxford University Press INC; 2003:13–19.

40. Emde RN: Early Narratives: A Window to the Child's inner world. In Revealing the inner world of young children. Edited by: Emde RN, Wolf DP, Oppenheim D. Oxford, New York: Oxford University Press INC; 2003:13–19.

41. Warren S, Oppenheim D, Emde RN: Can Emotions and Themes in chidren's play predict behavior problems? J Am Acad Child Adolesc Psychiatry 1996, 35:1331–1337.

42. Oppenheim D, Emde R, Hasson M, Warren S: Preschoolers face moral dilemmas: A longitudinal Study of Achmowledging and resolving internal conflict. Int J Psychoanal 1997, 78:943–57.

43. Robinson J, Mantz-Simmons L: The MacArthur Narrative Coding System: One Approach to Highlight Affective Meaning Making in the MacArthur Story Stem Battery. In Revealing the inner world of young children. Edited by: Emde RN, Wolf DP, Oppenheim D. Oxford, New York: Oxford University Press INC; 2003:13–19.

44. Robinson J, Mantz-Simmons L, MacFie J, the Mac Arthur Narrative Working Group: The MacArthur Narrative Coding System. In Unpublished document. University of Colorado Health Science Center, Denver; 1993.

45. Warren S: Narrative Emotion Coding System. In Revealing the inner world of young children. Edited by: Emde RN, Wolf DP, Oppenheim D. Oxford, New York: Oxford University Press INC; 2003:13–19.
46. Robinson J, Mantz-Simmons L: The MacArthur Narrative Coding System: One Approach to Highlight Affective Meaning Making in the MacArthur Story Stem Battery. In Revealing the inner world of young children. Edited by: Emde RN, Wolf DP, Oppenheim D. Oxford, New York: Oxford University Press INC; 2003:312–320.
47. Richardson DT: Second Step and Social/Emotional Learning Among African American and Latino Children. In Unpublished manuscript. University of California, San Francisco; 2003.
48. Björkqvist K, Österman K, Kaukiainen A: The development of direct and indirect aggressive stategies in males and females. In Of mice and women: Aspects of female aggression. Edited by: Björkqvist K, Niemelä P. San Diego, CA: Academic press; 1992:51–64.
49. Björkqvist K, Österman K, Kaukiainen A: Social Intelligence—Empathy = Aggression? Aggression and Violent Behavior 2000, 5:191–200.
50. Björkqvist K, Österman K, Lagerspetz KMJ: Sex differences in covert aggression among adults. Aggressive Behavior 1994, 20:27–33.
51. Knight GP, Guthrie IK, Page MC, Fabes RA: Emotional arousal and gender differences in aggression: A Meta-analysis. Aggressive Behavior 2002, 28:366–393.
52. Wittenberg RT: The moral dimension of children's and adolescents' conceptualisation of tolerance to human diversity. Journal of Moral Education 2007, 36:433–451.
53. Eisenberg N, Cumberland A, Guthrie IK, Murphy BC, Shepard SA: Age changes in prosocial responding and moral reasoning in adolescence and early adulthood. Journal of Research on Adolescence 2005, 15:235–260.
54. Kaukiainen A, Björkqvist K, Lagerspetz K, Österman K, Salmivalli C, Rothberg S, Ahlbom A: The Relationships Between Social Intelligence, Empathy, and Three Types of Aggression. Aggressive Behavior 1999, 25:81–89.

Adolescents' Experience of Comments about their Weight—Prevalence, Accuracy and Effects on Weight Misperception

Wing-Sze Lo, Sai-Yin Ho, Kwok-Kei Mak, Yuen-Kwan Lai and Tai-Hing Lam

ABSTRACT

Background

Weight comments are commonly received by adolescents, but the accuracy of the comments and their effects on weight misperception are unclear. We assessed the prevalence and accuracy of weight comments received by Chinese adolescents from different sources and their relation to weight misperception.

Methods

In the Hong Kong Student Obesity Surveillance (HKSOS) project 2006–07, 22612 students aged 11–18 (41.5% boys) completed a questionnaire on obesity. Students responded if family members, peers and professionals had seriously commented over the past 30 days that they were "too fat" or "too thin" in two separate questions. The accuracy of the comments was judged against the actual weight status derived from self-reported height and weight. Self-perceived weight status was also reported and any discordance with the actual weight status denoted weight misperception. Logistic regression yielded adjusted odd ratios for weight misperception by the type of weight comments received.

Results

One in three students received weight comments, and the mother was the most common source of weight comments. Health professional was the most accurate source of weight comments, yet less than half the comments were correct. Adolescents receiving incorrect comments had increased risk of having weight misperception in all weight status groups. Receiving conflicting comments was positively associated with weight misperception among normal weight adolescents. In contrast, underweight and overweight/obese adolescents receiving correct weight comments were less likely to have weight misperception.

Conclusion

Weight comments, mostly incorrect, were commonly received by Chinese adolescents in Hong Kong, and such incorrect comments were associated with weight misperception.

Background

It is well known that many adolescents misperceive their weight [1-8]. For example, Brener et al. [2] reported that half the normal weight high school students had weight misperception. About 30% of normal weight Hong Kong adolescents misperceive themselves as fat [9], and more girls than boys overestimate their weight [4,10-14]. The perception of suboptimal weight is associated with depressive symptoms and other psychological problems among adolescents in both cross-sectional and longitudinal studies [15-18].

The lack of knowledge about and access to growth charts has probably made it difficult for adolescents to evaluate their weight status objectively. Frequent exposure to the media's portrayals of thin ideal for females [19-23] and muscular physique for males [24-26] may therefore predispose adolescents to weight misperception [27-29]. The Tripartite Influence Model [30,31] also suggests that weight comments and opinions from parents and peers may influence adolescent weight perception.

Teasing about weight is common among adolescents [32-34]. Neumark-Sztainer et al. [32] reported in a large US study that 25% of secondary school students were teased about their weight several times in the past year. Adolescents are sensitive to weight-related influences, and may experience tremendous pressure from weight teasing. Cross-sectional and retrospective studies [30,32-36] have linked adolescent weight teasing to poor self-esteem and body image, unhealthy weight-control attempts, as well as eating disorders. There is also prospective evidence that weight teasing predicts psychological distress among adolescents [37].

Existing Western studies were implicit whether weight teasing was about being too fat or too thin, but given the high prevalence of obesity and the associated negative image, the former was much more likely. Although the effects of teasing about being too thin are uncertain, the perception of being too thin is associated with anxiety and depressive symptoms in cross-sectional studies [16,18]. Weight teasing about thinness would be more relevant in developing countries such as China, where underweight is common [38].

Most studies about adolescent weight teasing referred to that from peers and parents [32-35]. However, in Asian 3-generation families, grandparents may also exert great influence on adolescent eating patterns [39]. Grandparents, teachers, social workers and health professionals are all potential sources of adolescent weight comments although little is known about the prevalence, accuracy and effects of these comments.

Previous studies mainly focused on weight teasing, which included disparaging nicknames and making fun of others' weight and body shape [32-34,40]. However, weight-related comments could also be constructive and well-intentioned. For example, a caring mother could remind her teenage son of his bulging waistline and a family doctor could and should advise adolescents of their weight status. Moreover, to the best of our knowledge, no study has examined whether correct, incorrect or even conflicting (same person receiving opposite weight comments of being too fat and too thin) weight comments are independently associated with weight misperception in adolescents.

In the present study we extended existing research in three ways. First, we included parents, siblings and grandparents, as well as teachers, social workers and health professionals as separate sources of weight comments. Second, we examined the accuracy of those weight comments. Third, we investigated the effects of different types (correct, incorrect and conflicting) of weight comments on weight misperception. We hypothesized that adolescents who received incorrect and conflicting weight comments were more likely to have weight misperception, whereas those who received correct weight comments were less likely to have weight misperception, compared with adolescents who did not receive any weight comments over the past 30 days.

Methods

Data Collection and Subjects

The present study was part of a large population-based study, the Hong Kong Student Obesity Surveillance (HKSOS) project. Stratified cluster sampling was applied, and the schools were sampled with stratification by school district, source of funding, language of instruction (Chinese/English), religious background (Christian/Others/None) and single sex/co-education to represent all main stream non-international secondary schools in Hong Kong. Forty-two schools participated in this survey. All Form 1 to 7 students (equivalent to Grade 7–12 in US) in selected schools were invited to participate. We have obtained the consent of schools who acted in loco parentis for the students. Passive consent from the parents was obtained and all students participated on voluntary basis. Ethical approval was granted by the Institutional Review Board of the University of Hong Kong/Hospital Authority Hong Kong West Cluster.

In 2006–2007, the anonymous baseline survey was self-administered in classrooms under the supervision of trained researchers or teachers. In the present study 31603 students aged 11 to 18 were eligible. Of these, 2319 were excluded because self-reported height and weight data were missing. As the maximum reference values for height of the Hong Kong official weight-for-height (WFH) cut-offs were 175 cm in boys and 165 cm in girls [41], students who exceeded these height limits were excluded (n = 738). Also excluded were extreme body mass index (BMI) values beyond 10 (biological limit) and 50 (morbidly obese) (n = 4752) [42-44]. After further exclusion of 1182 questionnaires with incomplete data, 22612 (41.5% boys) remained for analyses. The test-retest reliability of measures used in this study was assessed with 1147 students (31.3% boys; mean age = 14.8 ± 1.6) from 3 of the participating schools over an interval of one month.

Measures

Actual Weight Status

Height (cm/inch) and weight (kg/lb) were self-reported by the participants to the nearest integer. Using sex-specific Hong Kong official weight-for-height (WFH) cutoffs [41], participants were defined as underweight (< 80% median weight-for-height), normal weight (80–120% median weight-for-height) and overweight/obese (> 120% median weight-for-height).

Weight Perception

Weight perception was measured using a standard question that asked students to describe their weight status as very thin, thin, just right, fat or very fat. This measure has been widely used in population surveys conducted in the US [26] and the UK [27], as well as in a large adolescent health behaviour study in Mainland China [4]. As relatively few students chose extreme categories (6.4% for very fat and 4.5% for very thin), the five categories were consolidated into three groups: thin (including very thin), just right and fat (including very fat). Overweight/obese, underweight and normal weight students who considered themselves other than fat, thin or just right, respectively, were classified as having "weight misperception" as opposed to "correct perception."

Prevalence of Weight Comments Received

To assess the prevalence of weight comments received by the students, two questions were used. The first question asked whether anyone had seriously commented over the past 30 days that the student was "too fat." A list of social contacts was provided as response options including: 1) family members (father, mother, siblings, grandfather, grandmother and other relatives); 2) peers (friends and classmates); 3) professionals (teachers, social workers and health professionals); and 4) others (domestic helpers and neighbors). There was also an option of "none" to indicate that no one had given such comments. The second question was identical except that "too fat" was replaced by "too thin."

Accuracy and Types of Weight Comments

The accuracy of the weight comments was assessed against the actual weight status. The weight comment of "too fat" was deemed correct to overweight/obese students but incorrect to normal or underweight students. Similarly, "too thin" was considered a correct comment to underweight students but incorrect to overweight/obese or normal weight students. Weight comments of "too fat" or "too thin" were considered incorrect to normal weight students. Therefore, according to the weight comments received and their accuracy, four categories were identified: 1) no weight comments, 2) correct weight comments, 3) incorrect weight comments and 4) conflicting weight comments (receiving both "too fat" and "too thin" comments regardless of actual weight status).

Data Analysis

Bivariate and Multivariate Analyses

Chi-square and Student's t-statistics were used to test the differences of basic characteristics and weight comments between boys and girls. The prevalence of weight comments from different sources was also compared between sexes by Chi-square statistics. In the bivariate analysis, the percentages of adolescents receiving correct, incorrect, conflicting or no weight comments were calculated, and Chi-square statistics were used to compare between correct perception and weight misperception.

In the multivariate logistic regression analyses, the binary outcome variable of weight misperception was regressed on the independent variable of weight comments taking the category "none" (no weight comments received) as the reference group. Odds ratios (ORs) and 95% confidence intervals (95% CI) for weight misperception were estimated, adjusting for age (as a continuous variables), BMI (weight (kg) divided by height in squared (m2)), and three socio-demographic factors, including the place of birth (Hong Kong, or elsewhere), the highest parental education (≤ primary, secondary, or ≥ tertiary), and perceived family affluence (relatively poor, medium, or relatively wealthy). The regression model was conducted with robust standard errors accounting for school clustering effect (design effect = 3.16). A CI range excluding 1 and $p < 0.05$ indicated that the OR was significant. Correlation coefficients between independent variables were examined ($r = 0.007$ to 0.44), and none of which exceeded 0.8 that indicates problems of multicollinearity [45]. All the main analyses were stratified by sex and actual weight status.

All statistical analyses were performed using STATA 9.0 (Stata Corporation, College Station, TX) with the significance level set at 5%. Multilevel logistic regression was not performed as we aimed to investigate population-averaged effects rather than school-specific effects.

Test-Retest Reliability

Intra-class correlation coefficients (ICC) were used to examine the test-retest reliability of continuous variables [46], while Kappa statistics (κ) was used for categorical variables. When there is a low prevalence of a particular response, or when the frequency is unbalanced, κ may be low [47], therefore percent agreement was also examined for categorical variables. The ICC for height (0.95) and weight (0.85) data in 1147 subjects were high and comparable to those of published reports [48]. Others have also found high correlations between self-reported and measured anthropometric data in adolescents [49]. Based on percent agreement, test-retest reliability of weight comments ranged from moderate to high (κ = 0.16—0.60; percent agreement = 74.3%—95.1%). Agreements of perceived weight status (κ = 0.68; percent agreement = 79.9%) was good. These test-retest reliability statistics were similar between boys and girls.

Handling of Missing Data

Missing values for the place of birth (0.6%), highest parental education (12.8%) and housing type (1.0%) were imputed using multiple imputations [50]. Five imputations were generated using the software program Amelia, based on a model that uses values from other variables to achieve optimal estimates [38]. Similar results were obtained using the imputed or original databases, so only the imputed database was used for its larger effective sample size.

Results

Basic Characteristics of Our Subjects

Table 1 summarizes the basic characteristics of the final sample (n = 22612; 41.5% boys). The sample was representative of Hong Kong adolescents in terms of sex, age, and residential district (all Cohen effect sizes [51] ≤ 0.2) (table not shown) despite some subjects were excluded due to missing data.

Table 1. Basic characteristics of participants (N = 22612).

Characteristics	All (N = 22612)	Boys (N = 9375)	Girls (N = 13237)	Test statistics
Age (years, mean, SD)	14.7 (1.70)	14.6 (1.70)	14.7 (1.70)	$t = -3.35$, $p = 0.019$
Form (%)				$\chi^2 = 27.52$, $p < 0.001$
Junior (F1–F3)[a]	59.6	61.7	58.2	
Senior (F4–F7)[a]	40.4	38.3	41.8	
BMI (kg/m², mean, SD)	19.2 (2.99)	19.5 (3.29)	18.9 (2.73)	$t = 13.29$, $p < 0.001$
Weight status by local references (%)				$\chi^2 = 204.15$, $p < 0.001$
Underweight	8.7	8.6	8.7	
Normal weight	79.0	75.3	81.6	
Overweight/Obese	12.4	16.1	9.7	
Parental education level (%)				$\chi^2 = 26.07$, $p < 0.001$
Primary or below	12.0	12.2	11.9	
Secondary	65.9	64.1	67.2	
Tertiary or above	22.1	23.7	21.0	
Place of birth (%)				$\chi^2 = 3.63$, $p = 0.06$
Hong Kong	73.9	73.6	74.6	
Other places[b]	26.1	26.4	25.4	
Family affluence (%)				$\chi^2 = 38.03$, $p < 0.001$
Relatively poor	36.4	38.6	34.9	
Medium	52.6	50.3	54.3	
Relatively wealthy	10.9	11.1	10.8	
Weight perception (%)				$\chi^2 = 177.35$, $p < 0.001$
Correct perception	47.6	52.9	43.9	
Weight misperception	52.4	47.1	56.1	
Received weight comments of being "too fat"	29.0	22.3	33.8	$\chi^2 = 350.00$, $p < 0.001$
Received weight comments of being "too thin"	26.9	28.4	25.8	$\chi^2 = 17.71$, $p < 0.001$

[a] F1–F3 is equivalent to grade 7 to grade 9; F4–F7 is equivalent to grade 10 to grade 12.
[b] Other places: Mainland China (majority), Macau, Western countries and others.

The prevalence rates of underweight and overweight/obese were 8.7% (8.6% for boys, 8.7% for girls) and 12.4% (16.1% for boys, 9.7% for girls), respectively. Boys and girls were significantly different in terms of age, form (school grade), BMI, weight status, highest parental education, place of birth, and perceived family affluence.

Weight misperception was identified in over half the subjects (52.4%) and was more common in girls (56.1%) than boys (47.1%) ($\chi 2 = 177.4$, $p < 0.001$). In the past 30 days, 29.0% and 26.9% of adolescents received weight comments of being "too fat" and "too thin," respectively. The comment of being "too fat" was more commonly received by girls (33.8%) than boys (22.3%) ($\chi 2 = 350.0$, $p < 0.001$), while that of "too thin" more commonly received by boys (28.4%) than girls (25.8%) ($\chi 2 = 17.7$, $p < 0.001$).

Sources of "Too Fat" and "Too Thin" Weight Comments by Sex

Among adolescents who received comments of being "too fat," 53.9% of the comments came from a single source, 24.7% from two sources and 10.6% from three sources. The corresponding figures for comments of being "too thin" were 54.2%, 22.7% and 10.2%.

Table 2 shows that the mother was the most common source of "too fat" comments for both boys and girls, followed by siblings and classmates. The comment of being "too fat" was more commonly received by girls than boys from all sources except the grandfather although more boys than girls were overweight/obese. The mother was also the most common source of "too thin" comments. Girls were nearly twice as likely as boys to receive comments of being "too thin" from classmates and friends. Grandparents, teachers, social workers and health professionals were more likely to give comments of being "too thin" than "too fat" to both boys and girls.

Table 2. Sources of weight comments received by boys and girls

	Too fat (%)				Too thin (%)				Ratio (Too fat/Too thin)	
	Boys	Girls	χ^2	p	Boys	Girls	χ^2	p	Boys	Girls
Family	14.4	26.1	446.21	< 0.001	22.2	18.9	37.35	< 0.001	0.65	1.38
Father	5.3	7.7	50.98	< 0.001	8.7	7.1	20.86	< 0.001	0.61	1.08
Mother	9.0	17.0	299.75	< 0.001	15.1	12.4	36.43	< 0.001	0.60	1.37
Siblings	4.1	10.6	312.18	< 0.001	3.9	2.8	20.41	< 0.001	1.05	3.79
Grandfather	1.0	0.7	5.09	0.024	2.0	1.3	18.53	< 0.001	0.50	0.54
Grandmother	0.6	1.5	41.33	< 0.001	2.5	2.7	0.56	0.45	0.24	0.56
Other relatives	1.2	4.4	185.95	< 0.001	3.6	4.4	6.88	0.009	0.33	1.00
Peers	7.5	13.5	204.17	< 0.001	6.5	11.4	153.34	< 0.001	1.15	1.18
Classmates	5.2	8.4	89.30	< 0.001	4.7	7.8	84.15	< 0.001	1.11	1.08
Friends	4.2	8.9	189.96	< 0.001	4.2	7.7	112.95	< 0.001	1.00	1.16
Professionals	2.1	2.2	0.62	0.50	2.8	2.8	0.007	0.97	0.75	0.79
Teachers	0.8	1.0	3.37	0.07	1.3	1.4	0.24	0.63	0.62	0.71
Social workers	0.4	0.4	0.09	0.77	0.5	0.6	0.75	0.39	0.80	0.67
Health professionals	0.9	0.9	0.19	0.89	1.2	1.2	0.03	0.87	0.75	0.75
Others[a]	5.5	4.7	5.17	0.012	6.5	4.5	42.99	< 0.001	0.85	1.04

[a] Others include neighbors and domestic helpers.

Prevalence of Correct Weight Comments from Different Sources

Figures 1 and 2 show the prevalence of correct weight comments from different sources. In general, health professionals and teachers were the two most accurate sources of "too fat" and "too thin" comments. The third most accurate source of "too fat" comments was the father, and of "too thin" comments classmates.

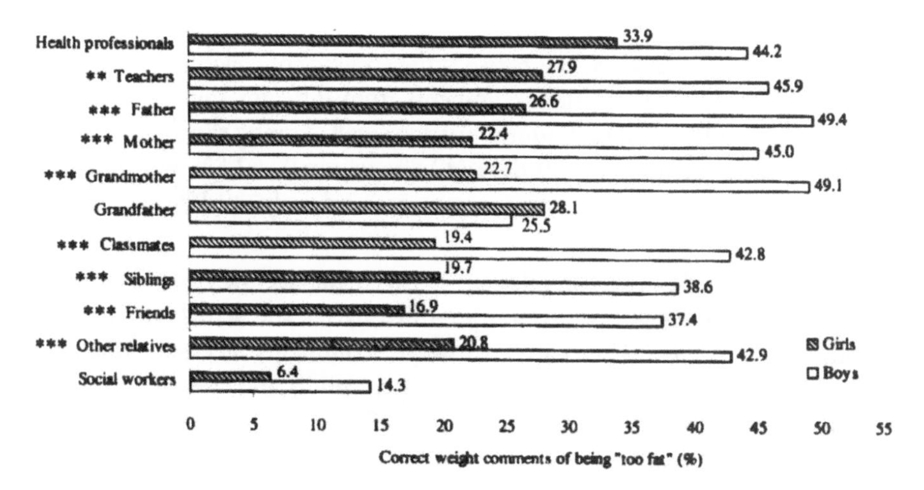

Figure 1. Prevalence of correct weight comments of being "too fat" by source and sex (descending order). Key: $^{*}p < 0.05$; $^{**}p < 0.01$; $^{***}p < 0.001$ between boys and girls.

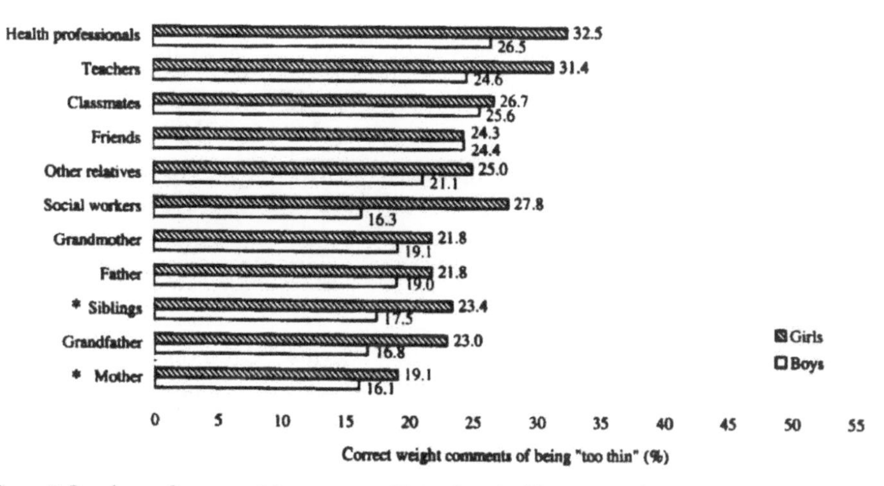

Figure 2. Prevalence of correct weight comments of being "too thin" by source and sex (descending order). Key: $^{*}p < 0.05$; $^{**}p < 0.01$; $^{***}p < 0.001$ between boys and girls.

Boys received more correct comments of being "too fat" than girls from all sources but the grandfather (28.1% girls vs 25.5% boys, $\chi 2 = 0.2$, $p = 0.69$). The accuracy of "too thin" comments from all sources was similar in boys and girls except that it was more accurate for girls than boys of comments from the mother (19.1% girls vs 16.1% boys, $\chi 2 = 4.8$, $p = 0.029$) and siblings (23.4% girls vs 17.5% boys, $\chi 2 = 3.9$, $p = 0.049$).

Bivariate Analyses

Before examining the associations between the accuracy of weight comments and weight misperception, the weight comments from different sources were combined as the sample size of each source was too small. Among those who received weight comments in the past 30 days (47.8% in total), 62.8% were incorrect comments, 17.1% were conflicting comments and only 20.1% were correct comments.

Table 3 shows the bivariate relations between the accuracy of weight comments received and weight perception, stratified by sex and actual weight status. In both boys and girls, the accuracy of weight comments received was significantly different between those with correct perception and weight misperception, among all weight status groups (all $p < 0.001$). As the accuracy of weight comments was significantly associated with weight perception, logistic regression models were then performed to estimate the OR and (95% CI) of having weight misperception, adjusting for potential confounders.

Table 3. The types of weight comments by weight perception in boys and girls of different weight statusa.

	Boys (n = 9375)				Girls (n = 13237)			
	Correct perception (n = 4958) %	Weight misperception (n = 4417) %	χ^2	p	Correct perception (n = 5812) %	Weight misperception (n = 7425) %	χ^2	p
UN	(n = 496)	(n = 311)	135.45	< 0.001	(n = 612)	(n = 539)	246.93	< 0.001
No comments	38.5	67.5			26.6	51.6		
Correct	54.6	15.1			65.7	22.3		
Incorrect	0	3.5			1.0	15.0		
Conflicting	6.9	13.8			6.7	11.1		
NW	(n = 3353)	(n = 3710)	545.23	< 0.001	(n = 4082)	(n = 6715)	600.24	< 0.001
No comments	75.7	50.3			65.5	41.6		
Correct	-	-			-	-		
Incorrect	16.2	40.6			27.8	50.3		
Conflicting	8.1	9.1			6.7	8.0		
OV/OB	(n = 1109)	(n = 396)	261.11	< 0.001	(n = 1118)	(n = 171)	228.84	< 0.001
No comments	37.9	62.1			30.5	53.8		
Correct	51.2	8.8			62.8	17.0		
Incorrect	2.0	13.9			1.0	19.3		
Conflicting	8.9	15.2			5.7	9.9		

[a] Using the sex-specific local weight-for-height (WFH) cutoffs, our subjects were defined as underweight (< 80% median WFH), normal weight (80% – 120% median WFH) and overweight/obese (> 120% median WFH).
UN = underweight; NW = normal weight; OV/OB = overweight/obese

Multivariate Analyses

Table 4 shows the adjusted ORs for weight misperception by the accuracy of weight comments received compared with not receiving any weight comments. ORs greater than 1 indicate that weight misperception is more likely. Incorrect weight comments were positively associated with weight misperception in normal weight and overweight/obese boys. ORs could not be calculated for underweight boys as none of them who had correct weight perception received incorrect weight comments. Conflicting comments were associated with an increased risk of weight misperception (OR = 1.73, 95% CI = 1.43–2.09, $p < 0.001$) among boys who were normal weight but not underweight or overweight/obese. Underweight and overweight/obese boys who received correct weight comments were 84% and 89%, respectively, less likely to have weight misperception, thus indicating a beneficial effect.

Table 4. Adjusted odds ratio (95% confidence interval) for weight misperception in relation to different types of weight comments received by sex and weight statusa (N = 22612).

		Boys		Girls	
	Types of weight comments received	Adjusted[b]	p	Adjusted[b]	p
UW	None	1		1	
	Correct	0.16 (0.11–0.22)	< 0.001	0.18 (0.14–0.23)	< 0.001
	Incorrect[c]	-	-	6.88 (2.68–17.68)	< 0.001
	Conflicting	1.25 (0.80–1.94)	0.3	0.74 (0.45–1.20)	0.2
NW	None	1		1	
	Correct	-	-	-	-
	Incorrect	3.73 (3.28–4.23)	< 0.001	2.89 (2.67–3.13)	< 0.001
	Conflicting	1.73 (1.43–2.09)	< 0.001	1.93 (1.67–2.23)	< 0.001
OV/OB	None	1		1	
	Correct	0.11 (0.08–0.15)	< 0.001	0.17 (0.11–0.26)	< 0.001
	Incorrect	4.75 (2.79–8.09)	< 0.001	12.7 (5.24–30.83)	< 0.001
	Conflicting	1.07 (0.80–1.44)	0.7	0.97 (0.53–1.79)	0.9

[a] Using the sex-specific local weight-for-height (WFH) cutoffs, our subjects were defined as underweight (< 80% median WFH), normal weight (80% – 120% median WFH) and overweight/obese (> 120% median WFH).
UW = Underweight; NW = Normal weight; OV/OB = Overweight/Obese.
[b] Adjusted for age, highest parental education, family affluence, place of birth, BMI and school effect.
[c] OR cannot be calculated due to the insufficient number in the group (Underweight boys with weight misperception did not receive any incorrect comments).

Similar results were observed in girls. Girls who received incorrect weight comments were more likely to have weight misperception regardless of their weight status. The small numbers of underweight and overweight/obese girls have resulted in the wide 95% CI in these subgroups (only 1% of girls with correct weight perception received incorrect weight comments, as shown in Table 3). Normal weight girls who received conflicting weight comments were also more likely to have weight misperception, with adjusted OR (95% CI) of 1.93 (1.67–2.23). However, conflicting comments in underweight and overweight/obese girls were not significantly associated with weight misperception. Consistent with the results in boys, underweight and overweight/obese girls who received correct weight comments had significantly lower risks of weight misperception, with adjusted odds ratio (95% CI) of 0.18 (0.14–0.23) and 0.17 (0.11–0.26), respectively.

Discussion

The most important finding of the present study is that despite the high rates of weight comments received by adolescent boys and girls, less than one-fifth of the comments were accurate. This is alarming as we also found incorrect weight comments were associated with weight misperception. Underweight and overweight/obese adolescents with weight misperception may be unaware of their weight problems [8,52], whereas normal weight adolescents with distorted perceptions of their weight may engage in unhealthy weight control behaviors [52]. Previous studies have found that normal weight adolescents with weight misperception were more likely to have psychosocial health problems and poor self-esteem [4,6,53].

Prevalence of Weight Comments from Different Sources

Our findings are in line with previous research that girls received significantly more weight comments of being fat from family members and peers than boys [32,34]. While previous research focused mainly on weight comments from the parents or the family as a group, we have specifically included the father, mother, grandfather, grandmother and siblings. We found that the comment of being "too fat" was most commonly made by the mother and siblings. It was suggested that parental weight comments would increase teasing by siblings [33] although this could not be examined using our data. We also found that girls were more likely than boys to receive comments of being "too fat" from siblings (Table 2). Although weight comments from grandparents were uncommon, they were interestingly dominated by the comment of being "too thin" (Table 2). As being fat is traditionally a sign of health and wealth among older Chinese people, many of whom have experienced poverty and hunger, grandparents may prefer children to eat and weigh more [39].

Our result that family members exceeded peers as a source of weight comments was in contrast to published findings [34]. This is probably because any well-intentioned comments, more likely from family members, were included in the present study, whereas only weight teasing was included in other studies.

Low Accuracy of Weight Comments by Health Professionals

As expected, health professionals were the most accurate source of weight comments, although less than half the comments were correct. In a US study, 76% of obese adolescents were correctly identified as obese by physicians [54]. A recent Canadian study also reported that about 60% of physicians could accurately estimate adolescent body size [55]. These findings may not be directly comparable due to differences in research methods and the proportion of borderline overweight and obese subjects whose weight status would be more difficult to determine [56]. Nevertheless, health professionals in Hong Kong should routinely measure the weight status of adolescents and advise accordingly despite the lack of consultation time, space and appropriate equipment [56]. The rapidly changing body dimensions during adolescence, and the lack of a well recognized weight status standard for children might have also contributed to the low accuracy of weight comments by health professionals.

Prevalence of Weight Misperception Among Hong Kong Adolescents

More than half our students had weight misperception (Table 3), which is consistent with previous studies in Western countries [2,3], Mainland China [4] and Hong Kong [9]. The lack of knowledge about and access to growth charts probably make it difficult for adolescents to evaluate their weight status objectively. As adolescents are susceptible to social influences [57-59], weight comments and opinions from the family and peers may also influence weight perception according to the Tripartite Influence Model [30,31]. However, our study found that most weight comments from parents and peers were inaccurate. In Hong Kong, one in three adults misperceived their own weight status [60] while in the UK and US, only 25–35% of parents could correctly identify the weight status of their obese adolescent child [49,61]. Whether parents who misperceived their own weight status are more likely to give incorrect weight comments to their children is still unclear.

Weight Misperception and the Accuracy of Weight Comments

We found that incorrect and conflicting weight comments were in general associated with weight misperception in adolescents. In contrast, correct weight comments were associated with correct weight perception. While weight teasing was linked to negative outcomes such as poorer self-esteem, unhealthy weight-control behaviors, depressive symptoms and even eating disorders [30,32-36], we found that accurate weight comments may help adolescents establish an appropriate weight perception. Further studies should examine whether correct weight perceptions are associated with effective weight control among adolescents.

Strengths and Limitations

Our large territory-wide sample allowed us to perform detailed analyses stratified by sex and weight status. We have also considered more sources of weight comments than previous studies and the comments were not limited to teasing only.

However, our study has several limitations. The anthropometric data and weight comments were self-reported although their test-retest reliability was good and comparable to that of other similar studies [2,49]. Weight comments were self-reported, but their association with weight misperception as expected supported the validity of the self-reported comments. To reduce any recall error of reported weight comments, a shorter time frame of the past 30 days was used. The frequency of weight comments was not assessed due to the length of the weight comments.

Our data were not sufficient to compare the impact of individual source of weight comments on weight misperception. According to Keery et al. [33], weight teasing from the father and elder brothers was associated with higher psychosocial distress than the mother and sisters. The effect of weight teasing or comments may also differ in male and female recipients. Therefore, more information about the relationship between the source and recipient of the weight comments is needed to understand the different effects. Future study should also investigate the feelings of adolescents after receiving weight comments. The association of incorrect weight comments and weight misperception with negative psychosocial health problems should be clarified by longitudinal studies.

Conclusion

Weight comments from family members and peers are commonly received by Chinese adolescents, yet most comments are inaccurate. Family members, peers and professionals should realize the potential adverse effects of incorrect weight comments, and adolescents should be taught how to correctly assess their weight status to establish correct weight perceptions. Health professionals should regularly give appropriate weight advice to adolescents based on objective measurements.

Competing Interests

We hereby declare that we do not have a financial association or other conflict of interest with the subjects mentioned in this manuscript.

Authors' Contributions

WSL contributed to study design and management, performed statistical analyses and drafted the manuscript; SYH is the principle investigator of the HKSOS project and critically revised the manuscript; KKM and YKL contributed to study design, coordination and revision of the manuscript; THL gave critical revision of the manuscript and supervision. All authors read and approved the final manuscript.

Acknowledgements

The study was supported by the University Research Committee, Strategic Research Theme on Public Health, The University of Hong Kong. We would like to thank the schools and students for their participation.

References

1. Holsen I, Kraft P, Roysamb E: The Relationship between Body Image and Depressed Mood in Adolescence: A 5-year Longitudinal Panel Study. Journal of Health Psychology 2001, 6(6):613–627.
2. Brener ND, Eaton DK, Lowry R, McManus T: The Association between Weight Perception and BMI among High School Students. Obesity Research 2004, 12(11):1866–1874.
3. Talamayan KS, Springer AE, Kelder SH, Gorospe EC, Joye KA, Talamayan KS, Springer AE, Kelder SH, Gorospe EC, Joye KA: Prevalence of overweight misperception and weight control behaviors among normal weight adolescents in the United States. Scientific World Journal 2006, 6:365–373.
4. Xie B, Liu C, Chou C-P, Xia J, Spruijt-Metz D, Gong J, Li Y, Wang H, Johnson CA: Weight perception and psychological factors in Chinese adolescents. Journal of Adolescent Health 2003, 33(3):202–210.
5. Xie B, Chou CP, Spruijt-Metz D, Reynolds K, Clark F, Palmer PH, Gallaher P, Sun P, Guo Q, Johnson CA: Weight perception and weight-related sociocul-

tural and behavioral factors in Chinese adolescents. Preventive Medicine 2006, 42(3):229–234.

6. Xie B, Chou CP, Spruijt-Metz D, Reynolds K, Clark F, Palmer PH, Gallaher P, Sun P, Guo Q, Johnson CA, et al.: Weight perception, academic performance, and psychological factors in Chinese adolescents. American Journal of Health Behavior 2006, 30(2):115–124.
7. Shi Z, Lien N, Nirmal Kumar B, Holmboe-Ottesen G, Shi Z, Lien N, Nirmal Kumar B, Holmboe-Ottesen G: Perceptions of weight and associated factors of adolescents in Jiangsu Province, China. Public Health Nutrition 2007, 10(3):298–305.
8. Standley R, Sullivan V, Wardle J: Self-perceived weight in adolescents: Over-estimation or under-estimation? Body Image 2009, 6(1):56–59.
9. Cheung C, Ip L, Lam S, Bibby H: A study on body weight perception and weight control behaviours among adolescents in Hong Kong. Hong Kong Medical Journal 2007, 13(1):16–21.
10. Maximova K, McGrath JJ, Barnett T, O'Loughlin J, Paradis G, Lambert M: Do you see what I see? Weight status misperception and exposure to obesity among children and adolescents. International Journal of Obesity 2008, 32(6):1008–1015.
11. Wardle J, Haase AM, Steptoe A: Body image and weight control in young adults: international comparisons in university students from 22 countries. Int J Obes (Lond). 2006, 30(4):644–651.
12. Prochaska JO, Velicer WF: The transtheoretical model of health behavior change. Am J Health Promot 1997, 12(1):38–48.
13. Prochaska JO, Velicer WF, Rossi JS, Goldstein MG, Marcus BH, Rakowski W, Fiore C, Harlow LL, Redding CA, Rosenbloom D: Stages of change and decisional balance for 12 problem behaviors. Health Psychology 1994, 13(1):39–46.
14. Rosenstock IM, Strecher VJ, Becker MH: Social learning theory and the Health Belief Model. Health Education Quarterly 1988, 15(2):175–183.
15. Al Mamun A, Cramb S, McDermott BM, O'Callaghan M, Najman JM, Williams GM: Adolescents' Perceived Weight Associated With Depression in Young Adulthood: A Longitudinal Study. Obesity 2007, 15(12):3097–3105.
16. Daniels J: Weight and weight concerns: Are they associated with reported depressive symptoms in adolescents? Journal of Pediatric Health Care 2005, 19(1):33–41.

17. Pesa JA, Syre TR, Jones E: Psychosocial differences associated with body weight among female adolescents: the importance of body image. Journal of Adolescent Health 2000, 26(5):330–337.
18. ter Bogt TFM, van Dorsselaer SAFM, Monshouwer K, Verdurmen JEE, Engels RCME, Vollebergh WAM: Body Mass Index and Body Weight Perception as Risk Factors for Internalizing and Externalizing Problem Behavior Among Adolescents. Journal of Adolescent Health 2006, 39(1):27–34.
19. Falkner NH, Neumark-Sztainer D, Story M, Jeffery RW, Beuhring T, Resnick MD, Falkner NH, Neumark-Sztainer D, Story M, Jeffery RW, et al.: Social, educational, and psychological correlates of weight status in adolescents. Obesity Research 2001, 9(1):32–42.
20. Pinhas-Hamiel O, Singer S, Pilpel N, Fradkin A, Modan D, Reichman B: Health-related quality of life among children and adolescents: associations with obesity. Int J Obes (Lond). 2006, 30(2):267–272.
21. Richardson LP, Garrison MM, Drangsholt M, Mancl L, LeResche L: Associations between depressive symptoms and obesity during puberty. General Hospital Psychiatry 2006, 28(4):313–320.
22. Swallen KC, Reither EN, Haas SA, Meier AM: Overweight, Obesity, and Health-Related Quality of Life Among Adolescents: The National Longitudinal Study of Adolescent Health. Pediatrics 2005, 115(2):340–347.
23. Viner RM, Haines MM, Taylor SJC, Head J, Booy R, Stansfeld S: Body mass, weight control behaviours, weight perception and emotional well being in a multiethnic sample of early adolescents. International Journal of Obesity 2006, 30(10):1514–1521.
24. Xie B, Chou CP, Spruijt-Metz D, Liu C, Xia J, Gong J, Li Y, Johnson CA: Effects of perceived peer isolation and social support availability on the relationship between body mass index and depressive symptoms. Int J Obes (Lond). 2005, 29(9):1137–1143.
25. Zhang HB, Tao FB, Zeng GY, Cao XQ, Gao M, Shao FQ: Effects of depression symptoms and other psychological factors on unhealthy weight reducing behaviors of adolescents. Chinese Journal of School Health 2000, 21:348–349.
26. Yu YZ, Hu YZ, Liang SY, Zhang L, Zhou LM, Zhang JZ: A case-control study on psychological behavioral characteristics in obese adolescents. Chinese Journal of Health Psychology 1996, 10:70–71.
27. Schwartz MB, Vartanian LR, Nosek BA, Brownell KD: The Influence of One's Own Body Weight on Implicit and Explicit Anti-fat Bias[ast]. Obesity 2006, 14(3):440–447.

28. Latner JD, Stunkard AJ: Getting Worse: The Stigmatization of Obese Children. Obes Res. 2003, 11(3):452–456.
29. Strauss RS, Pollack HA: Social Marginalization of Overweight Children. Arch Pediatr Adolesc Med. 2003, 157(8):746–752.
30. Keery H, Berg P, Thompson JK: An evaluation of the Tripartite Influence Model of body dissatisfaction and eating disturbance with adolescent girls. Body Image 2004, 1(3):237–251.
31. Thompson JK, Shroff H, Herbozo S, Cafri G, Rodriguez J, Rodriguez M: Relations Among Multiple Peer Influences, Body Dissatisfaction, Eating Disturbance, and Self-Esteem: A Comparison of Average Weight, At Risk of Overweight, and Overweight Adolescent Girls. Journal of Pediatric Psychology 2006, 32(1):24–29.
32. Neumark-Sztainer D, Falkner N, Story M, Perry C, Hannan PJ, Mulert S, Neumark-Sztainer D, Falkner N, Story M, Perry C, et al.: Weight-teasing among adolescents: correlations with weight status and disordered eating behaviors. Int J Obes Relat Metab Disord. 2002, 26(1):123–131.
33. Keery H, Boutelle K, Berg P, Thompson JK: The impact of appearance-related teasing by family members. Journal of Adolescent Health 2005, 37(2):120–127.
34. Eisenberg ME, Neumark-Sztainer D, Story M: Associations of Weight-Based Teasing and Emotional Well-being Among Adolescents. Arch Pediatr Adolesc Med. 2003, 157(8):733–738.
35. Lieberman M, Gauvin L, Bukowski WM, White DR: Interpersonal influence and disordered eating behaviors in adolescent girls: The role of peer modeling, social reinforcement, and body-related teasing. Eating Behaviors 2001, 2(3):215–236.
36. Wardle J, Robb KA, Johnson F, Griffith J, Brunner E, Power C, Tovee M, Wardle J, Robb KA, Johnson F, et al.: Socioeconomic variation in attitudes to eating and weight in female adolescents. Health Psychology 2004, 23(3):275–282.
37. Eisenberg ME, Neumark-Sztainer D, Haines J, Wall M: Weight-teasing and emotional well-being in adolescents: Longitudinal findings from Project EAT. Journal of Adolescent Health 2006, 38(6):675–683.
38. Wang Y, Monteiro C, Popkin BM: Trends of obesity and underweight in older children and adolescents in the United States, Brazil, China, and Russia. The American Journal of Clinical Nutrition 2002, 75(6):971–977.

39. Jingxiong J, Rosenqvist U, Huishan W, Greiner T, Guangli L, Sarkadi A: Influence of grandparents on eating behaviors of young children in Chinese three-generation families. Appetite 2007, 48(3):377–383.

40. Thompson JK, Cattarin J, Fowler B, Fisher E: The Perception of Teasing Scale (POTS): A Revision and Extension of the Physical Appearance Related Teasing Scale (PARTS). Journal of Personality Assessment 1995, 65(1):146–157.

41. Leung SS, Cole TJ, Tse LY, Lau JT, Leung SS, Cole TJ, Tse LY, Lau JT: Body mass index reference curves for Chinese children. Annals of Human Biology 1998, 25(2):169–174.

42. Wong JP, Ho SY, Lai MK, Leung GM, Stewart SM, Lam TH, Wong JPS, Ho SY, Lai MK, Leung GM, et al.: Overweight, obesity, weight-related concerns and behaviours in Hong Kong Chinese children and adolescents. Acta Paediatrica 2005, 94(5):595–601.

43. Grant JP: Duke procedure for super obesity: Preliminary report with 3.5-year follow-up. Surgery 1994, 115(6):718–726.

44. Shannon B, Smiciklas-Wright H, Min Qi W: Inaccuracies in self-reported weights and heights of a sample of sixth-grade children. Journal of the American Dietetic Association 1991, 91(6):675–678.

45. Tabachnick BG, Fidell LS: Using Multivariate Statistics. Needham Heights, MA: Allyn & Bacon; 2001.

46. Hume C, Ball K, Salmon J: Development and reliability of a self-report questionnaire to examine children's perceptions of the physical activity environment at home and in the neighbourhood. International Journal of Behavioral Nutrition and Physical Activity 2006., 3(16):

47. Chinn S, Burney PGJ: On Measuring Repeatability of Data from Self-Administered Questionnaires. International Journal of Epidemiology 1987, 16(1):121–127.

48. Brener ND, McManus T, Galuska DA, Lowry R, Wechsler H: Reliability and validity of self-reported height and weight among high school students. Journal of Adolescent Health 2003, 32(4):281–287.

49. Goodman E, Hinden BR, Khandelwal S, Goodman E, Hinden BR, Khandelwal S: Accuracy of teen and parental reports of obesity and body mass index. Pediatrics 2000, 106(1 Pt 1):52–58.

50. Harel O, Zhou XH, Harel O, Zhou X-H: Multiple imputation: review of theory, implementation and software. Statistics in Medicine 2007, 26(16):3057–3077.

51. Cohen J: Statistical power analysis for the behavioral sciences. Revised edition. New York: Academic Press; 1977.
52. Felts WM, Parrillo AV, Chenier T, Dunn P: Adolescents' perceptions of relative weight and self-reported weight-loss activities: Analysis of 1990 YRBS national data. Journal of Adolescent Health 1996, 18(1):20–26.
53. Jansen W, Looij-Jansen PM, de Wilde EJ, Brug J: Feeling Fat Rather than Being Fat May Be Associated with Psychological Well-Being in Young Dutch Adolescents. Journal of Adolescent Health 2008, 42(2):128–136.
54. O'Brien SH, Holubkov R, Reis EC: Identification, Evaluation, and Management of Obesity in an Academic Primary Care Center. Pediatrics 2004, 114(2):e154–159.
55. Chaimovitz R, Issenman R, Moffat T, Persad R: Body Perception: Do Parents, Their Children, and Their Children's Physicians Perceive Body Image Differently? J Pediatr Gastroenterol Nutr. 2008, 47(1):76–80.
56. Spurrier NJ, Magarey A, Wong C, Spurrier NJ, Magarey A, Wong C: Recognition and management of childhood overweight and obesity by clinicians. JJ Paediatr Child Health. 2006, 42(7-8):411–418.
57. Hutchinson DM, Rapee RM: Do friends share similar body image and eating problems? The role of social networks and peer influences in early adolescence. Behaviour Research and Therapy 2007, 45(7):1557–1577.
58. Presnell K, Bearman SK, Stice E: Risk factors for body dissatisfaction in adolescent boys and girls: A prospective study. International Journal of Eating Disorders 2004, 36(4):389–401.
59. Lee S, Lee AM: Disordered eating in three communities of China: A comparative study of female high school students in Hong Kong, Shenzhen, and rural Hunan, [http://www3.interscience.wiley.com/journal/70002627/abstract], International Journal of Eating Disorders 2000, 27(3):317–327.
60. Department of Health: Population Health Survey 2003/2004. Hong Kong 2005.
61. Jeffery AN, Voss LD, Metcalf BS, Alba S, Wilkin TJ: Parents' awareness of overweight in themselves and their children: cross sectional study within a cohort (EarlyBird 21). British Medical Journal 2005, 330(7481):23–24.

Children's Liking and Wanting of Snack Products: Influence of Shape and Flavour

Djin G. Liem and Liesbeth H. Zandstra

ABSTRACT

Background

Children's food choices are guided by their preferences. However, these preferences may change due to repeated exposure.

Methods

This study investigated children's (n = 242, 7–12 yrs-old) liking and wanting for snacks over 3 weeks of daily consumption. The snacks differed in size (small vs large) or flavour (sweet vs sweet-sour). Two conditions were designed: 1) a monotonous group in which children continuously consumed the same snack across the 3 weeks, and 2) a free choice group in which children were allowed to freely choose amongst 3 different flavours of the snack each day during 3 weeks.

Results

Shape influenced long-term liking, i.e. small shaped snacks remained stable in liking over repeated consumption, whereas large shaped snacks with the same flavour decreased in liking. Mean wanting ratings for all snack products decreased over 3 weeks daily consumption. Flavour did not significantly influence liking and wanting over time. The ability to freely choose amongst different flavours tended to decrease children's liking ($p < 0.1$) and wanting ($p < 0.001$) for these products. Changes in liking rather than initial liking was the best predictor of snack choice during the intervention.

Conclusion

Wanting rather than liking was most affected by repeated daily consumption of snack foods over three weeks. In order to increase the likelihood that children will repeatedly eat a food product, smaller sized healthy snacks are preferred to larger sized snacks. Future research should focus on stabilizing wanting over repeated consumption.

Background

Children's liking of the taste of a product has been identified as the most important determinant of children's food choice [1-3]. Several tests have been developed to measure children's liking (see [4] for review). Most of these tests use a one-off tasting to predict which products children like best. However, the taste children like (i.e. liking for sweet and dislike for bitter taste) changes during the life span [5-7] and across weeks, due to for example repeated exposure.

It has been argued that repeated exposure to a particular food can lead to an increase in liking of this food. It needs, however, to be noted that most of these studies were performed with foods which were quite different from each other (e.g. different vegetables or fruits) [8] or were novel to the child [9]. This makes it difficult to determine which specific product properties (e.g. flavour profile, appearance, size) are important for a change in liking after repeated exposure.

It could be that liking for some products increases after repeated exposure, whereas liking for other products remain stable or decreases after repeated exposure. Liem et al. [10] found that repeated exposure (i.e. 8 days, once a day) to a sweet drink increased children's liking for this drink. In contrast, repeated exposure to a sour drink, which was at the start similarly liked as the sweet drink, remained stable in liking. It remains to be investigated whether sweet-sour balance also plays a role in changes in liking for solid foods after a daily exposure during several weeks.

In adults changes in liking after repeated exposure has been investigated extensively. In these studies repeated exposure generally did not result in an increase in liking but rather a decrease in liking. This has been referred to as boredom or monotony, which can be defined as the lowered acceptance of a food as a function of the number of times a food is consumed (Sigel & Pelgrim, 1958 in [11]). Boredom can be caused by either neurophysiological responses, i.e. a decrease in actual liking caused by satiation with specific attributes of the consumed food, and/or cognitive response, i.e. a decrease wanting to eat the food [12,13]. These two causes have previously been set out as liking vs wanting. Liking can be defined as the pleasure derived from oro-sensory stimulation of food. Wanting can be defined as incentive salience, the motivation to engage in eating [14]. Extensive animals research by Berridge suggests that liking and wanting have separate neural substrates (i.e. dopamine vs opioid) and can act independently. This has been replicated in humans by using specific dopamine and opioid antagonists (see [15] for review). It has been suggested that liking and wanting play an important interdependent role in food choice and consumption in adults [14,16].

Studies which focussed on children's liking and wanting as separate pathways for food choice are scarce. Previous studies either measured liking and wanting as one concept [17,18], only measure one of the two pathways [19,20], or did not investigate which product properties are associated with a decrease in liking and wanting [21]. In order to investigate changes in liking and wanting for foods children have repeatedly been exposed to, we may learn from research conducted with adults.

It has been argued that the size of the food eaten plays an important role in the decrease of liking and wanting after repeated exposure. A recent study of Weijzen et al. suggested that after repeated consumption of small snack foods a statistically significant decrease in wanting but not in liking was observed. After a repeated consumption of large snack foods a statistically significant decrease in wanting and liking was observed [22]. They argued that the oral sensory stimulation that positively relates to the size of the food is related to liking and wanting. In this study, however, they investigated changes in liking and wanting within one meal consumption rather than over an extended period of time. It remains to be investigated whether size also influences liking and wanting after daily consumption of these foods for several weeks.

Furthermore, it has been suggested that a decrease in liking and wanting (measured as boredom) after repeated exposure can be minimized by giving adult [12,23] or children [21] a choice between different products. Hypothetically, in a choice situation participants have a larger feeling of control of what they eat, which decreases the perceived boredom [12].

The current study investigated three hypothesis related to children's change in liking and wanting. The first hypothesis concerned the influence of sweet-sour balance on children's liking and wanting. It was hypothesised that after daily consumption for three weeks, the liking and wanting of sweet snack foods would increase and the liking and wanting of sour snack foods would remain stable. This was tested by means of snack products which flavours were either Sweet or Sweet-Sour.

The second hypothesis concerned the influence of snack size on children's liking and wanting. It was hypothesised that Small sized snacks (e.g. nibbles) resulted in less decrease of liking but not wanting over daily consumption for three weeks than Large sized snack (e.g. bars). This was tested by means of snack products which differed in size.

The third hypothesis concerned the influence of choice on children's liking and wanting. It was hypothesised that children who could freely choose between snack products which differed in flavour and size would express a lower decrease in liking and wanting, over daily consumption for three weeks, than those who were not given a choice.

Methods

Participants

Children were recruited during door to door interviews in the Istanbul metro area in Turkey. Exclusion criteria were reported allergies for chocolate, polenta, sugar, dairy products, corn, corn oil, hazelnut or caramel. In addition, children were excluded from participation if they participated in any research concerning snack products in the past month. Initially 341 children started the study and 242 (n = 122, 7–9 yrs, n = 120, 10–12 yrs; 122 girls, 120 boys) completed the study. During the 3-week course of the study children dropped out of the study because of various reasons e.g. parents no longer gave permission or children did not want to participate any longer, failed to conduct the in-home liking test, or failed to give the products every day. The study was carried out according the ESOMAR ethical standards embodied in the ICC/ESOMAR Code of marketing and social research practise. Informed consent was obtain from the participants prior to participation

Stimuli

The stimuli comprised of 5 snacks. Two had the same size (Small) but were different in flavour (Sweet vs Sweet-Sour flavour). Two had the same flavour

(Chocolate Hazelnut) but were different in size (Small vs Large size). One snack was used as control-snack (Small size, caramel flavour). This product was only tasted during the baseline and end-measurement.

The Small sized snacks were on average 1.5 gram and 2.5 cm × 1.5 cm × 1.0 cm in size and presented in bags which contained 36 grams of snacks each (Unilever, Turkey). The Large size snack was on average 16.0 gram and 9.0 cm × 4.0 cm × 1.0 cm in size and presented in bags which contained 2 bars (Unilever, Israel). Both Small and Large snack foods comprised of a crunchy outer layer and a cream filling. The percentage cream relative to the weight of the snacks was kept constant. Per 100 gram the snacks contained 454 kcal- 63 carbohydrates, 9 gram protein and 16 gram fat. All stimuli were presented in non-labelled packs of aluminium foil which prevented light oxidation (see Table 1).

Table 1. Product characteristics (Flavour, maximum serving per day and abbreviation) Study design.

Format	Flavour	Maximum Serving per day*	Abbreviation
Small	Chocolate Hazelnut	1 bag = 32 gram	Small chocolate-hazelnut
Large	Chocolate Hazelnut	2 bars = 32 gram	Large chocolate-hazelnut
Small	Orange Bubble gum	1 bag = 32 gram	Small Sweet
Small	Orange Bubble gum with citric acid	1 bag = 32 gram	Small Sweet-Sour
Small	Caramel		Control

* each bag of nibbles contained about 20 nibbles

Overview

To test the hypotheses a between subject design was chosen. Children were asked to consume either Small chocolate-hazelnut, Large chocolate-hazelnut, Small sweet or Small sweet-sour snacks for a period of 3 consecutive weeks. Just before and at the end of the 3 weeks children's preference, liking and wanting for all snack foods were tested. After each week of exposure children's liking for snacks they consumed that week was tested (see Figure 1). Liking and wanting was individually measured in children's home by a trained interviewer. Parents were instructed to offer the snacks at the same time every day to minimize variation due to the time of day.

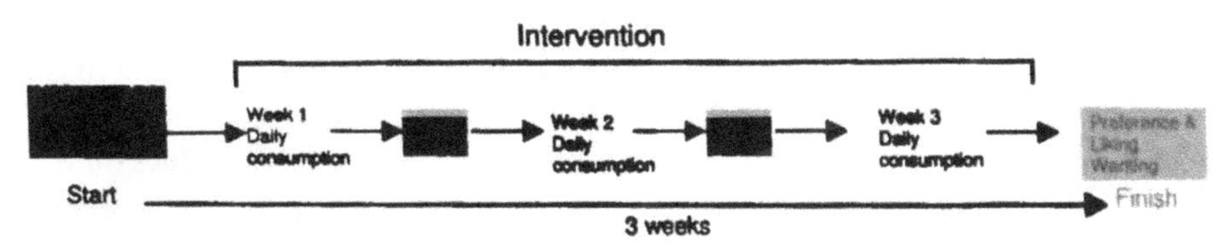

Figure 1. Schematic overview of the procedure. Measurements of liking (4 in total) and wanting (2 in total) are listed in grey blocks.

Group Composition

Because in a real life situation children would choose to eat products they like, children were grouped based on their initial liking for the products, which resulted in 6 groups. Children who preferred the Small snack with Chocolate Hazelnut flavour either as their most or second most favourite were placed in the group that received this snack daily for three weeks (Group Small chocolate-hazelnut, n = 41). In a similar way a Large chocolate-hazelnut group (n = 41), a Small Sweet group (n = 40), and a Small Sweet-Sour group (n = 40) were composed. A fifth group, whose children were randomly chosen from the previous groups (before the intervention started), was given a free choice of three different Small snacks (Small chocolate-hazelnut, Small Sweet, Small Sweet-Sour) (hereafter referred to as CHOICE group, n = 40). Every day children in the CHOICE group were presented with three bags of small snacks of which they could choose one to consume, after sampling one snack from each pack. The 6th group did not receive any snack foods other than at the beginning and the end of the intervention (hereafter referred to as CONTROL, n = 40).

Sensory Measures

Preference and Liking

Preference was measured by means of a rank-order method. Children were presented with all the different snacks. The interviewer asked the child to taste all the snacks and point to the one he or she liked best. This snack was removed from the table after which the procedure was repeated with the remaining snacks until all were place in a rank-order from most to least liked [24]. The least liked snack was assigned 1 point, the most liked was assigned 5 points. All other snacks where given points between 1 and 5 according their position in the preference rank-order.

Subsequently, the researcher showed the child pictures of 5 different drawings of faces representing 1) extremely liked, 2) liked, 3) maybe liked, maybe disliked, 4) disliked, 5) extremely disliked. The researcher explained the meaning of the 5 faces by saying: "This is the face that you make when you do not like something at all. This is a face you make when you do not like something. This is the face you make when you do not like it but also not dislike it. This is a face you make when you just like something. This is a face you make when you like something very much." Next, the child was asked to score the most preferred product on the 5-point facial scale. This procedure continued until all the stimuli were scored on liking. All children understood the procedure as suggested by the consistency between the ranking and scoring part. This procedure has been used and validated across different cultures such as France [25], UK [26], US [27].

Wanting

In order to obtain information about whether children's wanting for the product changed after three weeks of exposure, they were asked to taste the products and to rate them on how much they wanted to eat of it right now (i.e. really do not want to eat this-1 points; don't want to eat this-2 points; I do not know- 3 points; I want to eat it- 4 points; I really want to eat it- 5 points). Previous research used similar explicit measurements of wanting [19,20].

Amount Eaten

Children were free to consume any amount they wanted with a maximum of one bag of Small snacks, or two bars of the Large snack. Although snacks were offered every day children could decide not to consume the snacks at all. When children did not eat the whole bag or the two bars they were provided, parents were asked to save it. Each week the researchers collected and measured (grams) the left-overs. Due to logistic reasons the amounts consumed were measured on a group level rather than individual level.

Data Analyses

In order to determine significant differences between different products in initial liking and in initial wanting, Friedman analyses for ranks and post-hoc (Bonferroni) analyses were performed [28].

Changes in liking were analysed by comparing initial liking and wanting with the liking and wanting scores after the 3 week exposure. Paired sample t-tests were performed to investigate significant differences.

In order to investigate the association between food choice and liking, and food choice and wanting, two separate Anova's per snack food were carried out. Anova 1: independent variable = choice behaviour, dependent variables = liking before and liking after the intervention. Anova 2: independent variable = choice behaviour, dependent variables = wanting before and wanting after the intervention. Due to colinearity of liking and wanting, and the small number of subjects per group, measures of liking and wanting were not taken together in one Anova model (SPSS version 14). Choice behaviour in this matter was defined as: the number of times a particular snack was chosen during the 3-week intervention, by children in the CHOICE-group. P-values of less than 0.05 were considered statistically significant.

Results

Initial Preference, Liking and Wanting

Before the intervention the products were differently preferred ($F(4df) = 87.41$; $p < 0.0001$) and liked ($F(4df) = 65.10$; $p < 0.0001$). Post-hoc analyses suggested that the Large chocolate-hazelnut snack and the Control snack were significantly less liked than the remaining snacks (all above 4 on a 5-point liking scale) ($p < 0.05$). Furthermore, the Large chocolate-hazelnut and the Control snacks were significantly less wanted than the remaining products (all above 3.5) ($F(4df) = 101.35$; $p < 0.0001$; post-hoc analyses $p < 0.05$) (see Table 2)

Table 2. Initial liking and wanting scores of all children for Small chocolate-hazelnut, Large chocolate-hazelnut, Small Sweet and Small Sweet-sour snacks, n = 242 Consumption of products

Snack	Mean liking (± sem)	Mean wanting (± sem)
Small chocolate-hazelnut	4.04 ± 0.68	3.76 ± 0.8
Large chocolate-hazelnut	3.27 ± 0.09	3.07 ± 0.09
Small Sweet	4.06 ± 0.06	3.80 ± 0.07
Small Sweet-Sour	4.0 ± 0.06	3.81 ± 0.08
Control	3.60 ± 0.07	3.07 ± 0.09

On average children consumed between 89% and 97% of all the Small snacks (i.e. Small chocolate-hazelnut, Small Sweet and Small Sweet-Sour) they were offered during the three week intervention. Children who were asked to consume the Large chocolate-hazelnut snack for 3 weeks, ate between 67% and 85% of their daily servings, this was a statistically significant difference ($F(3df) = 3.9$; $p < 0.05$).

Difference in Liking Before and After Exposure

As shown in Figure 2 upper panel differences in flavour (Sweet vs Sweet-Sour) did not result in a different change in liking. Children who consumed the Small Sweet as well as children who consumed the Small Sweet-Sour both reported a stable liking for the snacks they consumed throughout the intervention.

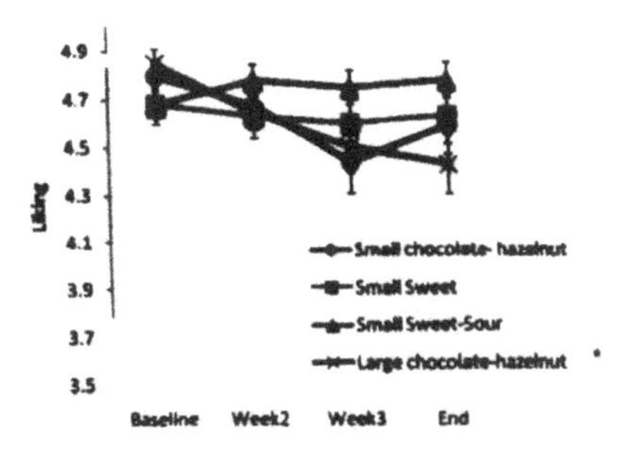

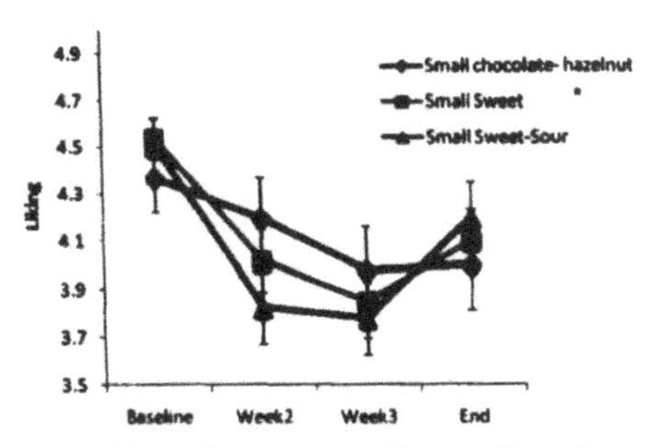

Figure 2. Mean (± sem)change in liking (from 1 = not liked at all, to 5 extremely liked) during a daily consumption of the Small chocolate-hazelnut, Small Sweet, Small Sweet-Sour or Large chocolate hazelnut snack. Shown for children who did not have a choice (upper panel) and children who could freely choose between the Small snacks (lower panel). * signifies significant decrease in liking from baseline to end $p < 0.05$.

Differences where, however, observed depending on size of the snack food. That is, children who consumed the Large chocolate-hazelnut snack on a daily basis for three weeks, significantly decreased their liking for this snack ($t(39df) = 3.19$; $p < 0.01$). In contrast with children who consumed the Small chocolate-hazelnut snack for three weeks. They did not decrease their liking for this product during the intervention ($t(40df) = 2.49$; $p = 0.10$).

Children in the CHOICE-group tended to report a decrease in liking for all products they could choose from during the exposure period. This, however, only reached significance for the Small chocolate-hazelnut snacks ($t(40df) = 2.88$; $p < 0.01$) (Figure 2, lower panel). Children who were not exposed to any experimental product (control group) did not change their liking for any of the snack products. Furthermore the Control snack which was only offered at baseline and the end of the intervention did not change in liking or wanting.

Difference in Wanting Before and After Exposure

In contrast to liking, after children ate specific snacks daily for three weeks their wanting to eat these products decreased. This was independent from flavour, size

(Figure 3, upper panel) or choice (Figure 3, lower panel) (all p-value's < 0.05). Children in the control group did not change their wanting for any of the products.

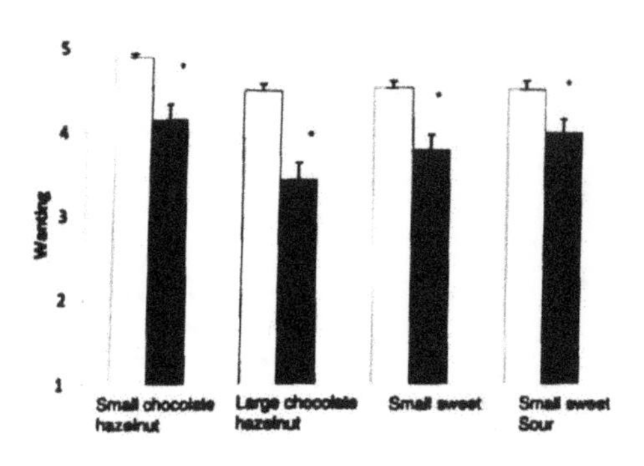

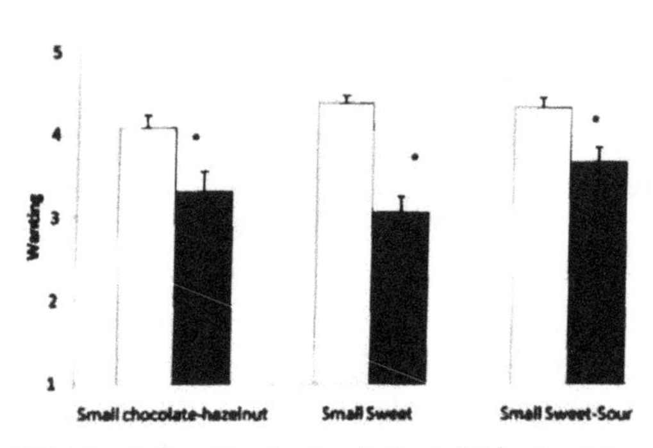

Figure 3. Mean (± sem) Wanting (before (clear bars) and after (solid bars) a daily consumption of either the Small chocolate-hazelnut (n = 41), Large chocolate-hazelnut (n = 41), Small Sweet (n = 40) and SmallSweet-Sour (n = 40). Shown for children who did not have a choice (upper panel) and children who could freely choose between the Small snacks (n = 40, lower panel).* signifies significant differences at P < 0.05

CHOICE Group: Free Choice During Three Weeks

Data from 4 subjects in the CHOICE group were incomplete, because parents failed to fill out which products were chosen each day. This resulted in 37 complete records of children in the CHOICE group. Only a few children (5.4%, n = 2) did not switch between products during the 3 weeks of the intervention and always choose the Small chocolate-hazelnut snack to consume. Most children (32.4%) switched between the three Small snack foods which were made available to them. As shown in Figure 4 upper panel, on average the Small chocolate-hazelnut snack was chosen the most often across the three weeks of intervention (on average 2.9 times out of 7). On average children's variety of snacks consumed was higher in the first week compared to the last week (F(3df) = 7.66; $p < 0.05$) (Figure 4 lower panel). Younger children (7 to 9 years of age) compared to older aged children (10 to 12 years of age), choose a large variety during the first week (t = 2.1, $p < 0.05$). No differences were observer for the second week (t = 0.95, p = 0.35), third week (t = -0.43, p = 0.67) or total variety across three week (t = 0.98, p = 0.42).

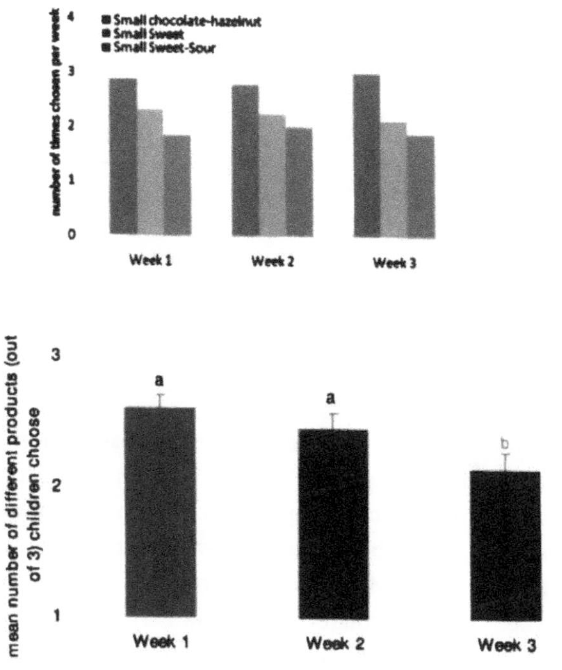

Figure 4. Upper panel- Mean number of times (± sem) children in the CHOICE group choose either Small chocolate-hazelnut, Small Sweet or Small Sweet-Sour snacks during the first, second and third week of the intervention (min = 0, max = 7). Lower panel- Mean number of different products (± sem) (out of 3: Small, Small Sweet, Small Sweet-Sour) children in the free CHOICE group (n = 37) choose during week 1, 2 and 3.
Liking and wanting as predictors for choice

As shown in Table 3, the liking after the 3 week intervention (with the liking-before-the intervention as covariate) rather than the liking before the intervention was a significant correlate of food choice for all three snack foods. The wanting after 3 week intervention (with the wanting-before-the intervention as covariate) showed to be a correlate for food choice for two out of the three snacks (Small chocolate hazelnut and Small sweet sour). When comparing the statistically significant effect sizes (B) of liking and wanting on food choice, it can be concluded that the effect size of liking on food choice is consistently larger than the effect size of wanting on food choice

Table 3. Anova 1-upper panel: association food choice (dependent variable) and liking (independent variables).

	model		Liking			
			before		after	
	F	p	B	95%CI	B	95%CI
Small chocolate hazelnut	21.0	<0.001	0.65	-1.4 – 2.8	3.8	2.5–5.0
Small sweet	3.3	<0.05	0.58	-1.8 – 3.0	2.0	0.33–3.6
Small sweet-sour	9.0	<0.01	0.18	-1.4 – 1.7	2.5	1.3 – 3.7

	model		Wanting			
			before		after	
	F	p	B	95%CI	B	95%CI
Small chocolate hazelnut	14.0	<0.001	1.4	-0.26 – 3.1	2.5	1.4–3.5
Small sweet	0.49	0.62	0.68	-2.0 – 3.3	0.58	-0.70 – 1.9
Small sweet-sour	8.2	<0.001	0.38	-1.3 – 2.0	2.1	1.1 – 3.2

Anova 2- lower panel: association food choice (dependent variable) and wanting (independent variables). Data for children in the CHOICE group (n = 37). Food choice is defined as the number of times children in the CHOICE group choose a particular snack during the 3 week intervention.

Discussion

This study investigated the influence of repeated consumption of snack foods on children's liking and wanting. This was tested with products which differed in Sweet-Sour balance, or size in two different conditions (monotonous and free choice condition). First, it will be discussed how repeated consumption of snacks influenced changes in liking and wanting in general. Subsequently, it will be discussed how size (i.e. Small chocolate-hazelnut vs Large chocolate-hazelnut), flavour profile (Sweet vs Sweet-Sour), and freedom of choice influenced changes in liking and wanting.

Previous studies showed that children increased their liking over repeated consumption of novel foods such as different cheeses [29] or novel vegetables [30]. In these studies initial liking for these novel foods was moderate to low, possibly due to children's food neophobic responses [31]. It has been shown that liking for novel foods increases after repeated consumption, because the novel foods become more familiar to children [32]. This contrasts findings of the present study as liking of the snack products remained relatively stable over time. Since these products were very recently introduced to the Turkish market, we may assume that children had no or little exposure to these foods prior to their participation in our study. Lack of increase in liking after repeated consumption might be due to a ceiling effect because most products were highly liked at the start of the intervention.

The present study hypothesised that small sized snacks resulted in less decrease of liking but not wanting, over daily consumption for three weeks, than large sized snacks. In the present study repeated consumption of small sized snack foods indeed seemed to show a less decrease in liking than a repeated consumption of large sized snack foods. Wanting, however, showed a similar decrease for small and large sized snack foods. At the same time children consumed more of the Small chocolate-hazelnut snacks during the three week intervention than of the Large chocolate-hazelnut snacks.

The latter finding is in contrast to Weijzen et al.'s study which focussed on sensory specific satiety in adults [22]. They suggested that adults consumed 12% less of small sized food than large sized foods when given the snacks during a one meal time occasion [22]. The present study and Weijzen's study are, however, different. It has been suggested that children, compared to adults show different chewing behaviour. That is, adults have a higher masticatory performance and a higher bite force than children [33]. Therefore, it may cost children more effort than adults to eat large snack foods compared with small snack foods. This may explain why children eat more of the small snack foods than of the large snack foods over the course of three weeks. Furthermore, Weijzen's study concerned sensory specific satiety which can be seen as an intra meal measurement of boredom. In the present study boredom and intake was measured over a course of three weeks. To date it is unclear, to our knowledge, whether decreases in liking during one meal occasions are related to decreases in liking after a prolonged exposure (i.e 3 weeks).

The effect of size on a decrease in liking might be related to the amount of oral stimulation. Children may eat small sized snacks faster than large sized snacks because they need less mastication and a lower bite force to breakdown the food before swallowing. It can be argued that when eating at a high rate, the food stays in the mouth during a shorter time than when eating slowly. Fast eating gives the sensory receptors in the mouth less time to interact with flavour and texture. This generates less sensory satiety of the sensory receptors [22,34]. This is still highly speculative because we did not measure speed of eating in the current study.

The present study hypothesised that after daily consumption for three weeks, the liking and wanting of sweet snack foods would increase and the liking and wanting of sour snack foods would remain stable. In the present study, Sweet-Sour balance did not influence long term liking, wanting or choice. A previous study with beverages which differed in sweet-sour balance found that repeated exposure to the sweet beverages increased liking for this beverage over time. Whereas repeated exposure to the sour beverages did not result in a different pattern of liking [10]. It could be that children were not able to taste differences between the sweet and sweet sour snack. However, a small pilot with adults showed that 6 out of 8 people reported the Sweet-Sour snack to be more sour than the Sweet snack. Children, however, might have been less sensitive to these small differences in taste than adults. An alternative explanation could be that the Sweet and the Sweet-Sour snacks were highly liked. In order for sweet-sour balance to be able to increase long term liking, the foods may need to be moderately liked as was the case in the previous study [10].

In adults it has generally been found that liking decreases after repeated exposure for a variety of foods (see [35] for review). In the present study most products, except for the Large snacks, remained stable in liking. This suggests that on average children's liking remains stable for at least three weeks. Potentially 3 weeks was not enough to show a decrease in liking. Le et al. [21], however, did also not observe a change in liking for noodle soup when children were exposed to this soup for 10 weeks. A decrease in liking for highly liked food might not be evident in children. This may prevent children from trying out foods they never tried, which could impact their dietary variety [36].

Wanting, however, decreased. This was most likely due to the repeated exposure. Recall that children who did not receive a repeated exposure to a particular snack food, did not change their wanting for these foods. The decrease in wanting was specific for flavour and size. For example, children who daily consumed the small snack with chocolate hazelnut flavour decreased their wanting for this particular snack but not for those snack with either a similar size or flavour. This suggests that in order to prevent a decrease in wanting, flavours and sizes may need to be rotated during the week. This does not mean that children should be given a wide variety of choice each day.

In the present study it was hypothesised that children who could freely choose between snack products which differed in flavour and size would express a lower decrease in liking and wanting, over daily consumption for three weeks, than those who were not given a choice. In contrast to our expectations, free choice did not prevent decrease in liking and wanting in children. Recall that children who were allowed to freely choose between 3 types of Small snacks tended to decrease their liking for these snacks during the course of three weeks. In adults it has generally be found that giving consumers a choice between different flavours prevented boredom with the products [12]. This is most likely due to consumers' feeling of control [23]. Possibly, children felt pressured by the given choice (i.e. you have to choose) which negatively impacted upon their liking and wanting. As pointed out by Schwartz, making a choice may make us realize that we missed out on the options we did not choose, which result in a lower satisfaction of the one we did choose [37]. In western societies children are overwhelmed by choice. Crisps, soda and many other products come in multiple flavours. Our research suggests that a large choice may have a negative effect on liking and wanting of any one specific product.

When children were given a choice we observed that after trying different flavours during the first week of exposure, children seem to pick their favourite and remain eating this snack for the remainder of the intervention. This suggests that initial success of different foods developed for children, might be misleading. Children may not continue eating a high variety of different foods but rather narrow down their choices. This might also be true for much younger children than we tested. Nicklaus and colleagues found that children decrease the number of different foods they eat between 2 and 3 years of age [38].

In the present study initial liking or wanting was not the strongest predictor of food choice. Food choice was best explained by the change in liking across the three weeks intervention. Furthermore, children who initially preferred the Large snacks tended to show a stronger decline in liking and wanting during the three week intervention than those who initially preferred the Small snacks. This suggests that consumer tests with children which select potential successful products based on initial liking may fail to select products which will be successful in the market in the long term. This may also be the case for adults [38].

Neither initial wanting nor the change in wanting played a significant role in food choice. This does not mean that wanting is irrelevant for children's food related behaviours. Recently it has been suggested that obese children showed a higher wanting for foods than lean peers [39]. Differences in children's liking for particular tastes are rarely observed [36]. Similar results have been obtained in adults [40]. It has previously been suggested that wanting depends on contextual factors such as the context in which a particular food is given (or not given), and the perceived appropriateness of consumption of particular foods [14]. Wanting can therefore fluctuate depending on the context in which a food is provided. Liking, as shown in the present study and previous studies [38], is more stable. It can be hypothesised that liking determine the range of food which are acceptable, whereas wanting plays a dominant role in which food will be eaten and in which quantity. This, however, needs to be tested in future research.

The present study had several limitations. In order to try to mimic a real life situation children were given the snacks they initially liked most. Regression to the mean effect could have accounted for differences observed between the first measurement of liking and subsequent measurements. After children were grouped based on their liking for the different snacks it seems that those in the Large-snack food group scored the large snack foods as extremely liked. Because of this, they are more likely to decrease their liking for this food than when the food had not been extremely liked, as was the case for the remaining groups. However, it needs to be noted that the present data suggest that the liking of the large snack foods gradually decreased over the course of three weeks. Children in the choice group were not selected based on their initial liking but rather comprised of a random sample of all children. Regression to the mean would therefore have little effect on changes in liking for this group. But it is this group which showed a decrease in liking for all products they were exposed to.

Consumption was not individually measured. Therefore we could not asses the relationship between liking & wanting and food consumption. Future studies should aim to investigate this relationship by measuring intake per individual rather than on a group level.

During the intervention period snacks were consumed at their children's home in their natural environment. By doing this we tried to minimize the impact of the researchers and lab-environment on children's rating of liking and wanting. Therefore we had no control about how the products were consumed (e.g whether children's played with their food, how long it took to eat). Parents were, however given strict instructions about how and when to offer the snacks to the children.

Furthermore, in order to investigate the intrinsic properties of the products (smell, taste, texture), products were provided unbranded. In real life, extrinsic product properties such as brand and nutritional messages are likely to have a large influence on children food choice [41-43]. Future studies should therefore focus on extrinsic properties as well.

Conclusion

The present study suggests that children's liking for large sized snack foods is more likely to decrease after a daily consumption than identically flavoured Small sized snacks. It needs to be investigated whether same principles hold true for foods which are not highly liked such as vegetables. We hypotheses that smaller sized foods encourage children to eat these food repeatedly due to the lower amount of effort involved than when eating large sized foods.

Decrease in liking during daily consumption of the same food was a better predictor of food choice than initial liking. Therefore a liking test with children based on a once off tasting may not represent market success. Wanting decreased more after daily consumption than liking. It remains to be determined how this decrease in wanting affects children's food consumption. Sensory testing with children should therefore not only focus on liking, but rather on liking & wanting.

Furthermore, choice appears to have a negative effect on liking. It is therefore recommended to offer children a limited choice rather than an unlimited choice.

Competing Interests

The authors declare that they have no competing interests.

Authors' Contributions

DGL designed the study, analyzed the data, and conceived and drafted the original manuscript. EHZ provided critical feedback on study design and drafts of the manuscript. All authors read and approved the final manuscript.

References

1. Ricketts CD: Fat preferences, dietary fat intake and body composition in children. Eur J Clin Nutr 1997, 51:778–781.
2. Perez-Rodrigo C, Ribas L, Serra-Majem L, Aranceta J: Food preferences of Spanish children and young people: the enKid study. Eur J Clin Nutr 2003, 57(Suppl 1):S45–S48.
3. Olson CM, Gemmill KP: Association of sweet preference and food selection among four to five year old children. Ecol Food Nutr 1981, 11:145–150.
4. Guinard JX: Sensory and consumer testing with children. Trends Food Sci Technol 2001, 11:273–283.
5. Desor JA, Greene LS, Maller O: Preferences for sweet and salty in 9- to 15-year-old and adult humans. Science 1975, 190:686–687.
6. Drewnowski A: Sensory control of energy density at different life stages. Proc Nutr Soc 2000, 59:239–244.
7. Mennella JA, Pepino MY, Reed DR: Genetic and environmental determinants of bitter perception and sweet preferences, sensory control of energy density at different life stages. Pediatrics 2005, 115:E216–E222.
8. Wardle J, Cooke LJ, Gibson EL, Sapochnik M, Sheiham A, Lawson M: Increasing children's acceptance of vegetables; a randomized trial of parent-led exposure. Appetite 2003, 40:155–162.
9. Cooke L: The importance of exposure for healthy eating in childhood: a review. J Hum Nutr Diet 2007, 20:294–301.
10. Liem DG, De Graaf C: Sweet and Sour Preferences in Young Children and Adults:Role of Repeated Exposure. Physiol Behav. 2004, 83(3):421–429.
11. Hetherington MM, Pirie LM, Nabb S: Stimulus satiation: effects of repeated exposure to foods on pleasantness and intake. Appetite 2002, 38:19–28.

12. Zandstra EH, De Graaf C, van Trijp HC: Effects of variety and repeated in-home consumption on product acceptance. Appetite 2000, 35:113–119.
13. Zandstra EH, Weegels MF, Van Spronsen AA, Klerk M: Scoring or boring? Predicting boredom through repeated in-home consumption. Food Qual Pref 2004, 15:549–557.
14. Mela DJ: Eating for pleasure or just wanting to eat? Reconsidering sensory hedonic responses as a driver of obesity. Appetite 2006, 47:10–17.
15. Berridge KC: Food reward: brain substrates of wanting and liking. Neurosci Biobehav Rev 1996, 20:1–25.
16. Finlayson G, King N, Blundell JE: Liking vs. wanting food: Importance for human appetite control and weight regulation. Neurosci Biobehav Rev 2007, 31:987–1002.
17. Fisher JO, Birch LL: Restricting access to foods and children's eating. Appetite 1999, 32:405–419.
18. Fisher JO, Birch LL: Restricting access to palatable foods affects children's behavioral response, food selection, and intake. Am J Clin Nutr 1999, 69:1264–1272.
19. Jansen E, Mulkens S, Jansen A: Do not eat the red food!: Prohibition of snacks leads to their relatively higher consumption in children. Appetite 2007, 49:572–577.
20. Jansen E, Mulkens S, Emond Y, JAnsen A: From the garden of Eden to the land of plenty: Restriction of fruit and sweet intake leads to increased fruit and sweet consumption in children. Appetite 2008, 51:570–575.
21. Le HT, Joosten M, Bijl-van-der J, Brouwer ID, Graaf C, Kok FJ: The effect of NaFeEDTA on sensory perception and long term acceptance of instant noodles by Vietnamese school children. Food Qual Pref 2007, 18:619–626.
22. Weijzen PLG, Liem DG, Zandstra EH, De Graaf C: Sensory specific satiety and intake: The difference between nibble- and bar-size snacks. Appetite 2008, 50:435–442.
23. Raynor HA, Niemeier HM, Wing RR: Effect of limiting snack food variety on long-term sensory-specific satiety and monotony during obesity treatment. Eat Behav. 2006, 7(1):1–14.
24. Birch LL: Dimensions of preschool children's food preferences. J Nutr Educ 1979, 11:77–80.
25. Leon F, Marcuz MC, Couronne T, Koster EP: Measuring food liking in children: a comparison of non-verbal methods. Food Qual Pref 1999, 10:93–100.

26. Guthrie CA, Rapoport L, Wardle J: Young children's food preferences: a comparison of three modalities of food stimuli. Appetite 2000, 35:73–77.

27. Birch LL, Sullivan SA: Measuring children's food preferences. J School Health 1991, 61:212–213.

28. Siegel S, Castellan NJ, Siegel S, Castellan NJ: The case of k related samples. In Nonparametric statistics. Volume 2. Boston: Mc Graw Hill; 1988:168–189.

29. Birch LL, Marlin DW: I don't like it; I never tried it: effects of exposure on two-year-old children's food preferences. Appetite 1982, 3:353–360.

30. Wardle J, Herrera ML, Cooke L, Gibson EL: Modifying children's food preferences: the effects of exposure and reward on acceptance of an unfamiliar vegetable. Eur J Clin Nutr 2003, 57:341–348.

31. Pliner P, Loewen ER: Temperament and food neophobia in children and their mothers. Appetite 1997, 28:239–254.

32. Birch LL, McPhee L, Shoba BC, Pirok E, Steinberg L: Looking vs. tasting. Appetite 1987, 9:171–178.

33. Julien KC, Buschang PH, Throckmorton GS, Dechow PC: Normal masticatory performance in young adults and children. Arch Oral Biol. 1996, 41(1):69–75.

34. Zijlstra N, Mars M, de Wijk RA, Westerterp-Plantenga MS, De Graaf C: The effect of viscosity on ad libitum food intake. Int J Obes (Lond). 2008, 32(4):676–683.

35. Sorensen LB, Moller P, Flint A, Martens M, Raben A: Effect of sensory perception of foods on appetite and food intake: a review of studies on humans. Int J Obes Relat Metab Disord 2003, 27:1152–1166.

36. Falciglia GA, Couch SC, Gribble LS, Pabst SM, Frank R: Food neophobia in childhood affects dietary variety. J Am Diet Assoc 2000, 100:1474–1481.

37. Schwartz B: The paradox of choice. Why less is more. New York: Harper Collins Publishers inc; 2004.

38. Nicklaus S, Boggio V, Issanchou S: Food choices at lunch during the third year of life: high selection of animal and starchy foods but avoidance of vegetables. Acta paediatrica (Oslo, Norway: 1992) 2005, 94:943–951.

39. Temple JL, Legierski CM, Giacomelli AM, Salvy SJ, Epstein LH: Overweight children find food more reinforcing and consume more energy than do non-overweight children. American Journal of Clinical Nutrition 2008, 87:1121–1127.

40. Saellens BE, Epstein LH: Reinforcing value for food in obese and non-pbese women. Appetite 1996, 27:41–50.

41. Robinson TN, Borzekowski DL, Matheson DM, Kraemer HC: Effects of fast food branding on young children's taste preferences. Arch Pediatr Adolesc Med. 2007, 161(8):792–797.
42. Bannon K, Schwartz MB: Impact of nutrition messages on children's food choice: Pilot study. Appetite 2006, 46:124–129.
43. King L, Hill AJ: Magazine adverts for healthy and less healthy foods: effects on recall but not hunger or food choice by pre-adolescent children. Appetite 2008, 51:194–197.

Preventing Delinquency Through Improved Child Protection Services

Richard Wiebush, Raelene Freitag and Christopher Baird

ABSTRACT

After increasing sharply in the early 1990s, rates of juvenile violent crime have been declining since 1993 (Snyder and Sickmund, 1999). Enhanced prevention and intervention efforts have had an impact. Yet, this encouraging trend should not invite tolerance of the status quo or inhibit efforts to drive rates still lower. This Bulletin examines a potentially powerful, yet often overlooked, delinquency prevention strategy: efforts to reduce the incidence of childhood maltreatment. The link between experiencing maltreatment as a child and committing offenses as a juvenile is profound. A substantial body of research (discussed more fully later) has shown that:

Maltreated children are significantly more likely than nonmaltreated children to become involved in delinquent and criminal behavior.

The prevalence of childhood abuse or neglect among delinquent and criminal populations is substantially greater than that in the general population.

Delinquent youth with a history of abuse or neglect are at higher risk of continuing their delinquent behavior than delinquents without such a history.

It follows, then, that if it were possible to reduce the incidence of children's maltreatment, delinquency rates would decline. Reducing the maltreatment of children is a goal best addressed on multiple levels—that is, through primary, secondary, and tertiary prevention efforts.1 Reducing childhood maltreatment also requires fundamental social changes in areas that contribute to increased rates of maltreatment (e.g., poverty). This Bulletin focuses on a tertiary prevention strategy that shows great promise in reducing subsequent maltreatment once a family has come to the attention of a child protection services (CPS) agency.

The Bulletin begins with a brief review of what is known about the link between childhood maltreatment and juvenile and adult offending. Next, it provides an overview of OJJDP's Comprehensive Strategy for Serious, Violent, and Chronic Juvenile Offenders and identifies where CPS tertiary prevention efforts belong within a community's continuum of delinquency prevention and intervention efforts. The main body of the Bulletin describes and assesses a particularly promising CPS tertiary prevention strategy known as Structured Decision Making (SDM).

Research on the Effects of Childhood Maltreatment

Childhood Maltreatment and Subsequent Offending

Researchers use two basic approaches to examine the possible link between childhood maltreatment and subsequent offending. The first approach is to sample maltreated children and follow them (retrospectively or prospectively) to observe rates of subsequent offending. The second approach is to sample juvenile or adult offenders and measure the rate at which they experienced maltreatment in childhood. Both approaches are strengthened when data on the maltreated sample group are compared with those for a nonmaltreated control group. Both approaches may use a variety of methods to define childhood maltreatment including self-reporting, referral to a CPS agency, substantiation of a CPS report, or court involvement related to childhood maltreatment victimization. Similarly, subsequent offending may be defined by self-report, arrest, or conviction. Some studies also examine the relationship between maltreatment and subsequent at-risk behaviors such as committing status offenses, becoming pregnant as a teenager, having a low grade point average (GPA), or experiencing mental health problems. The growing body of research on these issues uses a variety of methodologies but leads to a similar conclusion: "In general, people who experience any type of maltreatment in childhood... are more likely than people who were not maltreated to be arrested later in life" (Widom, 1995:4).

A closer examination of this body of research provides further details on the connections between abuse and delinquency. Several studies have examined the prevalence of abuse and neglect among delinquent and criminal populations and found that these populations have strikingly higher rates of childhood abuse and neglect than the general population. National estimates in 1997 indicated that approximately 4 percent of all children were reported to child welfare agencies as alleged victims of abuse or neglect (U.S. Department of Health and Human Services, Administration on Children, Youth and Families, 1999). In contrast: A study of court-referred juvenile offenders in Milwaukee County, WI, found that 66 percent of male offenders and 39 percent of female offenders previously had been victims in substantiated reports of abuse or neglect (Pawaserat, 1991).

A study of high-risk male juvenile parolees in three States revealed that the proportion of juveniles who had allegedly been victims of abuse or neglect ranged from 29 percent in Virginia to 45 percent in Colorado to 53 percent in Nevada (Wiebush, McNulty, and Le, 2000).

Findings are similar for adult offenders. For example, an estimated 5–8 percent of the general adult male population and 12–17 percent of the adult female population in the United States were physically or sexually abused as children (Gorey and Leslie, 1997). However, a recent study of adult offenders found that 16 percent of males and 57 percent of females in State prisons had experienced childhood physical or sexual abuse (Harlow, 1999).

Several empirically based risk assessment studies conducted by the National Council on Crime and Delinquency (NCCD) provide additional data linking childhood maltreatment and subsequent delinquency. These studies identify jurisdiction-specific risk factors that States use to assess and classify juvenile probationers and parolees according to their likelihood of committing subsequent offenses. In at least five States—Michigan, Nebraska, Rhode Island, Virginia, and Wisconsin (National Council on Crime and Delinquency, 1995a, 1995b, 1995c, 1997, 1999)—a prior allegation or confirmation of childhood abuse or neglect has been identified as an important risk factor for continuing delinquency. In Rhode Island, for example, juvenile probationers who were childhood victims of abuse or neglect recidivated at more than 1.5 times the rate of youth who had not been victimized (71 percent versus 46 percent).

Two conceptually and methodologically important research efforts have compared maltreated children with non-maltreated children to determine the impact of maltreatment on subsequent offending in general and violent offending in particular. First, in a study that was part of the OJJDP-funded longitudinal Rochester Youth Development Study, researchers examined the official and self-reported delinquency of a general population sample of 1,000 juveniles, 16 percent of whom had substantiated reports of abuse or neglect as children (Kelley, Thornberry, and Smith, 1997). The study compared the maltreated and nonmaltreated groups on the extent and frequency of their delinquent involvement. Among the study's key findings are the following:

Youth who had been victims of child abuse or neglect were significantly more likely than nonvictims to have an official record of delinquency (45 percent versus 32 percent).

Compared with youth who had not experienced childhood maltreatment, maltreated youth self-reported significantly greater involvement in delinquent behavior (79 percent versus 70 percent), serious delinquent behavior (42 percent versus 33 percent), and violent delinquent behavior (70 percent versus 56 percent).

The frequency of official and self-reported delinquent acts was significantly higher for maltreated youth than for youth who had not been maltreated.

As the frequency and severity of maltreatment increased, there were significant increases in the frequency of subsequent offending. In fact, the number of arrests among juveniles who had experienced multiple incidents of maltreatment (or multiple types of maltreatment or particularly severe maltreatment) as children was twice as high as the number among juveniles who had experienced less frequent or less severe maltreatment.

Second, Widom and her associates conducted a series of well-designed studies that compared the delinquent and criminal outcomes of a sample of maltreated children with those of a matched sample of nonmaltreated children (Widom, 1992; Widom, 1995; Maxfield and Widom, 1995). All the children (n=1,575) were followed through their teenage years into adulthood to determine the extent of their delinquent and criminal activity. Key findings include the following:

Juvenile arrest rates for the two groups differed significantly: 27 percent of the maltreated children (versus 17 percent of the nonmaltreated children) were arrested as juveniles. Moreover, the maltreated children had a higher average number of juvenile arrests than the nonmaltreated children (3.0 versus 2.4).

Adult arrest rates for the two groups were also significantly different: 42 percent of the maltreated children were arrested as adults, compared with 33 percent of the nonmaltreated children. In addition, the maltreated children had a higher average number of adult arrests (5.7 versus 4.2).

Both the maltreated and nonmaltreated groups included a substantial number of individuals who did not offend as juveniles but who were arrested as adults. The maltreated children, however, were significantly more likely than the nonmaltreated children to evidence this "adult onset" of criminal behavior (31 percent versus 26 percent). The maltreated children were also significantly more likely than the nonmaltreated children to commit violent offenses as teenagers and young adults.

Childhood Maltreatment and Other At-Risk Behaviors

Childhood maltreatment has been linked to a number of other adolescent problems. Compared with nonmaltreated matched control groups, abused or neglected children are significantly more likely to engage in violent behavior, become pregnant during adolescence, use drugs, have lower GPAs, and/or experience mental health problems (Kelley, Thornberry, and Smith, 1997). (See figure 1.)

Several other studies have reached similar conclusions, indicating that maltreated children are at increased risk of both delinquency (Bolton, Reich, and Gutierres, 1977; Alfaro, 1981; English, 1997) and other problems during adolescence (Dembo et al., 1992; Silverman, Reinherz, and Giaconia, 1996). As Kelley and colleagues have noted (1997:11):

Maltreatment diminishes the likelihood that children will come through adolescence with no serious problems. Moreover, a history of childhood maltreatment nearly doubles the risk that teenagers will experience multiple problems during adolescence.

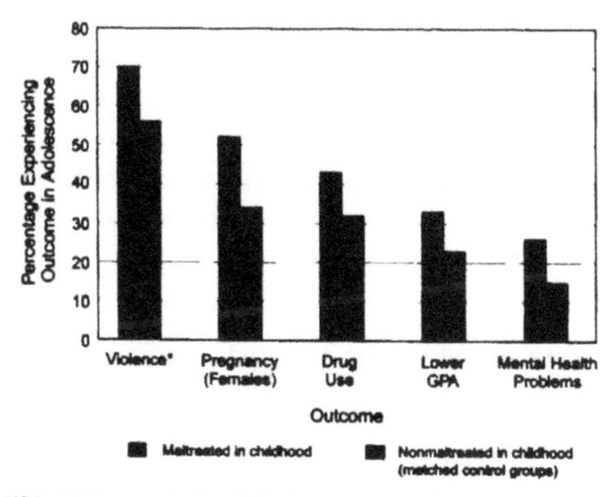

Figure 1: Relationship Between Child Maltreatment and Various Outcomes During Adolescence.

Still, most maltreated children do not become delinquents. What is it that keeps some maltreated children from being arrested as juveniles? Kelley and colleagues suggest that there may be "intervening factors, including the emergence of protective factors and the provision of effective services" (Kelley, Thorn-berry, and Smith, 1997:13). This Bulletin examines how CPS agencies can provide more "effective services" and potentially ward off subsequent delinquent/criminal involvement by abused and neglected children.

OJJDP's Comprehensive Strategy and the Role of CPS

In its 1995 Guide for Implementing the Comprehensive Strategy for Serious, Violent, and Chronic Juvenile Offenders, OJJDP emphasizes the need for delinquency prevention, stressing that it is the "most cost-effective approach to reducing juvenile delinquency" (Howell, 1995:7). In fact, prevention efforts constitute one of the two primary components of the Comprehensive Strategy (the other being reform of the juvenile justice system). The Comprehensive Strategy uses a risk-focused approach to prevention efforts—meaning that it requires careful attention to factors identified through research as precursors to delinquency, violence, and other problem behaviors. As shown in figure 2, the Comprehensive Strategy divides such risk factors into four basic domains: community, family, school, and individual/ peer. Central to the focus of this Bulletin are family risk factors. In defining family risk factors, the Comprehensive Strategy specifies that "family management problems" include failure to supervise and monitor children and excessively severe, harsh, or inconsistent punishment; "family conflict" includes domestic violence; and "family history of the problem behavior" includes caregiver substance abuse (Howell, 1995:20). Significantly, these family risk factors for delinquency and violence are also characteristics typically present in abusive or neglectful families. These identified risk factors, considered in conjunction with the clear link between childhood maltreatment and subsequent delinquency, strongly suggest that child welfare agencies and their efforts to reduce child abuse and neglect have a central role for delinquency prevention within the context of the Comprehensive Strategy.

Structured Decision Making: Background

The primary goals of the Structured Decision Making model are to (1) bring a greater degree of consistency, objectivity, and validity to child welfare case decisions and (2) help CPS agencies focus their limited resources on cases at the highest levels of risk and need. Structured assessment tools are used at various points in the case decision-making process (e.g., initial response to allegations, child removal, case opening/closing, and reunification). Each tool incorporates decision protocols—based directly on assessment results—to guide the agency's response to each family. One of the key assessment tools is a research-based risk assessment that classifies families according to their likelihood of continuing to abuse or neglect their children.

Risk Factor	Substance Abuse	Delinquency	Teenage Pregnancy	School Dropout	Violence
Community					
Availability of drugs	✓				✓
Availability of firearms		✓			✓
Community laws and norms favorable toward drug use, firearms, and crime	✓	✓			✓
Media portrayals of violence					✓
Transitions and mobility	✓	✓		✓	
Low neighborhood attachment and community organization	✓	✓			✓
Extreme economic deprivation	✓	✓	✓	✓	✓
Family					
Family history of problem behavior	✓	✓	✓	✓	✓
Family management problems	✓	✓	✓	✓	✓
Family conflict	✓	✓	✓	✓	✓
Favorable parental attitudes toward and involvement in the problem behavior	✓	✓			✓
School					
Early and persistent antisocial behavior	✓	✓	✓	✓	✓
Academic failure beginning in elementary school	✓	✓	✓	✓	✓
Lack of commitment to school	✓	✓	✓	✓	✓
Individual/Peer					
Rebelliousness	✓	✓		✓	
Friends who engage in the problem behavior	✓	✓	✓	✓	✓
Favorable attitudes toward the problem behavior	✓	✓	✓	✓	
Early initiation of the problem behavior	✓	✓	✓	✓	✓
Constitutional factors	✓	✓			✓

Source: Catalano and Hawkins, 1995; updated 1998–2000 by Developmental Research and Programs, Inc.

Figure 2: Risk Factors for Health and Behavior Problems.

SDM, then, focuses on how case management decisions are made and how agency resources can best be directed. Ultimately, SDM is a strategy designed to reduce subsequent maltreatment rates by improving both the efficiency and effectiveness of CPS agencies. To the extent that SDM accomplishes these goals—and available research indicates that it does—the system will have the added benefit of reducing the rate at which maltreated children subsequently become involved in the juvenile and/or criminal justice system.

Origins of SDM

In addition to its use in addressing the link between childhood maltreatment and subsequent delinquency, SDM has another tie to the juvenile justice system. A core component of SDM is the use of research-based risk assessment tools. The risk assessment methodology was originally developed to classify juvenile offenders according to their likelihood of committing additional offenses. In 1986 (in Alaska) and 1988 (in Michigan), the Children's Research Center (CRC), a division of NCCD, worked with State agencies to develop a research-based risk assessment tool for use with their delinquent populations. In both States, the agency that was responsible for juvenile justice also administered CPS. Knowing the value of risk assessment to decision making and resource utilization in the juvenile justice system, the Alaska and Michigan administrators raised the question of whether a similar tool could be used to identify which CPS families with at least one known incident of maltreatment were at highest risk of future maltreatment. Such a tool, the States reasoned, could help them make better decisions about which families to serve. In response, CRC staff conducted research in both States and found that fairly simple and highly effective tools could be constructed to identify families with low, moderate, and high risk of reabuse (Baird, Wagner, and Neuenfeldt, 1992).

Shortly before the CRC studies in Alaska and Michigan, Will Johnson conducted similar research in Alameda County, CA, and reached the same conclusion (Johnson and L'Esperance, 1984). It is not surprising that two independent studies confirmed the effectiveness of research-based risk assessment for CPS agencies. Although a substantial body of literature has demonstrated the effectiveness of empirical risk assessment in other human service fields, many in the child welfare field felt that the risk of childhood maltreatment was too complex an issue to be distilled to a short list of variables. However, it is precisely because human behavior is so complex that assessments based on a simple tool are generally more accurate than those based on clinical judgment (Dawes, Faust, and Meehl, 1989).

Since 1988, when CRC implemented the first research-based risk assessment scale for child welfare in Alaska's CPS system, it has conducted 11 risk studies in jurisdictions ranging from Rhode Island to California to South Australia. As the use of research-based risk assessment spread, it became evident that although risk assessment provided critical information for some CPS decisions, other key decision points required different types of structured assessments. Hence, the SDM model has been expanded to address most vital decisions, from the initial decision of whether to initiate an investigtion to the final decision of when to close a case. Not all of these decisions lend themselves to the use of research-based assessment methodologies; some require consensus-based approaches. However, all decision points in the child welfare case management process can use structured assessment tools that clearly identify what factors need to be considered by all staff and how those factors relate to the decision.

The Need for SDM in Child Welfare Services

The number of abuse and neglect allegations nationwide has risen dramatically during the past two decades. Most child welfare agencies have been hard pressed to respond effectively, as new demands have outpaced available resources. As a result, the Nation has seen class action lawsuits challenging the quality of services provided in more than 30 States, media exposés resulting from child deaths, increased concerns over worker and agency liability, and a continuous search for new strategies and resources to address the burgeoning problem.

Child welfare agencies' need for additional resources is obvious but not the only issue. Increasing pressures have highlighted a problem that has long plagued human services agencies in general and child welfare agencies in particular: the need for more efficient, consistent, and valid decision making. Child protection workers must make extremely difficult decisions. Yet, in many agencies, workers have widely different levels of training and experience. Consequently, decisions regarding case openings, child removal and reunification, and service provision have long been criticized as inappropriate and/or inconsistent. In fact, research has clearly demonstrated that decisions regarding the safety of children vary significantly from worker to worker, even among those considered to be child welfare experts (Rossi, Schuerman, and Budde, 1996). As the pressure to make critical decisions affecting children and families rises, so does the potential for error. Inappropriate decisions can be costly and may result in the overuse of out-of-home placements or, tragically, the injury or death of a child.

Problems of increasing referrals, limited resources, and liability exposure are inextricably linked with decision-making issues. Agencies overwhelmed by heavy workloads need to be able to consistently and accurately determine which cases should be investigated, which children need to be removed from their homes, and which families require the most intensive services. Clearly, new methods are required to help agencies and workers make decisions as efficiently and effectively as possible. Tools are also needed to help workers make accurate and reliable assessments of both immediate safety issues and long-term risk. Decision-making strategies are needed to help agencies focus limited resources on those families at the highest level of risk. These decision-making tools must be embedded in case management systems that incorporate clearly defined service standards, mechanisms for frequent reassessments, methods for measuring workload, and procedures for ensuring accountability and quality control.

How child welfare decisions are made and how agencies use resources are the key questions addressed by CRC's SDM model. The SDM model provides a comprehensive, systematic approach to case management, resource management, performance monitoring, quality assurance, and program and policy evaluation. The four principles that form the foundation of SDM are presented in the next section, followed by an indepth discussion of the major components of the model.

Principles of the SDM Model

The SDM model is based on four primary principles.

First, because decisions can be significantly improved when they are structured appropriately, every worker in every case must consider specific criteria through highly structured assessment procedures. Failure to clearly define decision-making criteria and identify how workers should apply these criteria results in inconsistency and, sometimes, inappropriate case actions.

Second, priorities assigned to cases and service plans (or responses) must correspond directly to results of the assessment process. The assessment process has little meaning unless its results lead directly to an appropriate decision. Decisions should be structured to ensure that the agency's highest priority is given to the most serious and/or the highest risk cases. Moreover, if agencies are to translate priority setting into practice, they must have clearly identified and consistently implemented service standards—differentiated by level of risk—for each type of case. Such differential service standards help focus how resources are used and provide a degree of accountability often missing in human services agencies. To ensure that the assessment process results in improved service, expectations for staff regarding the process must be clearly defined and practice standards must be readily measurable.

Third, virtually everything that an agency does—from providing services in an individual case to budgeting for treatment resources—should be a response to the assessment process. Risk and needs assessments, for example, should be linked directly to service plans. In the aggregate, assessment data also will help indicate the range and extent of service resources needed in a community. Similarly, assessment and case classification results are directly related to agency service standards, which in turn drive staff workload and budgeting requirements.

Fourth, a single, rigidly defined model cannot meet the needs of every agency. Not all State and county child welfare agencies are organized to deliver services in the same way. Nor do they always share similar service mandates. As a result, the CRC approach to designing an SDM system is collaborative and engages agencies in a joint development effort. Each system is built on a set of principles and components that are then adapted to local practices and mandates. The development process incorporates a great deal of input from local managers and staff. The result is a site-specific system that is "owned" by the agency and builds on its particular strengths as a service organization.

The SDM Model

The SDM model consists of a set of assessment instruments augmented by management components that provide accountability, quality assurance, and planning, budgeting, and evaluation data. Each assessment tool is designed specifically for use at a key decision point in the life of a CPS case. By focusing on particular decision points rather than attempting to address multiple issues with a single tool, the SDM model enhances clarity and allows agencies to more effectively monitor compliance with established policies and procedures. Although SDM tools identify the critical assessment factors for each decision point, the model also includes ways to account for unique case characteristics that may not have been captured on the assessment instrument. Most SDM tools incorporate an override provision that allows workers to change the assessment-indicated decision, when necessary. SDM does not replace worker judgment; instead, it provides an objective framework within which to articulate agency policy, thereby helping ensure that "best practices" are applied to all cases. The primary tools used in SDM sites are discussed below.

Response Priority Assessment

Most jurisdictions begin the SDM assessment processes after deciding to investigate a referral. At this point, a set of decision trees (see figure 3) guides caseworkers through key questions that allow them to determine how quickly to respond to the referral. For example, Cuyahoga County, OH, sorts all investigations into three groups: Priority 1 (those requiring a response within 1 hour), Priority 2 (response within 24 hours), and Priority 3 (response within 72 hours). Different decision trees are used to address different types of maltreatment (e.g., neglect, physical abuse, sexual abuse).

Figure 3 illustrates a response priority system for use in cases of alleged physical abuse. Under this system, the answer to each question directs the worker to the next question to be asked and, rather quickly, to a presumptive response level. Because of differences in State statutes and agency policy, response priority systems vary somewhat among jurisdictions. All systems, however, ensure that workers systematically apply certain key criteria to every case.

CRC researchers have assessed the effectiveness of response priority tools using SDM management data in California (Children's Research Center, 2000). The research questions were whether and to what extent the child removal (i.e., out-of home placement) rate was higher in cases that required an "immediate response" than the rate in cases that had been assigned a lower priority. Since the tools are designed to prioritize referrals based on the seriousness of the allegations, one would expect—if the tools are valid—to see a much higher rate of removal in the "immediate response" cases. The results provided strong support to the design of the response priority tools. They showed that over a 6-month period, the proportion of removals in the immediate response cases was four times higher than it was in the cases that were given lower priority (13 percent versus 3.2 percent).

Safety Assessment

When a CPS investigator first makes contact with a family, the worker must determine whether there are any immediate, pressing conditions that threaten the safety of the child. An SDM safety assessment generally consists of three parts. The first is a list of potential threats to children in the family—that is, conditions that would place a child in danger of immediate harm. Figure 4 shows an example of the first section of a safety assessment. The second section is an identification of the short-term interventions selected by the worker (e.g., monitoring by a neighbor or relative), which will constitute a safety plan, and the third is simply a record of the final decision.

Safety assessments should be completed during a CPS investigator's first face-to face contact with the family. If safety issues are present, workers are directed to consider a series of potential in-home interventions, beginning with the least restrictive. If in-home interventions are unavailable, refused, or insufficient to mitigate identified safety issues, placement emerges as the only alternative. All protective placements are based on the determination that available in-home interventions would fail to offer adequate protection for the child.

Although safety assessments may be characterized as simple checklists, their value cannot be overstated. Simplicity is, in fact, key to successful implementation, because CPS investigators are required to make decisions within very limited timeframes. By allowing investigators to focus on a relatively small set of important factors, safety assessments help investigators avoid mistakes and improve consistency.

Like the response priority decision trees discussed earlier in this Bulletin, safety assessments help ensure that CPS staff assess all cases based on a standardized set of issues. Safety assessments also require that agencies have a safety plan whenever any safety factor (i.e., a condition that threatens immediate harm) has been identified, thereby adding accountability to the process.

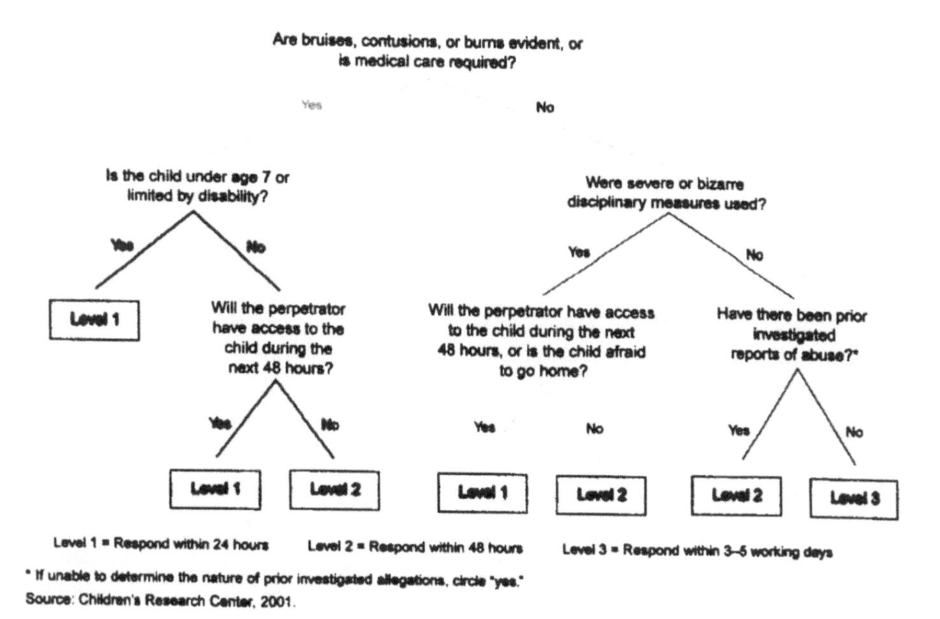

Figure 3: Response Priority Decision Tree: Physical Abuse Cases (Example)

Two studies conducted to date have identified a positive relationship between the safety-related issues typically incorporated in safety assessments and subsequent harm (Illinois Department of Children and Family Services, 1997; Wagner and Caskey, 1998). For example, researchers in Illinois used a pretest-posttest design to determine whether rates of child injuries that occur within 60 days of a CPS investigation declined after the Illinois Department of Child and Family Services began using its safety assessment. Compared with the period prior to implementation, use of the safety assessment appears to have reduced subsequent child injuries (Illinois Department of Children and Family Services, 1997).

Risk Assessment

The heart of the SDM model is its research-based risk assessment tool. Although other components of the model are based on a general consensus (often informed by available research) of what constitutes best practice, SDM risk assessment tools are based on the outcomes of actual cases. To develop a risk assessment tool, CRC and agency staff jointly identify a list of potential risk factors. These potential factors are based on literature, experience, and previous CRC research results. CRC researchers then review a large sample of case records (e.g., 1,000) and code them for the presence or absence of the factors, based on what was known about each family at the time of the sampled investigation. Case records are further explored to identify families that experienced reinvolvement with the agency after the sampled investigation. The definition of "reinvolvement" generally includes subsequent CPS referrals, subsequent substantiations of maltreatment, subsequent child injuries, and subsequent CPS placements. The research process then examines the statistical relationship between case characteristics and case outcomes to identify the variables that are most closely associated with risk. The set of risk factors that most effectively divides families into three or four different risk groups constitutes the risk assessment tool. Figure 5 presents a risk assessment scale that is used in California.

The test of any risk assessment tool is how well it classifies families based on their likelihood of subsequently maltreating their children. Families identified as "high risk," for example, should be expected to have re-referral (and/or resubstantiation) rates that are significantly higher than those in families that the tool classifies as "low risk." Figure 6 illustrates this principle using the results of the California risk scale. The data show a strong relationship between risk classification and outcomes. For example, after a 2-year followup period, the California families that were assessed as low risk had a resubstantiation rate of less than 8 percent. In contrast, among families classified as very high risk, the resubstantiation rate was 44 percent or more than five times the rate found for low-risk cases.

The 11 risk assessment studies that CRC has completed to date lead to the following conclusions:

- A child's risk of subsequent abuse and his or her risk of subsequent neglect are best approached as separate issues. Although some risk factors relate to both types of maltreatment (and, indeed, some families both abuse and neglect children), there are enough differences between the family characteristics and dynamics associated with abuse and those associated with neglect to warrant the use of separate scales to address each type of maltreatment.
- Risk instruments developed in one jurisdiction appear to transfer well to other locations. For example, the risk assessment instrument developed in Michigan has been shown to effectively classify families based on risk in California, Florida, and Missouri (Baird and Wagner, 2000), and a risk tool

developed in Wisconsin has proved effective with a predominantly Hispanic population in Texas (Wood, 1997). Still, when an agency adopts another jurisdiction's risk assessment instrument, it is critical that the agency routinely collect data to validate (and, if necessary, revise) the instrument within 2 years of implementation.

Family Case Name: ______________________ **Family Case #:** ____________

County Name: ____________ **County #:** ________ **Office:** ____________ **Worker #:** ________

CPS Referral Date: ___/___/___ **Current Date:** ___/___/___ **Assessment** (check one): __ **Initial** __ **Review**

Section 1: Safety Assessment

Part A. Safety Factor Identification

Directions: The factors in the following list are behaviors or conditions that may be associated with a child's being in immediate danger of serious harm. Identify the presence or absence of each factor by circling either "yes" or "no." **Note: The vulnerability of each child needs to be considered throughout the assessment. Children ages 0 through 6 cannot protect themselves. For older children, inability to protect themselves could result from diminished mental or physical capacity or repeated victimization.**

1. Yes No Caregiver's behavior is violent or out of control.
2. Yes No Caregiver describes or acts toward child in predominantly negative terms or has extremely unrealistic expectations.
3. Yes No Caregiver caused serious physical harm to the child or has made a plausible threat to cause serious physical harm.
4. Yes No The family refuses access to the child, or there is reason to believe that the family is about to flee, and/or the child's whereabouts cannot be ascertained.
5. Yes No Caregiver has not provided or will not provide supervision necessary to protect child from potentially serious harm.
6. Yes No Caregiver is unwilling or unable to meet the child's immediate needs for food, clothing, shelter, and/or medical or mental health care.
7. Yes No Caregiver has previously maltreated a child, and the severity of the maltreatment, or the caregiver's response to the previous incident(s), suggests that child safety may be an immediate concern.
8. Yes No Child is fearful of caregiver(s), other family members, or other people living in or having access to the home.
9. Yes No The child's physical living conditions are hazardous and immediately threatening.
10. Yes No Child sexual abuse is suspected, and circumstances suggest that child safety may be an immediate concern.
11. Yes No Caregiver's drug or alcohol use seriously affects his/her ability to supervise, protect, or care for the child.
12. Yes No Other (specify) ______________________

IF NO SAFETY FACTORS ARE PRESENT, GO TO SECTION 3: SAFETY DECISION

Part B. Safety Factor Description

Directions: For all safety factors that are marked "Yes," note the applicable safety factor number and then briefly describe the specific individuals, behaviors, conditions, and/or circumstances associated with that particular safety factor.

____

____

____

____

____

Source Children's Research Center, 2001

Figure 4: Section 1: Safety Assessment (Example).

Family Case Name: ______ Family Case #: ______

County Name: ______ County #: ______ Office: ______

Worker Name ______ Worker #: ______ CPS Referral Date: __/__/__ Assessment Date: __/__/__

Neglect — Score

N1. Current Referral Is for Neglect
- a. No — 0
- b. Yes — 1 ___

N2. Number of Prior Referrals
- a. None — 0
- b. One — 1
- c. Two or more — 2 ___

N3. Number of Children in the Home
- a. Two or fewer — 0
- b. Three or more — 1 ___

N4. Number of Adults in Home at Time of Referral
- a. Two or more — 0
- b. One/none — 1 ___

N5. Age of Primary Caregiver
- a. 30 or older — 0
- b. 29 or younger — 1 ___

N6. Characteristics of Primary Caregiver (check and add for score)
- a. Not applicable — 0
- b. ___ Parenting skills are a major problem — 1
- c. ___ Lacks self-esteem — 1
- d. ___ Apathetic or shows feelings of hopelessness — 1 ___

N7. Primary Caregiver Involved in Harmful Relationships
- a. No — 0
- b. Yes, but not a victim of domestic violence — 1
- c. Yes, as a victim of domestic violence — 2 ___

N8. Primary Caregiver Has a Current Substance Abuse Problem
- a. No — 0
- b. Alcohol only — 1
- c. Other drug(s) (with or without alcohol) — 3 ___

N9. Household Is Experiencing Severe Financial Difficulty
- a. No — 0
- b. Yes — 1 ___

N10. Primary Caregiver's Motivation To Improve Parenting Skills
- a. Motivated and realistic — 0
- b. Unmotivated — 1
- c. Motivated but unrealistic — 2 ___

N11. Response of Caregiver(s) to Investigation and Seriousness of Complaint
- a. Attitude is consistent with seriousness of allegation, and he or she has complied satisfactorily — 0
- b. Attitude not consistent with seriousness of allegation (minimizes) — 1
- c. Failed to comply satisfactorily — 2
- d. Both b and c — 3 ___

TOTAL NEGLECT RISK SCORE ___

INITIAL RISK LEVEL
Assign the family's risk level based on the highest score on either scale, using the following chart:

Neglect Score	Abuse Score	Risk Level
___ 0–4	___ 0–2	___ Low
___ 5–7	___ 3–5	___ Moderate
___ 8–12	___ 6–9	___ High
___ 13–20	___ 10–16	___ Intensive

FINAL RISK LEVEL: ___ Low ___ Moderate ___ High ___ Intensive

Abuse — Score

A1. Current Referral Is for Physical, Sexual, or Emotional Abuse
- a. No — 0
- b. Yes — 1 ___

A2. Prior Abuse Referrals
- a. None — 0
- b. Physical/emotional abuse referral(s) — 1
- c. Sexual abuse referral(s) — 2
- d. Both b and c — 3 ___

A3. Prior CPS Service History
- a. No — 0
- b. Yes — 1 ___

A4. Number of Children in the Home
- a. One — 0
- b. Two or more — 1 ___

A5. Caregiver(s) Abused as Child(ren)
- a. No — 0
- b. Yes — 1 ___

A6. Secondary Caregiver Has a Current Substance Abuse Problem
- a. No, or no secondary caregiver — 0
- b. Yes (check all that apply) — 1 ___
  - ___ Alcohol abuse problem
  - ___ Drug abuse problem

A7. Primary or Secondary Caregiver Employs Excessive and/or Inappropriate Discipline
- a. No — 0
- b. Yes — 2 ___

A8. History of Domestic Violence by Caregiver(s)
- a. No — 0
- b. Yes — 1 ___

A9. Caregiver(s) Over-Controlling
- a. No — 0
- b. Yes — 1 ___

A10. Child in the Home Has Special Needs or History of Delinquency
- a. No — 0 ___
- b. Yes (check all that apply) — 1 ___
  - ___ Diagnosed special needs
  - ___ History of delinquency

A11. Secondary Caregiver Motivated To Improve Parenting Skills
- a. Yes, or no secondary caregiver in home — 0
- b. No — 2 ___

A12. Primary Caregiver's Attitude Is Consistent With the Seriousness of the Allegation
- a. Yes — 0
- b. No — 1 ___

TOTAL ABUSE RISK SCORE ___

POLICY OVERRIDES
Policy: Override to Intensive. Check appropriate reason.
- ___ 1. Sexual abuse case and the perpetrator is likely to have access to the child victim.
- ___ 2. Nonaccidental physical injury to child under 2.
- ___ 3. Serious nonaccidental physical injury requiring hospital or medical treatment.
- ___ 4. Death (previous or current) of a child as a result of abuse or neglect
- ___ 5. Positive tox screen (any drug, including alcohol) of mother or child.

Discretionary Override to Risk Level
- ___ 6. Override and assign new risk level ______ __/__/__
  (Supervisor's initials as approval) (Date)

Discretionary override reason: ______

Source: Children's Research Center, 1996a.

Figure 5. California Family Risk Assessment Scale.

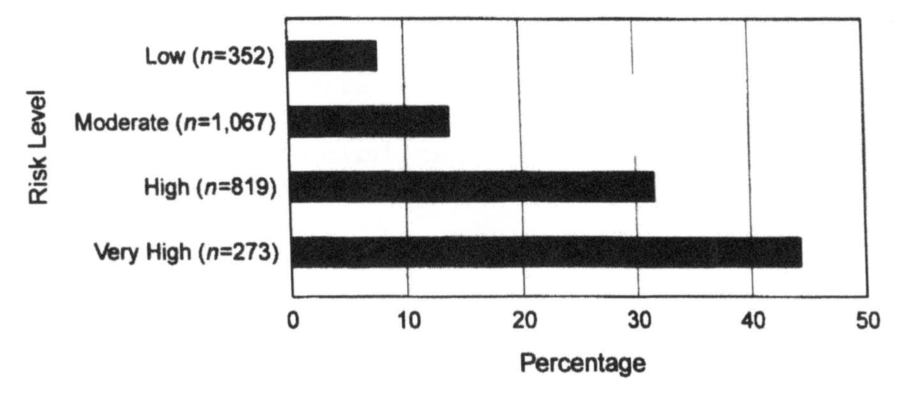

Source: Children's Research Center, 1998a.

Figure 6. Subsequent Substantiation Rates by Risk Classification: California (2-Year Followup).

Prediction Versus Classification

The risk level assigned to a case is not a prediction that a family will or will not maltreat a child in the future. Instead, a risk level designation simply denotes a case's inclusion in a group of families with relatively high or low historical rates of subsequent maltreatment. Accurate prediction in any field is difficult; prediction of human behavior is especially complex because many factors contribute to determining how individuals will act. Classification, on the other hand, is simply a systematic arrangement of clients into groups or categories according to established criteria. In the CPS context, classification is meant to assign cases to different risk categories based on observed outcome rates. Knowing that cases with certain similar characteristics have recidivism rates of 5 percent, 25 percent, or 50 percent helps social workers (and agencies) determine appropriate levels of intervention and allocate scarce resources in an effective manner.

Decisions Affected by Risk Assessment

Agencies typically use risk assessment results to guide decisions about whether families should have their cases opened for ongoing CPS services (e.g., moderate and high risk) or not (e.g., low risk). For cases that are opened, risk results are also typically used to determine the level of intervention required. The link between risk levels and service standards is discussed in greater detail later in this Bulletin.

Agencies should rely primarily on safety assessment results, rather than the risk assessment, to make protective placement decisions. Even among very high-risk families, about half will not have another substantiated incident of maltreatment. A policy that required protective placement of all very high-risk children, therefore, would lead to overuse of such placements, resulting in a crushing demand on scarce resources and unduly increasing the number of children and families who must endure the emotional impact that such placement brings.

Family Strengths and Needs Assessments

For families receiving ongoing CPS services, staff must decide precisely what services to provide and what the case plan objectives are. A family strengths and needs assessment (FSNA) covers a comprehensive array of critical domains of family life that affect the care of children (e.g., substance abuse, parenting skills, domestic violence). Families are rated on each domain along a continuum from strength to severe need. Definitions for each item and rating level help reduce subjectivity in these assessments. Items are weighted so that once completed, the assessment can identify a family's three most critical needs. Case plans, by policy, are to address those critical needs. With FSNAs, case plans are less likely to omit critical needs. Conversely, FSNAs help prevent case plans from including voluminous recommendations that are overwhelming for families.

More recent applications of the FSNA reflect the trend toward strength-based practice, requiring workers to identify a family's greatest strengths. Even though the literature has promoted strength-based practice for some time, the philosophy is rarely applied in the field. Including strengths in the SDM model helps make the practice more routine. In sum, FSNA instruments do the following:

Ensure that all CPS workers consistently consider each family's strengths and weaknesses in an objective format when assessing the family's need for services.

Provide workers and first-line supervisors an important case planning reference that eliminates long, disorganized case narratives and reduces paperwork.

Provide a basis for monitoring whether appropriate service referrals are made.

Through initial needs assessments followed by periodic reassessments, permit caseworkers and supervisors to easily evaluate change in family functioning and thus monitor the impact of services on the case.

Provide management with profiles containing aggregate information on the issues that client families face. These profiles can be used to develop resources to meet client needs.

One purpose of the FSNA is to allow caseworkers to consistently identify critical concerns facing families. Services provided to enhance child safety should not vary based on which worker has been assigned to a case. To what extent do these tools, in fact, promote consistency in the assessment process? Reliability testing has demonstrated a relatively high rate of inter-rater reliability for most items on SDM strengths and needs assessment instruments. Research conducted in California (Children's Research Center, 1998b) found that for caseworker assessments of whether a need existed, without regard to the severity of the need, most items had inter-rater reliability rates above 80 percent. For items that address some of the most critical issues facing CPS populations, such as substance abuse and mental health problems, rates of agreement were at or near 90 percent.

When the measure of agreement was based on workers' scoring of "needs" exactly the same along a four-point continuum (from "strength" to "severe need"), rates of agreement declined to some degree but remained high for key issues such as substance abuse (85 percent agreement), family relationships (70 percent), domestic violence (69 percent), and parenting skills (64 percent).

Risk and Needs Reassessment

Initial assessments of a family's risk level and service needs are followed by routine reassessments, conducted at established intervals (generally every 90 days) for as long as the case is open. Reassessment ensures that any potential changes in the family's risk level or service needs will be considered in subsequent stages of the service delivery process and that case decisions will be made accordingly. Case progress determines whether a lower or higher level of service is needed or whether the case can be closed. In most agencies, risk and needs assessment and reassessment instruments have become formal case planning documents and thus reduce the need for long case narratives and other paperwork.

Periodic reassessment also allows agencies to monitor important case outcomes on an ongoing basis. Such outcomes include new abuse or neglect incidents, changes in out-of-home placement status of children in the family, changes in a family's service utilization pattern, and changes in the severity of identified needs. In short, reassessing each family at fixed intervals provides direct service workers and their supervisors with an efficient mechanism for collecting and evaluating information necessary to effectively manage their cases.

Risk-Based Service Standards

Not all families that have been referred to CPS for child abuse or neglect require the same level of child welfare services. Yet, in terms of case assignment and resource allocation, many child welfare agencies treat all cases the same. Hence, agencies sometimes provide services to families that may not need them and at the same time fail to provide other, higher risk families with the resources needed to adequately protect children.

Risk assessment provides CPS agencies an objective framework within which to make service decisions. It also allows them to allocate service resources more efficiently. A primary mechanism for focusing resources is the use of differential service standards, under which the mandated frequency of caseworker-family contact is tied to the family's level of risk. Low-risk families do not need the same amount of agency resources (i.e., caseworker time) as high-risk families because the former are much less likely to maltreat their children again. When an agency establishes and uses differential service standards based on risk, existing service resources can reach farther and better results are possible. Figure 7 shows how the Michigan Family Service Agency has defined and differentiated service standards by risk level. Many other agencies have implemented similar standards.

SDM for Children in Out-of-Home Care

CRC also has applied the principles of standardized assessment and structured decision making to families that have children in foster care. SDM's foster care component is designed to ensure that State and Federal policies regarding reunification of families, permanency planning for children, and termination of parental rights are translated effectively into practice. To this end, the SDM model's presumptive guidelines for children in foster care are based on children's risk of future maltreatment, the safety of the home environment, and demonstrated parental interest and involvement in the lives of their children. The SDM foster care guidelines are a "best practice" tool that will facilitate implementation of new Federal legislation while leading to more consistent and appropriate decision making. Although every agency needs to modify this component of the SDM model to include its own assessment instruments, policies, and terminology, the overall logic of the component is universally applicable.

SDM guidelines governing children in out-of-home care are based on the following assumptions:

When a family reduces risk to an acceptable level and maintains appropriate visitation with a child, the child should be returned home if the home is judged to be safe.

When risk remains high, the home remains unsafe, or parents fail to meet their visitation responsibilities for a specified period (as set by Federal guidelines and/or agency policy), the goal for the case changes from returning the child home to developing another plan for permanency.

In SDM's foster care model, the initial risk level is established by using the research-based risk assessment instrument. The risk reassessment will reflect a reduced level of risk if the family has made significant progress toward treatment goals. However, the reassessment scoring system generally precludes a family from receiving a lower risk score if there has been any new substantiation of maltreatment of any child in the household since the previous assessment.

Case Type	Number of Required Face-to-Face Contacts per Month	Contact Level	Collateral Contacts
Low Risk	1	Face-to-face with child and/or parent/caretaker	1
Moderate Risk	2	Face-to-face with child and/or parent/caretaker	2
High Risk	3	Face-to-face with child and/or parent/caretaker	3
Intensive Risk	4	Face-to-face with child and/or parent/caretaker	4
Purchase of Service (POS) Through Private Agency	Varies, but minimum once per month	Worker contacts can be replaced on one-for-one basis by POS agency worker	2 Phone calls per month with agency worker
Families First/FTBS*	1	1 contact per month with Families First or FTBS worker	None required
Pending Adjudication	None required	Contacts as needed	None required

* Families Together Building Solutions.

Source: Michigan Family Independence Agency, 1996

Figure 7: Michigan Contact Standards for CPS Workers by Risk Level.

The reunification model consists of four assessment components:

- A structured risk reassessment.
- A structured evaluation of parental compliance with visitation schedules.
- A reunification safety assessment.
- Structured guidelines for changing the permanency planning goal.

As shown in figure 8 (presented as an example), results of the structured assessments (risk, visitation compliance, and safety) are considered jointly to guide decisions regarding a child's return to the home or changes in the permanency plan. In practice, CRC staff work with each agency to develop a protocol incorporating criteria that reflect key local policies and regulations.

Summary

The heart of SDM is a series of assessment tools and associated decision-making protocols that are designed to bring greater structure, objectivity, and consistency to child welfare practices. The model is also designed to assist agencies in systematically identifying the most problematic cases—and focusing resources on those families—in an effort to reduce the incidence of subsequent maltreatment. SDM assessment tools incorporate four qualities that are essential to improved decision making in child welfare: reliability, validity, equity, and utility. Reliability reduces the extent to which decisions vary simply because different workers bring different perspectives to CPS decision making. Validity helps ensure the accuracy of the decision-making process. Equity ensures that families are treated fairly, regardless of race or ethnicity. Finally, the tools must be useful both for workers making day-to-day decisions and, through the aggregate data generated by the system, for administrators making policy, program, and budget decisions. The utility of SDM for child welfare management practices is discussed in the following section.

Management Components of the SDM Model

In addition to helping agencies improve the consistency of their decision-making process and make more efficient use of their resources, the SDM model includes two components designed specifically to facilitate the management and administration of child welfare agencies. These components—workload measurement and management information reports—build on and help maximize the usefulness of the model's decision-making components.

Workload Measurement

The model's workload measurement component is based on the assumption that simple caseload counts do not adequately capture the amount of time—and therefore the number of staff members—needed to fulfill a child welfare agency's mandates. Moreover, given the SDM model's delineation of distinct case types and differential service standards, case-load counts are an ineffective measure for determining how workload should be distributed across work units or among individual staff members. Workload measurement translates a caseload into time requirements and, ultimately, into staffing needs. To establish a workload measurement system, agencies need to conduct a simple case-based time study and determine the amount of time that staff actually need to meet service standards for various types of cases. This information is used to calculate the agency's total workload "demand," which can then be compared with the agency's current "supply" of available staff. Knowing both the monthly time requirement for each type of case and the total workload demand allows an agency to:

Provide a rational, empirical basis for budget and staffing requests to external funding sources.

Develop an internal system for equalizing workload across work units or among staff members.

Estimate the impact of new service responsibilities or budget restrictions on the agency's delivery of services.

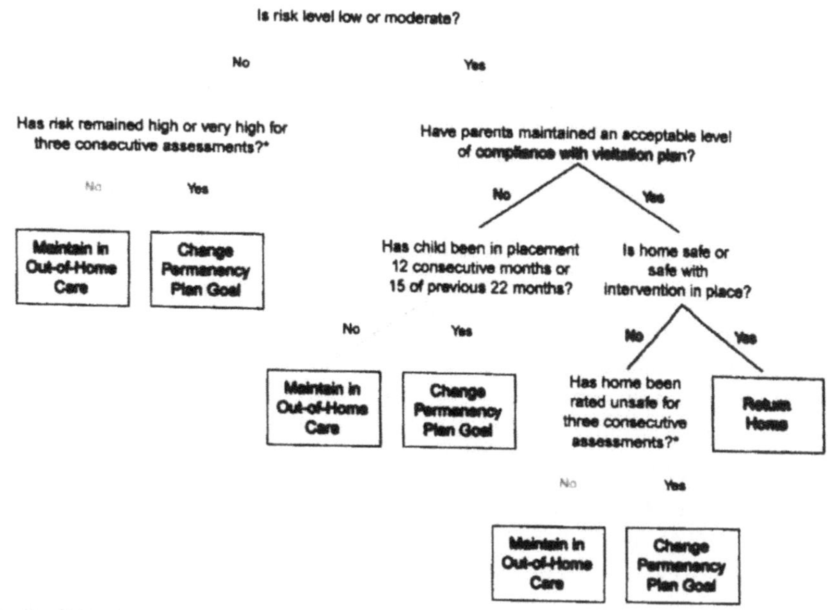

Figure 8: Placement/Permanency Plan Guidelines (Example).

Because a workload-based budget al.lows an agency to specify its case-related service standards and identify the number of staff members required to serve cases according to those standards, such a budget becomes, in essence, a service contract with funding sources. If, for example, a funding body agrees that high-risk cases should be seen by staff at least four times per month, it becomes the funding body's responsibility to provide a sufficient number of staff to allow the agency to meet that level of service. With a workload-based budget, funding bodies will know exactly what level of service will be provided based on the level of staff resources allocated. The effect of budget reductions on client service will be readily apparent, as will the effect of increases in resources. Workload measurement translates caseload into time requirements and, ultimately, staffing needs.

Quality Assurance and Evaluation Using SDM Management Information

Since the implementation of the Federal standards that accompany the Adoption and Safe Families Act of 1997, the emphasis on accountability in child welfare has reached new heights. Agencies now know that their decisions will be monitored for compliance with Federal mandates and State policy. An important feature of the SDM model, therefore, is that it provides information that allows agency management to routinely monitor compliance with standards, assess the impact of policy, identify service needs, and identify programs and intervention strategies that provide the best results for various types of cases. A basic premise underlying SDM is that the information needed to make good decisions at the individual case level (e.g., structured assessments of risk and service needs) is the same information needed in aggregate form by agency supervisors, analysts, and administrators. Figures 9 and 10 illustrate how agency managers can use aggregate information drawn from SDM records. Aggregate risk information can document changes in the nature of the client population. The data in figure 9, for example, reveal significant increases over a 5-year period in the proportion of substantiated cases identified as "high" and "very high" risk, clearly documenting changes in workload and indicating that new challenges face the agency. Figure 10 shows how managers can use needs and service referral data to monitor the extent to which clients are receiving services for identified problems and the effectiveness of those services in terms of reducing subsequent substantiated incidents of maltreatment.

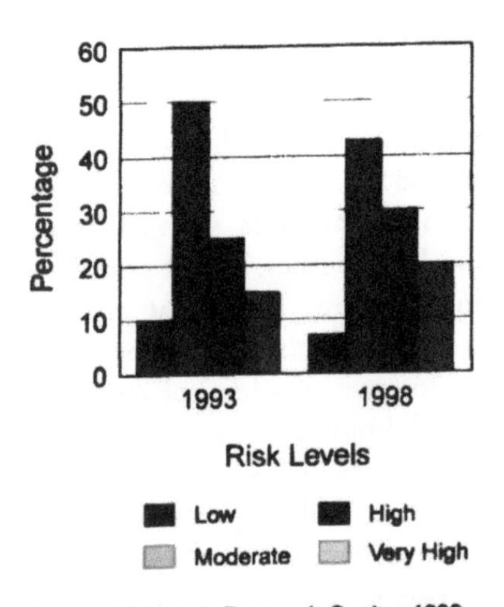

Figure 9: Changes in Initial Risk Levels of Substantiated Cases, 1993–98 (Example).

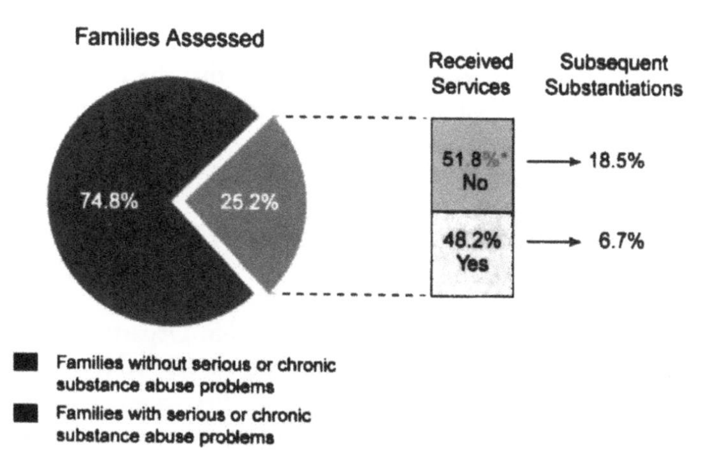

Figure 10: CPS Families With Serious or Chronic Substance Abuse Problems: Service.

Evaluations of the SDM Model

Few case management systems used in child welfare have been subjected to as much empirical scrutiny as the SDM model. This section summarizes some of the most salient evaluation research conducted in relation to SDM during the 12 years in which the model has been used in child welfare. The SDM system has been examined on two primary levels:

The extent to which SDM instruments add reliability, validity, and equity to decision making (i.e., whether they enhance consistency, accurately measure what they purport to measure, and treat races fairly).

The extent to which SDM is effective in reducing the subsequent maltreatment of children, as measured by outcome evaluations.

The next section presents the results of studies that have examined the reliability and validity of SDM risk assessment tools. This section is followed by a summary of impact evaluations conducted to date in Michigan.

Evaluation of Research-Based Risk Assessment

CPS agencies have traditionally relied on clinical judgment to establish the risk levels of families served by the system. However, recent research (Rossi, Schuerman, and Budde, 1996) has demonstrated that clinical decisions regarding the safety of children vary significantly from worker to worker, even among those considered to be child welfare experts. Moreover, although consensus-based risk assessment tools were developed to enhance consistency among caseworkers and improve decision making, recent studies indicate that the reliability and validity of these instruments are well below accepted standards (Baird et al., 1999; Baird and Wagner, 2000; Falco, in press).

Comparative Reliability and Validity of Different Risk Assessment Models

A study that CRC recently completed for the U.S. Department of Health and Human Services' Office of Child Abuse and Neglect (OCAN) compared the reliability and validity of three risk assessment tools. Two were consensus-based tools (the Washington State model and the Fresno, CA, risk assessment, which is a derivative of the Illinois CANTS system). The third was a research-based instrument (the Michigan version of SDM risk assessment).

To measure the reliability of the models, 80 randomly selected cases were assessed by 4 case readers using the Washington model, 4 using the Fresno model, and 4 others using the Michigan tool. Two measures of reliability were examined: the percentage of cases in which raters reached the same conclusion about risk level and Cohen's Kappa, a statistical measure of reliability. The results for the research-based Michigan risk tool showed that at least three of the four raters agreed on the risk level for 85 percent of the cases (Baird et al., 1999) (see figure 11). However, this level of agreement was obtained only for 45 percent of the cases with the California scale and 51 percent of the cases with the Washington risk tool. According to Cohen's Kappa, which was computed for each set of raters, the SDM system (Michigan) was again deemed far more reliable than the two consensus-based systems (see figure 12).

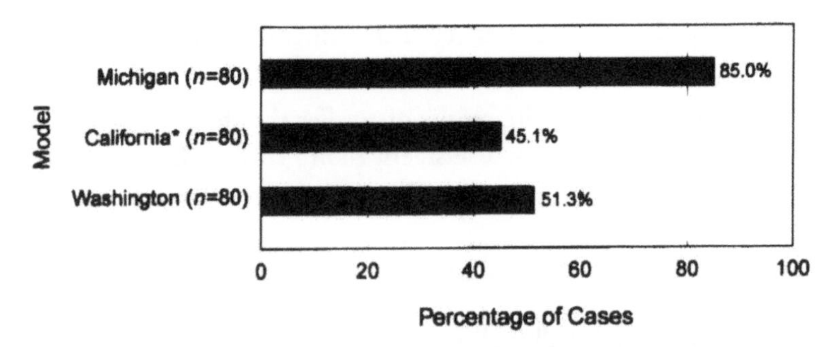

* The California risk assessment involved in this study was a consensus-based tool that predated California's implementation of SDM and research-based risk assessment.

Source: Baird et al., 1999.

Figure 11: OCAN Risk Assessment Reliability Study: Percentage of Cases With at Least 75-Percent Rater Agreement for Overall Risk.

Because risk assessment tools attempt to classify families according to the likelihood of future maltreatment, the validity of these tools can be demonstrated by showing a significant increase in subsequent maltreatment for every increase in risk level. According to this criterion, the OCAN study found superior validity in the research-based tool. In a sample of more than 1,400 cases from California, Florida, Michigan, and Missouri, the SDM research-based risk assessment tool categorized families into groups with significantly different risk levels. The families classified as higher risk had many more subsequent substantiations of maltreatment than the families classified as lower risk. The consensus-based tools, on the other hand, sorted families into risk levels that had little correlation with actual outcomes (Baird et al., 1999). These findings are shown in Table 1. For each risk model, the table shows the percentage of cases at each risk level that had subsequent investigations and substantiations during an 18-month followup. The data show that for both outcome measures there was little difference between families classified as moderate risk and families classified as high risk by the California model. Further, there was little difference in subsequent substantiations for the families classified at each risk level on the Washington model. In contrast, there were significant differences in outcomes by risk classification when the Michigan model was used to assess the families.

Table 1: OCAN Risk Assessment Validation: Percentage of Cases With Subsequent Investigations and Substantiations, by Risk Level (18-Month Followup).

	California Model* (n=876)		Michigan Model (n=929)		Washington Model (n=[illegible])	
Risk Level	**Subsequent Investigations**	**Subsequent Substantiations**	**Subsequent Investigations**	**Subsequent Substantiations**	**Subsequent Investigations**	**Subsequent Substantiations**
Low	28%	15%	16%	7%	25%	16%
Moderate	38	18	32	15	35	16
High	38	18	46	28	39	21

* The California risk assessment involved in this study was a consensus-based tool that predated California's implementation of SDM and research-based risk assessment.

Source: Baird and Wagner, 2000.

Research-Based Risk Assessment and Equity

Disproportionate numbers of minority children, particularly African Americans, are placed in foster care, and minority children spend more time in placement than their Caucasian counterparts (Hill, 2001). This disproportionality was the case long before SDM was introduced and remains a prevalent pattern, raising the issue of equity in CPS decision making. Because empirically based risk assessment tools use information related to poverty and other social conditions, some practitioners have questioned whether the instruments contribute to racial bias. Under the SDM system, however, foster care placement is guided by safety assessment, not risk. Risk level guides case openings and intensity of services decisions.

Disproportionate representations of minority youth in foster care, therefore, should not be attributed to the use of research-based risk assessment tools.

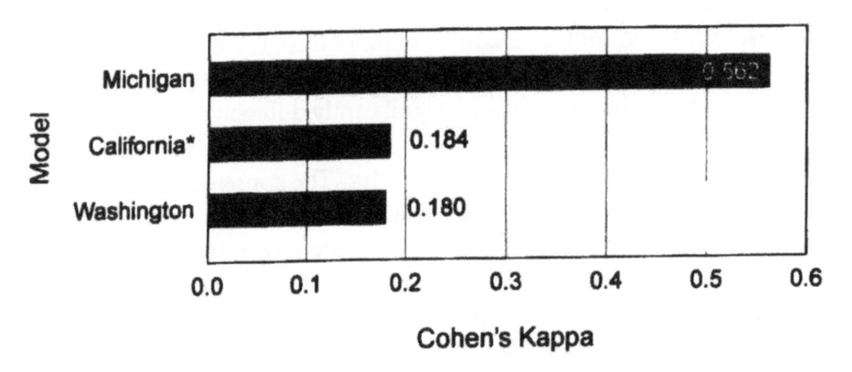

* The California risk assessment involved in this study was a consensus-based tool that predated California's implementation of SDM and research-based risk assessment.

Source: Baird et al., 1999.

Figure 12: OCAN Risk Assessment Reliability Study: Cohen's Kappa Among Raters for Overall Risk.

Nevertheless, it is important to measure how SDM risk assessments perform across racial and ethnic groups. Because equity is a key principle of SDM development efforts, every SDM risk tool validated to date has been subjected to an examination of its validity within racial and ethnic populations. These tests have shown that the use of SDM instruments results in virtually equal assignment of all races and ethnicities to each risk level. Table 2 presents data from Michigan as an example of the level of equity SDM has attained. These data on more than 6,500 white and 5,000 African American families show that there is no disproportionate representation at any risk level (Baird, Ereth, and Wagner, 1999). The California Department of Social Services also conducted an independent, detailed analysis of the individual items incorporated in that State's new research-based tool and an overall assessment of the tool's equity. The analysis found no bias in any item or in the instrument as a whole (Johnson, 1999).

Equally important findings relevant to the issue of equity have come from Baird and his colleagues. They have found that the rate of subsequent maltreatment observed within racial and ethnic groups increases with each incremental rise in risk level and that maltreatment rates within each risk category are similar among all groups (Baird, Ereth, and Wagner 1999).

Evaluation of the Michigan SDM System

Between 1989 and 1992, CRC and Michigan child welfare staff worked together to design an SDM system for CPS cases (Baird et al., 1995). When initially implemented, the system consisted of risk and needs assessment instruments, case planning and reassessment tools, and differentiated service standards. System implementation began in 13 pilot counties in 1992.

Michigan's phased implementation schedule for the system presented an opportunity to formally evaluate the impact of SDM by comparing outcomes in the 13 pilot counties with those in a matched sample of 11 counties still operating under the traditional system. The evaluation sample included all cases with substantiated reports of abuse or neglect between September 1992 and October 1993. The SDM and comparison study samples each consisted of approximately 900 families. Outcome measures included new referrals, investigations, and substantiations during a 12-month followup period.

The evaluation revealed several important differences in decision making and case processing in the SDM and comparison counties. These findings are summarized in the sections that follow.

Case Closing Decisions

The SDM counties were significantly more likely than non-SDM counties to close low- and moderate-risk cases following substantiation, and the non-SDM counties were more likely than SDM counties to close high- and intensive-risk cases. Moreover, cases that were closed without services in the SDM counties had significantly lower re-referral rates than those closed without services in the comparison group. This finding indicates that the use of risk assessment led to improved decisions in the SDM counties regarding which cases could be safely closed at the completion of the investigation.

Program Participation

Service program participation by families was significantly higher in the SDM counties than in the comparison counties—particularly among high- and intensive-risk families. For example, high-risk families in SDM counties were more likely than those in non-SDM counties to receive parenting skills training, substance abuse treatment, family counseling, and mental health services (see figure 13). This outcome is likely a result of the clear identification (via the risk assessment) of these families as being more likely to reabuse or reneglect their children and the more consistent identification of existing problems (via the SDM strengths and needs assessment).

Table 2: Distribution of Families by Race and Risk Level: Michigan Risk Assessment.

Risk Level	White (*n*=6,651)	African American (*n*=5,296)
Low	10.5%	11.3%
Moderate	30.7	30.0
High	45.1	46.0
Very high	13.7	12.7

Source: Baird, Ereth, and Wagner, 1999.

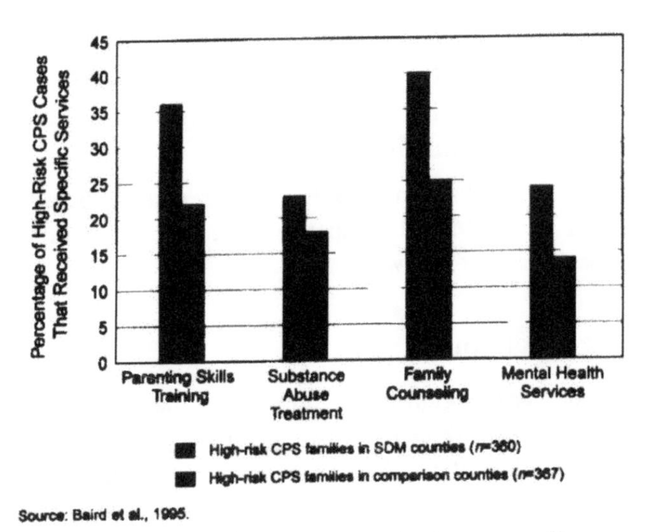

Source: Baird et al., 1995.

Figure 13: Michigan SDM Evaluation Results: High-Risk CPS Families' Receipt of Services.

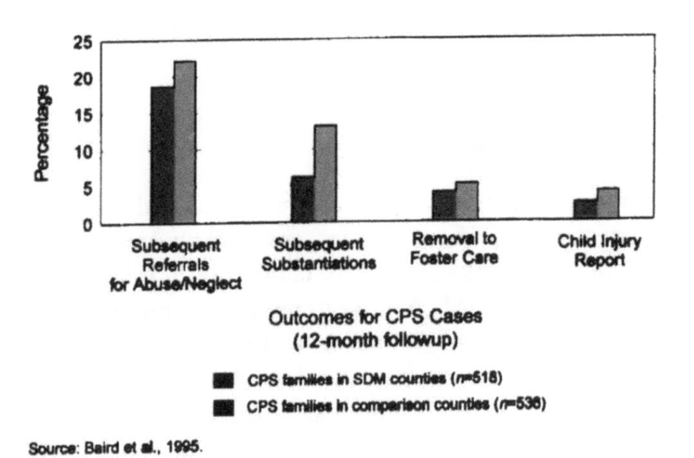

Source: Baird et al., 1995.

Figure 14. Michigan SDM Evaluation Results: CPS Case Results.

Outcomes

The evaluation also examined whether implementation of the SDM system resulted in a better overall system of child protection-and, in particular, lower rates of subsequent maltreatment. Families in the SDM and comparison counties were followed for 12 months to determine whether use of the SDM system resulted in lower rates of re-referral or resubstantiation. Figure 14 compares results for CPS cases in SDM counties with those in non-SDM counties. For every outcome measure, families in the SDM counties had better results than families in the comparison counties. The greatest dufferebce was ub rates of subsequent maltreatment substantiations: that rate was 50 percent lower in SDM counties than in non-SDM counties (6.2 percent versus 13.2 percent).

An analysis of outcomes by risk group also showed positive results for the Michigan SDM system. For example, high risk CPS cases handled in the SDM counties had fewer subsequent referrals for maltreatment and fewer subsequent child injuries than high-risk cases in the non-SDM counties. They also had lower rates of subsequent placement in foster care and were only half as likely to have a subsequent maltreatment substantiation.

Summary

The results of this carefully controlled evaluation show not only that SDM resulted in important changes in decision making and service provision for child welfare cases, but as anticipated, that it ultimately had a positive impact on the protection of Michigan's children. Although the rigor of the Michigan study has not been duplicated in other agencies, data from Wisconsin counties seem to support the Michigan findings. In Wisconsin, high and very high-risk cases that were opened for services had much lower rates of subsequent reports of maltreatment than cases at similar levels of risk that did not receive child protection services. As table 3 illustrates, high levels of intervention in these cases lowered the rate of subsequent reports dramatically. At the same time, services had negligible effects on low- and moderate-risk families (Wagner and Bell, 1998). Thus, data from both Michigan and Wisconsin indicate that accurate identification of families with the greatest potential for subsequent maltreatment, together with appropriate allocation of resources, can play a significant role in protecting children from harm.

Conclusion

Although a large proportion of children involved in the child welfare system subsequently become involved in the juvenile justice system, statistics alone do not adequately tell the story of these children. To fully understand the toll exacted by child abuse and neglect, policymakers must read the files of incarcerated youth. Many of these youth come from abusive homes and often have moved in and out of foster care for years before ending up in the juvenile correctional system. Few have ever known a truly stable home environment.

Any program that effectively reduces abuse and neglect can serve as a prevention strategy for juvenile delinquency. Given the firmly established relationship between abuse/neglect and subsequent delinquency and criminality, it seems imperative that policymakers embrace emerging technologies that significantly improve decision making and help communities devote resources to children and families most at risk. It is clearly time to resolve age-old conflicts between clinical judgment and structured decision making. In particular, the use of empirically based risk assessment is not a question of replacing professional judgment with statistical inference; it is simply a matter of using the best information available to protect children from harm.

Table 3: Percentage of 1995 Investigations Re-Referred During 2-Year Followup, by Case Opening Status and Risk Level: Wisconsin

	Subsequent Referrals	
Risk Level at 1995 Investigation	**Cases Closed After 1995 Investigation (n=1,014)**	**Cases Opened for Services After 1995 Investigation (n=216)**
Low/moderate	14.1%	14.6%
High	27.7	15.2
Very high	44.8	23.6

Source: Wagner and Bell, 1998.

As demonstrated by the OCAN and Michigan evaluation research, structured decision making represents a practical and efficient way to improve the Nation's child welfare systems. By reducing the extent of maltreatment experienced by children, the SDM model can make a significant contribution to breaking the link between abuse and delinquency.

Acknowledgements

Richard Wiebush and Raelene Freitag are Senior Researchers with the Children's Research Center, National Council on Crime and Delinquency. Christopher Baird is Senior Vice-President of the National Council on Crime and Delinquency.

References

1. Alfaro, J.D. 1981. Report on the relationship between child abuse and neglect and later socially deviant behavior. In Exploring the Relationship Between Child Abuse and Delinquency, edited by R.J. Hunter and Y.E. Walker. Montclair, NJ: Allenheld, Osmun.
2. Baird, S.C., Ereth, J., and Wagner, D. 1999. Research-Based Risk Assessment: Adding Equity to CPS Decision Making. Madison, WI: National Council on Crime and Delinquency, Children's Research Center.
3. Baird, S.C., and Wagner, D. 2000. The relative validity of actuarial and consensus-based risk assessment systems. Children and Youth Services Review 22(11/12): 839–871.
4. Baird, S.C., Wagner, D., Caskey, R., and Neuenfeldt, D. 1995. Michigan Department of Social Services Structured Decision Making System: An Evaluation of Its Impact on Child Protection Services. Madison, WI: National Council on Crime and Delinquency, Children's Research Center.
5. Baird, S.C., Wagner, D., Healy, T., and Johnson, K. 1999. Risk assessment in child protective services: Consensus and actuarial model reliability. Child Welfare 78(6):723–748.
6. Baird, S.C., Wagner, D., and Neuenfeldt, D. 1992. Protecting children: The Michigan model. Focus (March):1–7.
7. Bolton, F.G., Reich, J.W., and Gutierres, S.E. 1977. Delinquency patterns in maltreated children and siblings. Victimology 2:349–357.
8. Catalano, R.F., and Hawkins, J.D. 1995. Risk-Focused Prevention Using the Social Development Strategy. Seattle, WA: Developmental Research and Programs, Inc.
9. Children's Research Center. 1998a. California Preliminary Risk Assessment. Madison, WI: National Council on Crime and Delinquency, Children's Research Center.
10. Children's Research Center. 1998b. California Structured Decision Making System, Field Test Report. Madison, WI: National Council on Crime and Delinquency, Children's Research Center.
11. Children's Research Center. 1999. A New Approach to Child Protective Services: Structured Decision Making. Madison, WI: National Council on Crime and Delinquency, Children's Research Center.
12. Children's Research Center. 2000. Combined California Counties: Structured Decision Making Case Management Reports, January–June, 2000. Madison, WI: National Council on Crime and Delinquency, Children's Research Center.

13. Children's Research Center. 2001. Structured Decision Making in CPS: Generic Policy and Procedures Manual. Madison, WI: National Council on Crime and Delinquency, Children's Research Center.
14. Dawes, R.M., Faust, D., and Meehl, P. 1989. Clinical versus actuarial judgement.Science 243:1668–1674.
15. Dembo, R., Williams, L., Wothke, W., Schmeidler, J., and Brown, C. 1992. The role of family factors, physical abuse and sexual victimization experiences in high risk youths' alcohol and other drug use and delinquency: A longitudinal model. Violence and Victims 7(3):245–246.
16. English, D.J. 1997. Current knowledge about CPS decision making. In Decision Making in Children's Protective Services: Advancing the State of the Art, edited by T.D. Morton and W. Holder. Atlanta, GA: National Resource Center on Child Maltreatment.
17. Falco, G. In press. Clinical vs. Actuarial Risk Assessment: Results From New York State. Albany, NY: Office of Program Evaluation for the New York State Department of Social Services.
18. Gorey, K.M., and Leslie, D.R. 1997. The prevalence of child sexual abuse: Integrative review adjustment for potential response and measurement biases. Child Abuse and Neglect 21:391–398.
19. Harlow, C.W. 1999. Prior abuse reported by inmates and probationers. In Bureau of Justice Statistics Selected Findings (April).
20. Washington, DC: U.S. Department of Justice, Office of Justice Programs, Bureau of Justice Statistics.
21. Hill, R.B. 2001. The Role of Race in Foster Care Placements: An Outline Summary. Rockville, MD: Westat, Inc.
22. Howell, J.C., ed. 1995. Guide for Implementing the Comprehensive Strategy for Serious, Violent, and Chronic Juvenile Offenders. Washington, DC: U.S. Department of Justice, Office of Justice Programs, Office of Juvenile Justice and Delinquency Prevention.
23. Illinois Department of Children and Family Services. 1997. Illinois Child Endangerment Risk Assessment Protocol: A Report to the General Assembly Concerning the Implementation and Validation of the Protocol. Springfield, IL: Illinois Department of Children and Family Services.
24. Johnson, W. 1999. Race and the California family risk assessment. Paper presented at the 13th Annual CPS Risk Assessment Roundtable, San Francisco, CA.

25. Johnson, W., and L'Esperance, J. 1984. Predicting the recurrence of child abuse.
26. Social Work Research and Abstracts 20(2):21–26.
27. Kelley, B.T., Thornberry, T.P., and Smith, C.A. 1997. In the Wake of Childhood Maltreatment. Bulletin. Washington, DC: U.S. Department of Justice, Office of Justice Programs, Office of Juvenile Justice and Delinquency Prevention.
28. Maxfield, M.G., and Widom, C.S. 1995. Childhood victimization and patterns of offending through the life cycle: Early onset and continuation. Paper presented at the American Society of Criminology, Boston, MA.
29. Michigan Family Independence Agency. 1996. CPS Policy and Procedures Manual. Lansing, MI: Family Independence Agency.
30. National Council on Crime and Delinquency. 1995a. Assessing the Risk of Juvenile Offenders in Nebraska. Madison, WI: National Council on Crime and Delinquency.
31. National Council on Crime and Delinquency. 1995b. Revalidation of the Michigan Juvenile Risk Assessment. Madison, WI: National Council on Crime and Delinquency.
32. National Council on Crime and Delinquency. 1995c. Rhode Island Juvenile Risk Assessment Findings. Madison, WI: National Council on Crime and Delinquency.
33. National Council on Crime and Delinquency. 1997. Wisconsin Juvenile Offender Classification Study: County Risk Assessment Revalidation Report. Madison, WI: National Council on Crime and Delinquency.
34. National Council on Crime and Delinquency. 1999. Development of an Empirically Based Risk Assessment Instrument for the Virginia Department of Juvenile Justice: Final Report. Madison, WI: National Council on Crime and Delinquency.
35. Pawaserat, J. 1991. Identifying Milwaukee Youth in Critical Need of Intervention: Lessons From the Past, Measures for the Future. Milwaukee, WI: University of Wisconsin Employment and Training Institute, pp. 1–12.
36. Rossi, P., Schuerman, J., and Budde, S. 1996. Understanding Child Maltreatment Decisions and Those Who Make Them. Chicago, IL: Chapin Hall Center for Children, University of Chicago.
37. Silverman, A.B., Reinherz, H., and Giaconia, R. 1996. Long-term sequelae of child and adolescent abuse: A longitudinal community study. Child Abuse and Neglect 20(8):709–723.

38. Snyder, H., and Sickmund, M. 1999. Juvenile Offenders and Victims: 1999 National Report. Report. Washington, DC: U.S. Department of Justice, Office of Justice Programs, Office of Juvenile Justice and Delinquency Prevention.

39. U.S. Department of Health and Human Services, Administration on Children, Youth and Families. 1999. Child Maltreatment 1997: Reports From the States to the National Child Abuse and Neglect Data System. Washington, DC: U.S. Government Printing Office.

40. Wagner, D., and Bell, P. 1998. The Use of Risk Assessment To Evaluate the Impact of Intensive Protective Service Intervention in a Practice Setting. Madison, WI: National Council on Crime and Delinquency, Children's Research Center.

41. Wagner, D., and Caskey, R. 1998. Safety Assessment Validation Report for the Michigan Family Independence Agency. Madison, WI: National Council on Crime and Delinquency.

42. Widom, C.S. 1992. The Cycle of Violence. Research in Brief. Washington, DC: U.S. Department of Justice, Office of Justice Programs, National Institute of Justice.

43. Widom, C.S. 1995. Victims of Childhood Sexual Abuse—Later Criminal Consequences. Research in Brief. Washington, DC: U.S. Department of Justice, Office of Justice Programs, National Institute of Justice.

44. Wiebush, R., McNulty, B., and Le, T. 2000. Implementation of the Intensive Community-Based Aftercare Program. Bulletin. Washington, DC: U.S. Department of Justice, Office of Justice Programs, Office of Juvenile Justice and Delinquency Prevention.

45. Wood, J.M. 1997. Risk predictors for re-abuse or re-neglect in a predominantly Hispanic population. Child Abuse and Neglect 21(4):379–389.

Parents' Assessment of Parent-Child Interaction Interventions – A Longitudinal Study in 101 Families

Kerstin Neander and Ingemar Engström

ABSTRACT

Background

The aim of the study was to describe families with small children who participated in parent-child interaction interventions at four centres in Sweden, and to examine long term and short term changes regarding the parents' experience of parental stress, parental attachment patterns, the parents' mental health and life satisfaction, the parents' social support and the children's problems.

Methods

In this longitudinal study a consecutive sample of 101 families (94 mothers and 54 fathers) with 118 children (median age 3 years) was assessed, using self-reports, at the outset of the treatment (T1), six months later (T2) and 18 months after the beginning of treatment (T3). Analysis of the observed differences was carried out using Wilcoxon's Signed-Rank test and Cohen's d.

Results

The results from commencement of treatment showed that the parents had considerable problems in all areas examined. At the outset of treatment (T1) the mothers showed a higher level of problem load than the fathers on almost all scales. In the families where the children's problems have also been measured (children from the age of four) it appeared that they had problems of a nature and degree otherwise found in psychiatric populations. We found a clear general trend towards a positive development from T1 to T2 and this development was also reinforced from T2 to T3. Aggression in the child was one of the most common causes for contact. There were few undesired or unplanned interruptions of the treatment, and the attrition from the study was low.

Conclusion

This study has shown that it is possible to reach mothers as well as fathers with parenting problems and to create an intervention program with very low dropout levels—which is of special importance for families with small children displaying aggressive behaviour. The parents taking part in this study showed clear improvement trends after six months and this development was reinforced a year later. This study suggests the necessity of clinical development and future research concerning the role of fathers in parent-child interaction interventions.

Background

Parent-Child Intervention

Finding ways to prevent mental health problems is perceived as an important task within child psychiatry, in concurrence with other authorities and organizations striving to promote the course of children's development. Since the 1960s the arena of early childhood interventions has been transformed from a modest collection of pilot projects to a multidimensional domain of theory, research, practice and policy [1]. Such interventions were previously directed towards the children themselves—specifically targeting the needs of disabled children and children growing up in poverty [2]. The scope and the target group for these interventions have since then broadened and may now include mental health problems at large. As research in the field of child development has grown, the proliferation of parent-child and family interventions have reflected our increased understanding of the critical and determinative nature of parent-child interaction [2]. Early childhood intervention has thus experienced a paradigm shift from a child-oriented to a family-oriented approach [3].

The main theoretical basis generally applied for this type of intervention is attachment theory [4,5] which emphasizes the importance of the quality of early relationships [2]. A core feature of this theory is the importance for a child to experience everyday interaction with a reasonably sensitive and sufficiently predictable parent able to provide a "secure base" [6] from which the child can comfortably engage with the world, balancing inquisitiveness with a need for security.

This theory is often complemented by the ecological perspective [7], which highlights both the interaction of the child as a biological organism within its immediate social environment in terms of processes, events and relationships and the interaction of social systems in the child's social environment [8]. Within the transactional model [9] the development of the child is seen as a product of continuous dynamic interactions between the child and his or her family and social context. In this web of transactional processes, of which the child and his/her parents form part, researchers have been able to empirically identify a number of aspects that have proved to be important for a positive development of the child; parental stress [10], parental patterns of attachment [11,12], parents' mental health and well being [13], parents' access to a social network [14], and the possibilities of obtaining social support [15].

Among the seminal contributions to the fields of infant development and parent-child treatment, the writings of Daniel Stern [16-18] have offered critical and highly influential new theoretical perspectives. Stern describes the clinical system shaped during parent-child interventions and emphasizes that the interaction includes the inner representations of the child and the parent as well as their observable behaviour. These aspects constantly influence each other and the intervention can therefore choose different ports of entry to achieve change—for example the parent's inner images of the child, the representations of himself/herself as a parent, or the observable interaction. Stern [19] stresses the fact that the therapeutic alliance in parent-child treatment must be far more positive and validating than in a traditional psychodynamic therapeutic context.

Studies on the efficacy of interventions

The first systematic survey of interventions specifically directed towards the parent-child interaction, based upon attachment theory, was undertaken by van Ijzendorn et al. [20]. This survey, including twelve mother-child interventions, supported the theory that such interventions increased the mothers' sensitivity, but the effect on the children's attachment was surprisingly weak. This result indicated the influence of parental attachment representation on children's attachment through mechanisms other than responsiveness; referred to as "the transmission gap" [21]. A narrative review by Egeland et al. [22] of 15 attachment-based interventions pointed out that there are many factors at different ecological levels that may interfere with successful intervention. The source of obstacles to a secure parent-child attachment may be found in the child, the caregiver, the care-giving environment, or a combination of all these. In order to meet the participants' needs, the authors recommend flexible broad-based interventions—particularly for high-risk samples, where the parents are often dealing with multiple challenges and barriers in their own lives. Such comprehensive interventions should be designed to make services available that can meet both the attachment-related and other needs of high risk families; e.g. enhancing parental well being and providing and promoting social support.

A different conclusion was reached by Bakermans-Kranenburg et al. [23] in a meta-analysis of interventions with the purpose of enhancing parental sensitivity and/or child attachment security. This review comprises 70 studies where the intervention started at an average child age of below 54 months. The intervention studies were not restricted to a specific population: both middle-class samples with healthy children, at risk populations, and clinical samples were included. The analysis revealed that the interventions had an impact both on the mothers' sensitivity and—to a lesser degree—on the children's attachment. Interventions with video feedback were found to be more effective than those without. The most effective interventions used a moderate number of sessions and focused on sensitivity in families with, as well as without, multiple problems. These findings were summarized in the title of the article: Less Is More. Only three of the studies included fathers and these studies are all fairly old [24-26] but the conclusion in the review was that interventions including fathers appeared to be significantly more effective than interventions focusing on mothers only.

It has thus been shown that early interventions directed towards parent-child interaction may have a positive effect upon parenting [23], but whether "less is more" or "more is better" is an issue that can only be resolved through further studies [27].

A critical analysis of interventions based on attachment theory, limited to research that has been peer-reviewed, paid special attention to methodological aspects of the primary studies [28]. The conclusions, based upon 15 prevention studies published between 1988 and 2005, revealed that attachment interventions produce on average weak to moderate effects across caregiver and child outcomes. In only one of the studies were fathers involved. The authors emphasize that data on treatment integrity or social validity—if the interventions are accepted by key agents e.g. parents, children and intervention agents—are essentially nonexistent in the literature. This is significant since an intervention must be accepted by important participants in order to have high effectiveness under real-world conditions—and not only high efficacy under tightly controlled research conditions. Naturalistic studies, i.e. studies carried out under real-world conditions have a special value in so far as they can provide answers concerning treatment acceptability by giving information about dropout from treatment, which may be seen as a proxy for acceptance of treatment. Egeland et al. [22] ask for more research on interventions based upon the ecological model taking into account such factors as social support and parents' emotional health and well-being. Bakermans-Kranenburg et al. [23] stress the need for long-term follow-up studies, since sleeper effects—effects that emerge a long time after the intervention—on for example attachment security might otherwise remain undetected.

Cultural Considerations

It is also of great importance to study parent-child interventions within various cultural contexts. Even though the development of such interventions has been considerable for the last thirty years in Sweden as well, only a small number of these have been assessed with regard to outcome [29,30]. There are cultural variations with regard to children's mental health. Heiervang et al. [31] have shown that the Norwegian prevalence of externalising disorders (behavioural and hyperactivity) was about half that found in Britain, whereas rates of emotional disorders were similar. Differences like this offer a rationale for the study of parent-child interventions in different cultural contexts. Research results from the Nordic countries—with their resources in the field of mother and child health care, parental leave, and a well-developed pre-school—may be of specific interest to complement and enhance knowledge about various conditions for these interventions. The most obvious deficit in this research field hitherto is, however, the almost complete lack of intervention studies that include fathers.

A Swedish Example of Parent-Infant Intervention Approaches

This study is based on an intervention programme that has been developed during the last two decades in Sweden. Attachment theory [4,5] along with an ecological, transactional perspective [7,9] and Stern's theories of development in infancy [16] and of preconditions for treatment [17,18] provide the theoretical foundation employed at these centres. Attachment theory, which is usually associated with infants and small children, is also relevant for families with children in their middle childhood (7–12), when attachment to the parent(s) is still salient and important [32] though with a somewhat altered goal: from proximity of the attachment figure in early childhood to his/her availability in middle childhood according to Bowlby [33]. This gradual development is taken into consideration in the therapeutic work. A salutogenetic [34] therapeutic approach implies a focus on factors that support a positive development and not only an interest in factors that cause problems.

The Work Assignment

The linchpin of the therapeutic work is the collaborative relationship between the parent(s) and the therapist. A basic principle is that the goals of intervention should be established through a dialogue between the parents and the therapist based on the parents' own descriptions of the problem with the changes they desire being crucial. Priority is given to the parents' interpretation of the problem. This means that even though both the person referring the family and the therapist may suggest themes to work with, it is always the parents who decide what problem areas are ultimately selected as the focus of the treatment, as long as this is in accordance with the therapist's competence and role. The interventions may concentrate on outer, observable behaviour and/or on the inner images the parent has of his or her child and him or herself. The dialogue leads to the agreement upon a work assignment, which also entails clarification of the roles of the practitioners and the parents. On the basis of these discussions the professionals endeavour to shape the treatment according to the pronounced needs of each family.

Elements in the Program

The intervention comprises a number of elements combined on the basis of the needs of the family in conformity with the ideas behind stepped care, which refers to the practice of beginning therapeutic measures with the least extensive intervention possible and moving on to more extensive interventions only if deemed necessary in order to achieve a desired therapeutic goal [35]. The first step—which is always involved but which never constitutes the entire intervention—is parental counselling. The next step—which comprises the main element of the intervention—is interaction treatment which can be carried out in different forms as described below; "in video," "in vivo" (live), and "in verbis" (verbally). A combination of these three forms is most often used. When required, collaboration with the family's social network forms yet another step.

Interaction Treatment "In Video" – Marte Meo

Marte Meo was developed in the Netherlands by Maria Aarts in the 1980s [36], and may be regarded as an application of modern developmental psychology [16]. The starting point in the Marte Meo intervention is the question raised by the parent. The therapist makes a short video recording (3–7 minutes) of the child interacting with his/her parent(s) and analyses it, using a number of basic principles for a natural supportive dialogue. The principles the therapist is looking for are whether and how (1) the child's focus of attention is recognized by the parent, (2) the child's states, initiatives and feelings are acknowledged by the parent, (3) the child is given the time and space to react, (4) the child's ongoing actions, experiences and feelings are interpreted, punctuated and named by the parent, (5) the child is assisted to experience structure and predictability, (6) the child is guided by well-adjusted information and gets approving confirmation when a desirable behaviour is emerging, (7) the child is assisted through inevitable unpleasantness, (8) the child is encouraged to take an interest in other persons and their actions and feelings/sentiments, and (9) the child is helped to start and close an activity or a dialogue [37]. The therapist then chooses sequences to review with the parent, to create a link between the parent's initial question and the therapist's idea of what kind of support the child needs. The basic purpose is to afford an opportunity for joint observation and reflection on the child and his/her needs. The sequences selected are preferably ones that contain "moments of solutions" where the child is provided with the support he/she needs and the parent thus becomes his/her own model. The second best choice is where the needs of the child are displayed. The parent becomes an active, reflective participant in the work of developing his/her interaction with the child, and the child is mentalized instead of problemized [37]. The parent is encouraged to practise in everyday situations, and the process continues with new recordings, analyses and joint reflections.

Interaction Treatment "In Vivo"

Modern developmental psychology and attachment theory emphasize the quality of the everyday interaction for the development of the child. In interaction treatment "in vivo" the therapist and the parent use ordinary everyday life situations as points of departure. The work is framed by the work assignment and the situations can be planned by the therapist and the parent(s) together or utilized as they arise. Interaction treatment "in vivo" always includes the child and can take place in the homes of the families or/and at the centres, in a group setting or with only one family and the therapist partaking.

Interaction treatment "in vivo" is guided by the same understanding of a child's need for dialogue as Marte Meo. Since the structure is less well-defined "in vivo," the therapist faces other challenges, e.g. not to make up for the support the child needs but is not given by his/her parent. The parent is encouraged to become more attentive to the focus of attention of the child, his/her initiation of dialogue, expressions of emotions, rhythm and the child's need of assertion, guidance and protection. The aim of this part of the treatment is to enhance the parent's own ability to mentalize [38], i.e. to imagine how the world is conceived from the child's perspective, which may be of crucial significance in parenthood. Moments of intersubjectivity—the sharing of lived experience—are considered indispensable both for the therapeutic relationship and for the child's development [18].

In accordance with attachment theory, special attention is given to those factors which, alongside sensitive attunement, are thought to be of the greatest importance in helping the child to experience that his/her parent is providing a secure base and a safe haven. This must be communicated to the child through the parent's behaviour and includes for instance that the parent is not perceived as frightened/frightening, that he/she is not explicitly hostile, that the parent shows a fundamental willingness to soothe and comfort in times of fear and distress [39] and that he/she is predictable in his/her reactions and actions.

Interaction treatment "in vivo" involves the joint reflection of therapist and parent and the child may also take part if that is felt to be appropriate with regard to age and other circumstances.

Interaction Treatment "In Verbis" (Verbally)

The port of entry in interaction treatment "in verbis" is the parent's representations, e.g. his/her inner pictures of the child or of himself/herself as a parent. There may also be focus on the parent's own attachment history. It might for example be of help for parents to reflect upon how their own avoidant attachment behaviour was quite an appropriate strategy when they were children, but that the situation is now different, with new possibilities both in relation to their partners and in their ways of meeting their own children's needs of a secure base. Parents may also have a strong wish not to repeat their own parents' way of bringing up children—for example by using threats or violence—but realize that they lack alternative models.

Obstacles in the parent's history are often referred to as "the ghosts in the nursery" [40], but together with the exploration of painful memories it can be valuable to identify "the angels in the nursery," i.e. the beneficial experiences [41].

Collaboration with the Families' Social Network

In accordance with the ecological perspective, collaboration with the families' private and professional network is also often taken into account. The aim may be to give the family access to resources from other micro-systems; to develop connections fraught with conflict between micro-systems (e.g. the family and the childcare); or to coordinate multiple micro-systems involved in network meetings.

Aims of the Current Study

This longitudinal multi-centre study includes fathers, mothers and children in parent-child interaction interventions at four treatment centres in Sweden. Since one of the fundamental principles behind these interventions is that the parents have the right to define the problems and to take an active part in planning the intervention, it is logical to focus on the parents' experience of change. The self-report measurements used in this study cover those areas, presented earlier in the text, that have been shown to be of importance for good parenting and child development.

The aim of this study was

- to describe families—where difficulties in the interaction between parents and children have led to participation in parent-child interaction interventions at four centres in Sweden—with respect to social characteristics and psychological aspects of scientifically proven importance. These aspects were: the parents' experience of parental stress, parental attachment patterns, the parents' mental health and life satisfaction, the parents' social support and the children's problems at the outset of the treatment (T1)
- to examine long term changes (18 months after beginning of treatment (T3)) and short term changes (6 months after beginning of treatment (T2)) regarding the same aspects as those assessed at the outset of the treatment.

Ethical Approval

This study has been approved by the Research Ethics Committee of Orebro # 319/02.

Methods

The Four Centres for Parent-Child Intervention

The families included in this study have participated in treatment at one of the following four centres for parent-child intervention in Sweden: Gryningen in Karlskoga (ages 0—6), Lindan in Lindesberg (ages 0—5), Lundvivegården in Skövde (ages 0—12) and Björkdungen's family centre in Örebro (ages 0—12). Gryningen is run by the Department of Child and Adolescent Psychiatry in collaboration with the Social Welfare authorities, Lindan by the Department of Child and Adolescent Psychiatry while Lundvivegården and Björkdungen fall under the auspices of the Social Welfare authorities. They are all outpatient departments. Treatment is voluntary, but some parents may nevertheless feel themselves coerced into complying with the wishes of social authorities for them to participate in the intervention.

The therapists at the centres all have degrees (e.g. social workers, preschool teachers) and have been trained in the Marte Meo method. Some of the therapists have acquired additional qualifications in, for instance, cognitive psychotherapy and family therapy.

In spite of organizational differences at the centres, the shared theoretical foundation, essential features in their therapeutic approach and the elements in the intervention programme (described above) justify the idea of including them all in a multi-centre study.

Subjects

This study is based on a consecutive sample of all parents who commenced treatment during three years at one of these four centres (Figure 1). The study excluded parents displaying substantially impaired cognitive capacity due to acute and serious mental reactions. Of the five families excluded for that reason, four were refugees seeking political asylum. In all, 154 parents (94 mothers and 60 fathers) in 101 families agreed to participate in the study. In the 54 two-parent families all of the mothers and 45 (83%) of the fathers participated in treatment.

Altogether the 101 families had 118 children taking part in the treatment (Table 1). Forty-four (37%) of these were girls and 74 (63%) were boys. The children's ages varied from unborn (the treatment started towards the end of pregnancy) up to 12-year-olds, with a median age of 3. The parents' ages varied between 18 and 49 with a median age of 31.

Of the 154 parents (94 ♀; 60 ♂) in the study 131 (77 ♀; 54 ♂) were born in Sweden. There were 10 foreign-born parents (7 ♀; 3 ♂) from European countries and 11 parents (10 ♀; 1 ♂) from countries outside Europe (data is lacking for two of the fathers). This means that Swedish-born parents were somewhat overrepresented in the study compared to society as a whole, but the parents born abroad dominated among the parents excluded for reasons of health. One-third of the parents taking part in the study were either unemployed or on sick leave, which constitutes a considerably higher proportion than in the population as a whole.

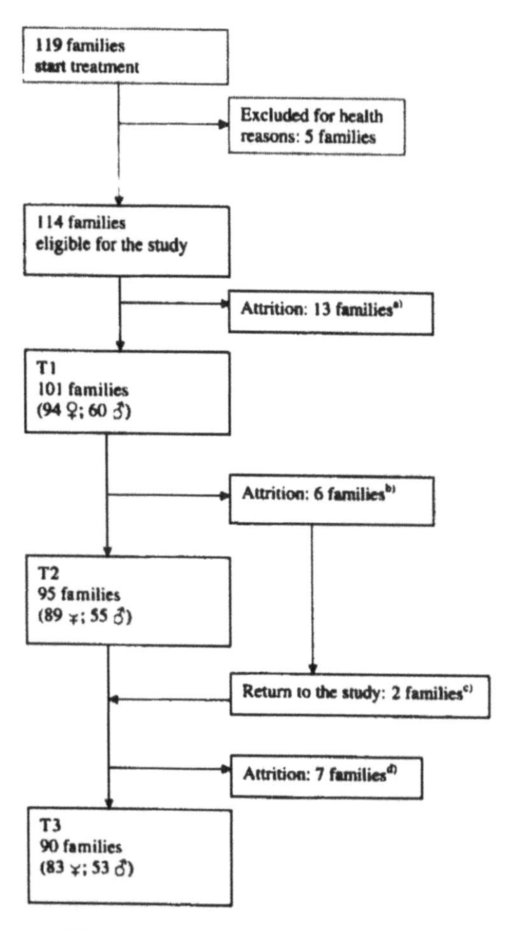

a) 10 parent related; too burdened (5), hesitant about their own ability to answer (3), wished to protect private life (1), had not time (1) & 3 staff related: oblivion (1) uncertainty about the families' intention (2)

b) answered too late (2), declined (2), expelled from the country (1), staff did not manage to establish contact (1)

c) 2 families not present at T2 (answered too late (1) staff did not manage to establish contact (1)) returned to the study at T3

d) declined (1), hidden because of threat of expulsion (1), ill-health (1), left the country (1), staff did not manage to establish contact (1), information of cause missing (2)

Figure 1. Study flowchart.

Table 1. Subjects & contact initiators

			n
Children's age (n = 118 children; 44 girls & 74 boys)	♀	♂	
Unborn	0	4	4
0 – 11 months	10	6	16
1 year	1	6	7
2 years	7	11	18
3 years	6	10	16
4 years	4	10	14
5 years	4	7	11
6 years	1	6	7
7 years	2	4	6
8 years	1	5	6
9 years	2	1	3
10 years	4	1	5
11 years	0	3	3
12 years	2	0	2
Child's residence (n = 101 families)			
Mother & Father			54
Single Mother			26
Mother & Stepfather			9
Alternating residence (at least 10 days a month with each parent)			6
Single Father			4
Father & Stepmother			1
Foster home			1
Parents' occupation (n = 154 parents, 94 ♀; 60 ♂)	♀	♂	
Employed	35	41	76
Unemployed/employment measures	15	11	26
Long-term sick-leave/temporary disability pension/pension	24	2	26
Student	13	3	16
Seeking political asylum	3	3	6
Working in the home	3	0	3
Information missing	1	0	1
Initiating contact (≥ 1 per case; 124 contact initiators in 101 families)			
Social services			48
Parents			37
Adult psychiatry			12
Child health service			11
Paediatric clinic			4
Preschool			3
Child psychiatry			3
Maternity welfare			2
Other			4

Contact Initiators and Contact Causes

The parents may themselves contact the centres or be referred to them by child health care, social services, preschools or some other body (Table 1). Contact cause (Table 2) is always related to the interaction between the parent and the child. When, for example, a parent's poor self-esteem is indicated as the cause of contact, it is therefore its impact on the parent-child relationship that is the reason for contact. Contact causes shown in table 2 refer to what was indicated when the parents applied to the centres or were referred to them. It is not, therefore, an assessment made by the staff at the centres. Dysfunction in parent-child interaction was the most common reason for seeking treatment. The predominant cause with reference to the children was externalizing behaviour and it is worth noting that aggression was by far the most frequent cause for contact. These are examples of how the parents expressed their goals for the treatment: "to put an end to Oscar's biting and fighting," "to feel confident as a mother of my baby," "to help Anna to concentrate on one thing" or "to be able to communicate with Alan without constant trouble."

Table 2. Contact cause (≥ 1 per case)

	n
Interaction between parent/parents – child (174 causes stated in 91 families)	
Need for support in the parent role	58
Interaction difficulties	53
Boundary setting problems	44
Attachment difficulties	14
Suspected abuse	4
Other	1
Child (142 causes in 78 children in 75 families)	
Externalizing problems	81
Aggressiveness (37), Hyperactivity & concentration problems (31), Cannot/Does not want to listen/obey (7), Troublemaking/Obstinacy/Acting out (6)	
Regulation problems	31
Sleeping (17), Feeding (7), Screaming (5), Toilet training (2)	
Contact difficulties	5
Interaction difficulties with siblings and/or other children	5
Internalizing problems	4
Delayed development	6
Handicap/illness	6
Trauma	3
Other	1
Parents (89 causes in 70 parents in 55 families)	
Mental problems/mental illness	35
Insecurity/low self-esteem/immaturity/very young	38
Worn-out & tired	6
Abuse	4
Feeling of loneliness	2
Somatic illness	2
Assaulted others	1
Other	1
Relationship between the parents/step-parents (47 causes in 35 families)	
Conflict or crisis with the partner/the other parent	18
Separation	17
Violence or threat of violence	3
Death	3
Other	6
Social network (32 causes in 27 families)	
Insufficient network	12
Conflict filled network	18
Other	2
Social situation (25 causes in 23 families)	
Burdened social situation	17
Strains in connection with refugee situation	5
Other	3

Treatment, Duration, Compliance, and Termination

The interaction treatment consisted of various combinations of the three modalities "in video," "in vivo" and "in verbis" (Table 3). Collaboration with the families' social network was reported for 60% of the families, most frequently with child-care and school followed by social services and relatives.

Table 3. Interventions in 101 families

	Families	Number of sessions		
	n	Md	Mean	Sd
Interaction treatment	101			
"In vivo"	88	21.5	32.3	28.5
At a centre – family & therapist in a group setting	56	33.5	39.1	28.9
At a centre – family & therapist exclusively	22	9.0	12.8	12.4
At home	57	4.0	6.9	7.9
"In video" Marte Meo	83	6.0	6.1	3.6
Reviews with one parent	67	4.0	4.7	3.2
Reviews with two parents	44	4.0	4.5	2.7
"In verbis"	95	9.0	13.0	12.3
Number of sessions with one parent	74	6.5	8.9	8.5
Number of sessions with two parents	60	6.0	9.3	8.7
Combinations of treatment modalities				
"In vivo" & "In video" & "In verbis"	72			
"In vivo" & "In verbis"	13			
"In video" & "In verbis"	5			
"In vivo" & "In video"	3			
"In verbis"	5			
"In video"	3			
"In vivo"	0			
Collaboration with the families' social network	61			
Child care (24) & School (8)	32			
Social services	31			
Relatives	24			
Psychiatry (adults)	8			
Child health care	7			
Child psychiatry	6			
Friends	3			
Maternal health care	1			
Network meeting	9			

If a family or a member of a family was receiving services at the outset of treatment from e.g. a psychiatric outpatient unit, these services generally continued during the intervention time since the centres have no wish to act as a substitute for other agencies.

After six months (T2) 74 of the 101 families were still in treatment, and when the final assessment (T3) took place—18 months after the outset—treatment was still under way for 19 families (Table 4). For the families that had completed treatment at T2 or T3, the time of treatment varied from 1 to 18 months. The median treatment period for all 101 families was 10 months. Slightly more than a third of the families attended treatment once a week, half of them more often (maximum three days a week) and the rest less frequently. Failure to attend treatment was low for almost three-quarters of the families (≤ 15% of planned treatment sessions).

Out of the 101 families taking part in the study, treatment was interrupted for a total of ten families: three families moved from the neighbourhood, two families seeking political asylum were expelled from the country, two families were subject to child welfare assessments by the social services and finally there were three families whose treatment was interrupted because of staff reasons: sick leave or retirement. The median length of treatment for these ten families was eight months. There were no other dropouts from the treatment.

Table 4. Treatment duration, compliance and termination

	Families	Number of months		
	n	Md	Mean	Sd
Treatment duration				
Length of treatment for all 101 families	101	10		
Treatment completed	*72*	*8*	*8.9*	*4.60*
Interrupted treatment	*10*	*8*	*9.5*	*6.12*
Still in treatment at T3	*19*			
Treatment completion at T2 (6 m) & T3 (18 m)				
Treatment completed at T2	24			
Treatment interrupted at T2	4			
Treatment completed at T3 *(another 48 families)*	72			
Treatment interrupted at T3 *(another 6 families)*	10			
Proportion of failure to attend treatment (101 families)				
≤ 15% of planned treatment sessions	73			
16 – 25%	9			
26 – 50%	14			
51 – 75%	1			
≥ 76%	2			
Missing data	2			
			After n months	
Treatment interruption	10			
Families moved from the neighbourhood	3		*6; 7; 19 months*	
Asylum seeking families expelled from the country	2		*1; 11 months*	
Investigation by social services	2		*4; 9 months*	
Staff reasons: sick leave or end of service	3		*5; 15; 18 months*	

Measures

The Parents' Experience of Parental Stress

The Swedish Parenthood Stress Questionnaire (SPSQ) [42] is based on the Parent Domain of the Parenting Stress Index [43]. This instrument comprises of five subscales: incompetence, role restriction, social isolation, spouse relationship problems, and health problems. The total experience of stress is measured by a general parenting stress scale consisting of all items. The instrument has been used in several studies and has displayed good psychometric properties [42]. Since about half of the families seeking help at the four centres are single parents a special "single version" was designed for them in which the questions regarding the sub-scale on spouse relationship problems had been removed.

The Parents' Patterns of Attachment

The Relationship Questionnaire (RQ) [44] is a self-report instrument designed to measure four categories of attachment (avoidant/dismissive; secure/autonomous; ambivalent/preoccupied and disorganized/fearful), using combinations of a person's self-image (positive or negative) and image of others (positive or negative).

On the RQ the respondent is asked to rate, on 7-point scales, how well he/she feels the description of the four patterns apply to their own experiences. The psychometric properties of the Swedish version have proved to be satisfactory [45].

The Parents' Mental Health

The instrument used to measure psychological health was the General Health Questionnaire 12 (GHQ12) [46], a questionnaire with 12 questions. The index can vary between the values 0 and 12, with a low value indicating good psychological health. The threshold value for poor psychological health is 3 [47]. The instrument has displayed good psychometric properties [46].

The Parents' Present and Expected Life Satisfaction

Cantril's Self-Anchoring Ladder of Life Satisfaction [48] is a measure of an individual's overall assessment of life satisfaction. Subjects are asked to evaluate their life at the present time, one year ago and one year from now on a ladder, with the bottom (0) representing the worst possible life and the top (10) the best possible life. The Cantril Ladder has been reported to have good validity and stability and reasonable reliability [49].

The Parents' Social Support

In order to obtain a measure of perceived availability and adequacy of support from intimates and the wider social network we used a brief version of The Interview Schedule for Social Interaction [50]. The Swedish version [51] consists of 30 items measuring both the availability and the adequacy of attachment and social interaction and is divided into four subscales. The maximum obtainable scores are: for Availability of Social Integration (AVSI) 6 points, Adequacy of Social Integration (ADSI) 8 points, Availability of Attachment (AVAT) 6 points, and Adequacy of Attachment (ADAT) 10 points, 1 for each item. The ISSI has displayed good psychometric properties [52].

The Children's Strengths and Difficulties

The Strengths and Difficulties Questionnaire (SDQ) [53] is a brief behavioural screening questionnaire concerning 3–16 year olds. It exists in several versions: the versions used in this study were questionnaires for completion by the parents of 4–16 year olds. In this study there were 50 families with children 4 years or older. All versions of the SDQ incorporate statements regarding 25 attributes, some positive and others negative. These 25 items are divided into 5 sub-scales:

emotional symptoms; conduct problems; hyperactivity/inattention; peer relationship problems and prosocial behaviour. The first four sub-scales produce a total difficulties score. The SDQ also includes an impact supplement. The instrument has been translated into Swedish and its psychometric properties are considered good [54,55].

Procedure

The first point of assessment called T1 took place at the outset of the treatment, the second assessment (T2) six months later and the third point (T3) 18 months after treatment began. In order to minimize attrition, members of the staff contacted the families and asked them to come to the centres to fill in the questionnaires if they were no longer undergoing treatment at T2 and T3. If this was not possible, the questionnaires were sent home to the family. There was no loss of data from the great majority of informants. The exact number of persons completing each questionnaire is indicated in tables 5, 6 and 7. The staff at the four centres supplied information for the Background data (at T1) and a Treatment Journal (at T2 & T3) with data concerning the intervention.

Statistical Analysis

The results of the assessments made by the parents at the outset of treatment (T1) were compared with available community and clinical samples. No individual data were accessible from these studies, which ruled out the possibility of using non-parametric tests. The accessible studies were mostly based on reports of means and standard deviations. Student's t-test was therefore carried out to analyse the statistical significance of differences. A chi-square test for non-parametric data was used to determine the significance of differences in proportions.

The long term changes (T1 → T3) and short term changes (T1 → T2) were analysed using Wilcoxon's Signed-Rank test. To complete the description of this study and to enable comparison with other intervention studies Cohen's d [56] was also used, with the definitions small (0.20–0.49), moderate (0.50—0.79) and large effect size (≥ 0.80).

Since a relatively large number of statistical tests were performed, the possibility of the random significance of some results cannot be ruled out. A threshold p value of 0.01 was therefore deemed statistically significant.

Results

The Families' Problems at the Outset of the Treatment (T1)

The design of the study did not include a control group that could serve as a comparison at the outset. In order to give an idea of the occurrence and the extent of problems—whether they should be labelled "everyday problems" or could be considered to be of clinical significance—in the families participating in the study, the results have been compared to data from available community and clinical samples, preferably Swedish ones (Tables 5 and 6).

Table 5. Intervention mothers at the outset (T1) and comparative data.

Scale	Intervention mothers			Community samples ♀				Clinical samples ♂			
	n	Mean	sd	n	Mean	sd	p	n	Mean	sd	p
SPSQ[a] (couples)	66	3.11	.58	1081	2.52	.56	<.001	75	2.81	.59	.003
incompetence	66	3.11	.78		2.27	.68	<.001		2.57	.84	<.001
role	66	3.84	.84		3.42	.82	<.001		3.88	.75	.766
isolation	66	2.65	.84		2.05	72	<.001		2.21	.82	.002
spouse	66	2.68	1.03		2.25	.94	<.001		2.29	1.07	.030
health	66	3.17	.83		2.61	.88	<.001		3.09	.88	.581
SPSQ (single)	24	3.36	.56								
incompetence	24	3.46	.60								
role	24	3.72	.98								
isolation	24	2.93	.92								
health	23	3.22	.69								
RQ[b] (B)	92	3.95	1.88	211	5.02	1.50	<.001				
RQ (D)	91	3.74	2.33	209	2.40	1.70	<.001				
LoL[c] past	92	4.67	2.47					103	4.4	2.3	.430
L-o-L present	92	5.22	2.07	2032	7.30	1.48	<.001	103	5.2	2.0	.945
L-o-L future	89	7.92	1.81					102	7.7	2.0	.429
								Clinical sample ♀ & ♂			
ISS[d] total	93	16.06	7.98					103	16.3	6.2	.813
AVAT	92	4.64	1.72					103	4.4	1.6	.314
ADAT	92	5.08	3.20					103	5.9	2.9	.062
AVSI	93	2.04	1.80					103	2.0	1.7	.873
ADSI	93	4.37	2.77					103	4.4	2.5	.936
SDQ[e] total	37	19.24	5.75	260	6.15	5.24	<.001	62	16.71	7.23	.073
SDQ impact	37	3.54	2.40		0.34	1.16	<.001		3.14	2.76	.466
emotional	37	4.00	2.15		1.60	1.84	<.001		4.50	2.60	.327
conduct	37	4.86	2.00		1.09	1.29	<.001		3.23	2.23	<.001
hyperactivity	37	6.65	3.09		2.38	2.18	<.001		6.00	2.83	.288
peer	37	3.73	2.12		1.15	1.90	<.001		3.03	2.40	.146
prosocial	37	6.62	2.13		8.62	1.50	<.001		7.00	2.20	.402
GHQ 12[f] *Prop. of poor psychol. health*	93	78.3%		8792	25.6%		<.001				

Student's t-test; statistical significance set at $p < .01$

[a] The Swedish Parenthood Stress Questionnaire: *Incompetence; Role restriction; Social isolation; Spouse relationship problems; Health problems*. Low values are desirable.

[b] The Relationship Questionnaire: B – High values are desirable; D – Low values are desirable.

[c] Cantril's Self-Anchoring Ladder of Life Satisfaction. High values are desirable.

[d] The Interview Schedule for Social Interaction:(AVAT) *Availability of attachment; (ADAT) Adequacy of attachment; (AVSI) Availability of social integration; (ADSI) Adequacy of social integration*. High values are desirable.

[e] The Strengths and Difficulties Questionnaire: *Emotional symptoms; Conduct problems; Hyperactivity; Peer problems; Prosocial behaviour*. Low values are desirable except for prosocial behaviour where high values are desirable.

[f] The General Health Questionnaire.

The mothers participating in the study showed a statistical significant higher degree of parental stress as measured by SPSQ compared to a community sample formed by 1500 randomly selected mothers with children aged from 6 months up to 3 years. Both fathers and mothers displayed significantly higher degrees of stress than a clinical sample consisting of 104 families seeking help for their children from a Specialist Child Health Centre [57]. The single parents showed even higher degrees of parental stress. The parents' attachment patterns differed from those of a community sample of 500 randomly selected families with children up to 6 years of age from the western region of Sweden [58]. The RQ results showed that the parents in this study had a significantly lower degree of secure attachment B than parents in the community sample and the mothers showed a higher degree of disorganized attachment D than mothers in the community sample.

Table 6. Intervention fathers at the outset (T1) and comparative data.

Scale	Intervention Fathers			Community samples ♂				Clinical samples ♂			
	n	Mean	Sd	n	Mean	sd	P	n	Mean	sd	p
SPSQ (couple)	51	2.72	.59					65	2.39	.50	.002
incompetence	51	2.53	.72						2.02	.57	<.001
role	51	3.29	.81						3.23	.86	.703
isolation	51	2.61	.66						2.18	.73	.001
spouse	50	2.48	.81						1.98	.79	.001
health	51	2.70	.85						2.57	.81	.403
SPSQ (single)	8	2.81	.78								
incompetence	8	2.81	.87								
role	8	3.06	1.18								
isolation	8	2.85	.75								
health	8	2.31	.97								
RQ (B)	60	4.13	1.71	192	4.88	1.48	.001				
RQ (D)	60	2.95	2.06	188	2.57	1.67	.149				
LoL past	59	5.20	2.23					47	5.2	2.1	1.000
L-o-L present	59	5.95	1.92					47	5.5	1.9	.231
L-o-L future	59	7.58	1.78					46	7.4	1.8	.610
								Clinical sample ♀ & ♂			
ISSI total	60	18.73	7.64					103	16.3	6.2	.028
AVAT	60	4.73	1.53	83	5.1	1.4	.136	103	4.4	1.6	199
ADAT	60	6.15	3.09	83	7.6	2.6	.003	103	5.9	2.9	.605
AVSI	60	2.58	1.86	83	3.0	1 7	.163	103	2.0	1.7	.044
ADSI	60	5.27	2.64	83	6.5	1.8	.199	103	4.4	2.5	.038
SDQ total	25	17.92	6.47								
SDQ impact	25	2.76	2.89								
emotional	25	3.56	2.36								
conduct	25	4.32	1.91								
hyperactivity	25	6.84	2.94								
peer	25	3.20	1.80								
prosocial	25	6.36	2.61								
GHQ 12 Prop. of poor psychol. health	60	43.3%		7126	18.6%		<.001				

Student's t-test; statistical significance set at $p < .01$

The parents' mental health as measured with GHQ12 differed significantly ($p < .001$) from that of a sample of 7126 men and 8792 women aged 16—44 in an annual, national public health survey conducted by the Swedish National Institute of Public Health [59]. With a cut-off value of 3, 78.3% of the mothers and 43.3% of the fathers reported poor psychological health versus 25.6% for women and 18.6% for men in the community sample. There are no available data from Swedish community samples concerning the parents' present and expected life satisfaction as measured with Cantril's ladder. The instrument has, however, recently been used in a Dutch study [60] on a sample of 2032 mothers with children aged 1–3 years, recruited from community records of several cities and towns in the western region of the Netherlands. The mothers in our study made a significantly lower assessment of their current life satisfaction. The levels in our study are consistent with data from a Swedish study [61] comprising parents of children aged 3—9 who had been clinically assessed by professionals as displaying behaviour management problems.

Data from a Swedish sample concerning social support as measured with ISSI were based on 83 middle-aged men [51], and indicated on all four subscales a more favourable result than those of the fathers in our study, but only the differences in adequacy of attachment is statistically significant. In a recent Swedish study [52], data were presented from three psychiatric samples. The parents (results from both fathers and mothers) in our study are comparable with a sample consisting of patients aged 18—55 years (both men and women) from an outpatient unit for people with long-term mental illness, mainly psychosis.

The children's problems, as measured with the SDQ, deviated considerably from a Swedish community sample, consisting of the parents of 450 children, 5–14 years old, randomly selected from the population register [62]. The clinical comparison sample consists of children from four child psychiatric outpatient clinics in Sweden, with a mean age of 10 years. The children in our study displayed more severe problems in every subscale except emotional symptoms. The difference in conduct problems was statistically significant. The average scores were above cut-off scores for psychiatric cases [55] on the total score, the impact score and all of the sub-scales except for the prosocial scale where they were even.

At the outset of treatment (T1) the mothers showed a higher degree of problem load than the fathers on almost all scales. The only exceptions consisted of the mothers' somewhat more positive rating of the future than the fathers, and the fathers' higher rating of hyperactivity problems in the children and their lower rating of prosocial behaviour.

To sum up the results from commencement of treatment, the parents in this study had considerable problems in all areas examined. In the families where the children's problems have also been measured (children from the age of four) it appeared that the children undergoing treatment had problems of a nature and degree otherwise found in psychiatric populations.

Long Term Changes (after 18 Months (T3)) and Short Term Changes (after 6 Months (T2))

We found a clear general trend towards a positive development from T1 to T2 and this development was also reinforced from T2 to T3 (Tables 7 and 8). This trend was stronger for mothers than for fathers. The gender differences will—for space reasons—be further analyzed and discussed in a forthcoming article.

Table 7. Parents' assessments at T1, T2 & T3

	T1			T2			T3		
Scale	*n*	Mean	sd	*n*	Mean	sd	*n*	Mean	sd
SPSQ (couples) total stress	117	2.94	.61	108	2.78	.55	103	2.67	.59
incompetence	*117*	2.86	*.81*	*108*	2.70	.75	*103*	2.53	.74
role restriction	*117*	3.60	*.87*	*108*	3.37	.84	*104*	3.28	.89
isolation	*117*	2.63	.77	*108*	2.48	.74	*103*	2.38	.79
spouse	*116*	2.60	.94	*109*	2.52	.88	94	2.53	.95
health	*117*	2.97	*.87*	*109*	2.81	*.81*	*103*	2.64	.86
SPSQ (single parents) total stress	32	3.22	.66	35	2.93	.68	32	2.75	.63
incompetence	32	3.29	.72	35	2.99	.77	32	2.79	.74
role restriction	*32*	3.56	*1.05*	35	3.37	*1.04*	32	3.22	.97
isolation	*32*	2.91	*.87*	35	2.56	.88	32	2.33	.92
health	*31*	2.98	.86	35	2.68	.76	32	2.53	.90
RQ (B)	152	4.02	1.81	143	4.20	1.78	135	4.50	1.67
RQ (D)	151	3.42	2.26	143	3.01	2.09	135	2.84	2.00
Cantril's L-o-L present	151	5.50	2.04	142	6.30	2.05	131	6.99	1.64
Cantril's L-o-L future	148	7.78	1.79	139	7.96	1.80	131	8.28	1.33
GHQ12	153	4.46	3.37	144	3.42	3.30	136	2.70	3.00
ISSI	153	17.11	7.93	143	17.99	7.25	136	19.36	7.03
AVAT	*152*	4.68	*1.64*	*143*	4.72	*1.56*	*135*	*5.16*	*1.28*
ADAT	*152*	*5.50*	*3.19*	*143*	5.89	3.00	*136*	6.29	2.96
AVSI	*153*	2.25	*1.84*	*143*	2.22	*1.79*	*136*	2.44	*1.83*
ADSI	*153*	4.72	2.75	*143*	*5.15*	2.69	*136*	5.49	2.68
SDQ total difficulties	62	18.71	6.03	59	15.92	6.74	56	14.21	7.37
SDQ impact	62	3.23	2.61	59	1.53	2.32	56	1.50	2.54
emotional symptoms	62	3.82	2.23	59	3.47	2.48	56	2.71	2.10
conduct problems	62	4.65	1.97	59	3.78	2.04	56	3.36	2.34
hyperactivity	62	6.73	3.01	59	5.93	2.91	56	5.43	2.96
peer problems	62	3.52	2.00	59	2.73	2.26	56	2.71	2.08
prosocial behaviour	62	6.52	2.32	59	6.69	2.19	56	7.27	2.33

Table 8. Parents' long term changes T1→T3 and short term changes T1→T2

	Long term T1→T3				Short term T1→T2			
Scale	d	Z	p		d	Z	p	
SPSQ (couples) total stress	.45	-4.539	<.001	***	.28	-3.643	<.001	***
incompetence	42	-4.678	<.001	***	.20	-2.812	.005	**
role restriction	.36	-3.964	<.001	***	27	-2.809	.005	**
social isolation	.33	-2.678	007	**	.20	-1.945	.052	
spouse relationship problems	07	-.586	.558		09	-.292	770	
health problems	.38	-3.795	<.001	***	.19	-1.890	.059	
SPSQ (single) total stress	.73	-3.375	.001	**	.43	-3.015	.003	**
incompetence	.69	-3.084	.002	**	41	-2.440	.015	
role restriction	.34	-2.469	.014		.18	-.931	.352	
social isolation	65	-2.611	009	**	40	-2.973	003	**
health problems	.51	-1.732	.083		.37	-1 948	.051	
RQ (B)	.28	-2.851	.004		10	-.972	.331	
RQ (D)	.27	-3.426	.001	**	.19	-2.202	.028	
Cantril's L-o-L present	.80	-6.335	<.001	***	.39	-4.093	<.001	***
Cantril's L-o-L future	.31	-3.090	.002	**	10	-1.606	.108	
GHQ12	.55	-5.466	<.001	***	.31	-4.051	<.001	***
ISSI	.30	-2.636	008	**	12	-1.271	204	
AVAT	.33	-2.361	.018		.03	-.522[a)]	.602	
ADAT	.26	-2.187	.029		.14	-1.166	.244	
AVSI	10	-1.018	.308		-.02[a)]	-.585[a)]	.559	
ADSI	.28	-3.400	.001	**	16	-2.313	021	
SDQ total difficulties	68	-4.254	<.001	***	.44	-4.167	<.001	***
SDQ impact	.67	-4.342	<.001	***	.69	-4.790	<.001	***
emotional symptoms	.51	-2.764	.006	**	.15	-1.217	.224	
conduct problems	.60	-4.466	<.001	***	43	-3.578	<.001	***
hyperactivity	.44	-3.199	.001	**	.27	-3.392	.001	**
peer problems	.40	-2.168	.030		.37	-2.920	.004	**
prosocial behaviour	.32	-2.718	.007	**	.08	-1.164	.244	

d Cohen's d; effect size small 0.20 – 0.49, moderate 0.50 – 0.79, large ≥ 0.80.
Z Wilcoxon Signed Ranks test: $^{**}p < 0.01$, $^{***}p < 0.001$. Statistical significance in this study set at $p < 0.01$.
[a)] changes in an unfavourable direction

Reduced Experience of Parental Stress

The experience of parental stress was reduced from T1 to T2, and the stress continued to diminish from T2 to T3. The change from T1 to T3 was statistically significant for spouses ($p < .001$) as well as for single parents ($p = .001$) and the effect size (Cohen's d) was moderate for spouses ($d = 0.45$) and moderate to large for single parents ($d = 0.73$).

Changes in Parental Attachment

The outcomes considered of special importance concerning the patterns of attachment were changes regarding pattern B (secure attachment), where an increase is desirable, and for pattern D (fearful or disorganized), where, instead, a decrease is desirable.

The parents showed a certain development towards the desirable pattern of attachment B from T1 to T2, and a stronger reinforcement from T2 to T3. The change from T1 to T3 was significant ($p = .004$), but the effect size according to Cohen's d was small ($d = 0.28$). The negative pattern of attachment D decreased from T1 to T2, a trend that also continued between T2 and T3, but the effect size was still small ($d = 0.27$).

Improved Mental Health

The parents' improved mental health expressed as an average value improved considerably from T1 to T2, as well as from T2 to T3. The change was highly significant statistically ($p < .001$) and the effect size was considered to be medium ($d = 0.55$). The proportion of persons with good mental health (cut off = 3) altered significantly ($p < .001$) from 35.3% at T1 to 52.1% at T2 and 61% at T3.

Improved Present and Expected Life Satisfaction

The parents' present life satisfaction was significantly improved from T1 to T3, ($p < .001$) and their expected life satisfaction also improved considerably ($p = .002$). The effect size was large concerning present life satisfaction ($d = 0.80$) and small ($d = 0.31$) with regard to the future.

More Satisfactory Social Support

A certain short-term improvement took place from T1 to T2 and a more marked change was visible from T2 to T3. Wilcoxon's test showed a significant change from T1 to T3 ($p = .008$). The effect size was small as measured with Cohen's d ($d = 0.30$). On the sub-scales the effect size was next to non-existent ($d = 0.10$) for access to a social network, but significant and clear, albeit small, to adequacy of attachment.

Problem Reduction with the Children and Reduced Impact of the Problems

The total symptom charge was significantly reduced from T1 to T3 ($p < .001$) and the effect size was of medium size ($d = 0.68$). The effect of the problems in the lives of the children and the families was also significantly reduced ($p < .001$), with a medium effect size ($d = 0.67$). The most important changes concerned conduct problems, which corresponds well with the problem description given by the parents at the outset.

When calculating the effect size concerning SDQ a measure called added value is sometimes used which takes into account a certain amount of "self-healing." In this study the measure of added value is 2.56, which would give an effect size of 0.51.

Summary

The results of the study showed that the subjective assessment of parents partaking in parent-child interventions was that less parental stress was experienced after six months, with the exception of factors concerning the way in which the spouse relationship had been influenced. The parents' ways of relating to other people (patterns of attachment) had developed in a positive direction: their mental health had improved, as had their present and expected life satisfaction. The possibility of obtaining social support had increased—not primarily through a larger network but through experiencing the existing network as being more adequate. Finally the children's problems—especially conduct problems—had decreased, as had their effect in their daily life. The positive development in all these areas had continued and been reinforced eighteen months after the outset of the treatment. As can be seen from Tables 7 and 8, the variables under examination exhibited different patterns of improvement: there are "quick starters," which are more evident during the six first months (e.g. SDQ Impact); "slow starters" that improve over time (e.g. aspects of perceived social support) and others where the development seems to have taken a more even course (e.g. life satisfaction).

Discussion

Positive Impact of a Multi-Modal Approach to Parent-Child Intervention

The main result of this study is that the families experienced a manifest improvement during the period of intervention. This improvement concerned all the aspects studied and led to an experience of increased mental wellbeing, increased faith in the future, reduced parental stress, greater possibilities of obtaining social support, positive changes in the way of relating to other people and a reduction of the impact of the problems pertaining to the children on everyday life. A clear pattern was visible: there was improvement after six months in all the areas studied and a continued and reinforced development was observed a year later.

With regard to the discussion of whether "less is more" or "more is better" the centres in this study endeavour to match the extent of the intervention to the needs of each family and there is a readiness to meet needs on different ecological levels and to choose different ports of entry in the interaction treatment. This approach supports the standpoint that "less is more" is relevant for some whereas "more is better" is more relevant for others [63]. There are families whose treatment may be restricted for instance to a limited number of Marte Meo-sessions with a narrow focus, but there are also families with a long history of mistrust of authorities to surmount before a collaborative relationship can be established and treatment can start. The differential patterns of improvement described above may reflect the variation of needs—the immediate impact of a child's behavioural problems can change rapidly whereas the parent's way of relating to other people seem to alter more slowly. At this stage the present study cannot claim to add much evidence on the question of "less" or/and "more." Further analysis of the dataset will, however, shed light on this issue with respect to the families in this study.

The tendencies were, with a few exceptions, similar for mothers and fathers but improvement was considerably stronger for mothers. The manifest and intriguing gender differences with regard to problem weight at the outset and improvement during the intervention will—as already noted—be further addressed in a forthcoming article.

One interesting result was that the level of dropout from treatment was low. There were only ten undesired or unplanned interruptions of the treatment, and when they occurred they were related to external circumstances. This result was unexpected since problems with high levels of dropout often are encountered in the literature concerning interventions in early childhood [64], and several studies have shown an attrition of 40–60% in children and families who began outpatient treatment services [65,66]. Attention has therefore been drawn to the need for interventions designed to improve commitment and decrease attrition [67], and Staudt [68] emphasizes, as did Cook [28], that research on interventions must include their acceptability to clients and their potential to reach and engage the families of at-risk children.

This raises questions about which aspects of the intervention in this study contributed to the low dropout levels. Successful negotiation and acceptance by the therapist and client of the goals, tasks and techniques have been found to increase engagement and hope [69]. In their research concerning barriers to treatment participation Kazdin and Wassell [70] point out the importance of the parents' perceived relevance of treatment. The principle adopted by the centres in this study that the goals and means of the intervention should be established through a dialogue between the parents and the therapist might therefore be a vital element. This is corroborated by a Swedish study [71] with 4–12-year-old children who displayed externalizing behaviour problems—using partly the same therapeutic approach—where there were no dropouts in the intervention group after the intervention had begun.

Another reason for the low number of dropouts from treatment might be that the intervention is adapted to the needs of each family. A mismatch—either way—between a family's assistance needs and the extent of the intervention can jeopardize the families' motivation to participate. The low number of dropouts from treatment has led to a very limited attrition from the study, which is a major strength since it implies that the results we have obtained have a high validity. As the study was a naturalistic one, it is the effectiveness of the centres' everyday practice that we are measuring. There is, therefore, no need to fear that the results depend upon special conditions during the intervention period. Another essential merit in this study lies in the fact that the change has been measured both in a short term and a long-term perspective. The long-term improvements in this study raise questions about what happens in an even longer perspective, especially since the results suggest that the notion of sleeper effects is of relevance in this kind of intervention programme.

Variables of Clinical Importance

One of the most important reasons for seeking help was aggression in the child. This is of great interest as aggression and other anti-social behaviour—especially in children below 12—is one of the main predictors for continued negative development [72]. Since a meta-analysis [73] has shown that aggressive behaviour tends to remain stable in all age groups when untreated, it is of utmost importance to provide effective treatment programmes for families.

Most of the results in this study, however, relate to improvements in the parents and an important question concerns which of these aspects may be considered important from a clinical perspective. A secure attachment is an important protective factor for a child growing up in a risk environment [74] and a disorganized attachment is a serious risk factor for externalizing problems [75]. Within attachment research, questions concerning the stability of patterns of attachment over time are studied and discussed as well as to what extent patterns of attachment are "inherited" by one generation from another. There is clear evidence of the importance of the parents' own attachment patterns for the child's possibility to develop a secure attachment [76]. This could imply that even small changes in a positive direction—an increased proportion of secure attachment and a reduced proportion of fearful/disorganized attachment—might be of significant importance for the children's development.

There is also strong evidence [13] indicating that the mother's mental health and well-being affect the child's development. Improved mental well-being should therefore be of vital importance. Likewise, the experience of parental stress is important. Anderson [77] has shown that the experience of stress is associated with a heightened risk of anxiety in the child, which indicates that stress reduction is clinically relevant.

In a study [78] comprising 152 infant parents there was an association between social support and increasingly positive parent-child activities over time, but this effect was mediated by mothers' attachment styles. It is considered important to reduce the feelings of relationship anxiety, and the authors consider that parenting interventions can achieve this by actively building on parents' successful social experiences within the framework of the intervention. This concurs with the emphasis of the centres on the therapist-parent relationship [79].

Limitations

A limitation of this present study is its lack of a control group. For ethical and practical reasons it was not possible to create one and we cannot therefore say with certainty what the development would have been like for these families had they not received help. A crucial question is whether results corresponding to those displayed by the families in this study could be obtained through spontaneous improvement. In a prospective study [80] 2587 children were followed up 3 years after the original survey for a sub-sample of the 1999 British Child and Adolescent Mental Health Survey. Latent mental health scores (i.e. combined information from multiple informants) showed strong stability over time ($r = 0.71$). A poorer outcome was associated for instance with externalizing as opposed to emotional symptoms and after exposure to parental mental illness. The authors conclude that there is a need for effective intervention with children with impairing psychopathology, since they are unlikely to improve spontaneously. The predictors of change in mental health were closely comparable across the range of initial SDQ scores, suggesting that they operated in a similar manner regardless of the initial level of (mal)adjustment.

A control group of "community families" would have enabled better comparison with respect to the burden of problems at the outset of treatment. Though the comparative data presented do not offer a perfect match—for more detailed information about the samples see references—they do contribute to the description of the subjects in the study.

Zaslow et al. [81] have shown that self-reports have a predictive value and that they are an appropriate choice when budgets or time are limited. It was logical to prioritize the parents' subjective perspective in this study, but we realize that deeper knowledge could be attained if supplemented by observations/assessments; e.g. parent-child interaction, the children's attachment, health data and how the children function in day care.

Conclusions and Directions For Future Research

This study has shown that it is possible to reach mothers and fathers with parenting problems, and to create an intervention program with very low dropout levels. This is of special importance since aggressive behaviour by the children was one of the most important reasons for seeking help in this study. Aggression in childhood has been shown to be a serious risk factor for further negative development, and families facing these problems have often displayed high levels of dropout.

The role of fathers in parent-child interaction interventions remains unexplored. Future research regarding fathers in parent-child interventions is of special importance so that the continued development of these interventions will be tailored to the needs of the fathers as well as these of the mothers. Clinicians also need more empirical knowledge on the question of "less" or/and "more," and this will be the focus of another forthcoming article from this study.

Another important and neglected aspect is the children's own experiences of participation in parent-child interventions. We have addressed parents' subjective accounts of participating in treatment at the four centres in a previous study [79], but this perspective should be complemented by assessments of the parents, the children and the interaction from a third-person perspective.

Since living conditions in different cultures may create different problems there is a demand for further knowledge about parent-child interventions with various designs from various cultural contexts. There is therefore a need to deepen our understanding of what support society should offer to vulnerable fathers and mothers in order to help them to provide "a secure base" for their children.

Competing Interests

The authors declare that they have no competing interests.

Authors' Contributions

KN conceived the study, shared responsibility for the design and was responsible for the data collection, performed the statistical analysis and drafted the manuscript. IE shared responsibility for the design and helped to draft the manuscript. Both authors read and approved the final manuscript.

Acknowledgements

This research was supported by grants from the Allmanna Barnhuset Foundation in Stockholm, whose support we gratefully acknowledge. We would like to thank all the staff at the four centres for their engagement in data gathering.

References

1. Shonkoff JP, Meisels SJ: Preface. In Handbook of early childhood intervention. 2nd edition. Edited by: Shonkoff JP, Meisels SJ. New York: Cambridge University Press; 2000:XVII–XVIII.
2. Meisels SJ, Shonkoff JP: Early Childhood Intervention: A Continuing Evolution. In Handbook of early childhood intervention. 2nd edition. Edited by: Shonkoff JP, Meisels SJ. New York: Cambridge University Press; 2000:3–31.
3. Peterander F: Preparing practitioners to work with families in early childhood intervention. Educational and Child Psychology 2004, 21:89–101.
4. Bowlby J: Attachment and loss: Attachment. Volume 1. New York: Basic Books; 1969.
5. Ainsworth MDS: Object relations dependency and attachment: A theoretical review of the mother-infant relationship. Child Dev 1969, 40:969–1025.
6. Ainsworth MDS: Infancy in Uganda: infant care and the growth of attachment. Baltimore: Johns Hopkins University Press; 1967.

7. Bronfenbrenner U: Ecology of the family as a context for human development: Research perspectives. Dev Psychol 1986, 22:723–742.
8. Garbarino J, Ganzel B: The human ecology of early risk. In Handbook of early childhood intervention. 2nd edition. Edited by: Shonkoff JPE, Meisels SJE. New York, NY, US: Cambridge University Press; 2000:76–93.
9. Sameroff AJ: Ports of Entry and the Dynamics of Mother-Infant Interventions. In Treating parent-infant relationship problems Strategies for intervention. Edited by: Sameroff AJE, McDonough SCE, Rosenblum KLE. New York, NY, US: Guilford Press; 2004:3–28.
10. Hadadian A, Merbler J: Mother's stress: Implications for attachment relationships. Early Child Dev Care 1996, 125:59–66.
11. Benoit D, Parker KC: Stability and transmission of attachment across three generations. Child Dev 1994, 65:1444–1456.
12. Belsky J: The Developmental and Evolutionary Psychology of Intergenerational Transmission of Attachment. In Attachment and bonding A new synthesis. Edited by: Carter CS, Grossmann KE, Hrdy SB, Lamb ME. Cambridge, MA, US: MIT Press; 2005:169–198.
13. Speranza AM, Ammaniti M, Trentini C: An overview of maternal depression, infant reactions and intervention programmes. Clinical Neuropsychiatry: Journal of Treatment Evaluation 2006, 3:57–68.
14. Cochran M, Walker SK: Parenting and Personal Social Networks. In Parenting: An ecological perspective. Edited by: Luster T, Okagaki L, Mahwah NJ. US: Lawrence Erlbaum Associates Publishers; 2005:235–273.
15. Osofsky JD, Thompson MD: Adaptive and Maladaptive Parenting: Perspectives on Risk and Protective Factors. In Handbook of early childhood intervention. 2nd edition. Edited by: Shonkoff JP, Meisels SJ. New York Cambridge University Press; 2000:54–75.
16. Stern DN: The Interpersonal World of the Infant: A view from Psychoanalysis and Developmental Psychology. New York: Basic Books; 1985.
17. Stern DN: The Motherhood Constellation: A Unified View of Parent-Infant Psychotherapy. New York: Basic Books; 1995.
18. Stern DN: The Present Moment in Psychotherapy and Everyday Life. New York: Norton; 2004.
19. Stern DN: The Motherhood Constellation: Therapeutic Approaches to Early Relational Problems. In Treating parent-infant relationship problems Strategies for intervention. Edited by: Sameroff AJ, McDonough SC, Rosenblum KL. New York: Guilford Press; 2004:29–42.

20. van Ijzendoorn MH, Juffer F, Duyvesteyn M: Breaking the intergenerational cycle of insecure attachment: a review of the effects of attachment-based interventions on maternal sensitivity and infant security. J Child Psychol Psychiatry 1995, 36:225–248.
21. van Ijzendoorn M: Adult attachment representations, parental responsiveness, and infant attachment: A meta-analysis on the predictive validity of the Adult Attachment Interview. Psychol Bull 1995, 117:387–403.
22. Egeland B, Weinfeld NS, Bosquet M, Cheng VK: Remembering, repeating, and working through: Lessons from attachment-based interventions. In Handbook of infant mental health Infant mental health in groups at high risk. Volume 4. Edited by: Osofsky JD, Fitzgerald HE. New York: Wiley; 2000:35–89.
23. Bakermans-Kranenburg MJ, van Ijzendoorn MH, Juffer F: Less is more: Meta-analyses of sensitivity and attachment interventions in early childhood. Psychol Bull 2003, 129:195–215.
24. Dickie JR, Gerber SC: Training in social competence: The effect on mothers, fathers, and infants. Child Dev 1980, 51:1248–1251.
25. Metzl MN: Teaching parents a strategy for enhancing infant development. Child Dev 1980, 51:583–586.
26. Scholz K, Samuels CA: Neonatal bathing and massage intervention with fathers, behavioural effects 12 weeks after birth of the first baby: The Sunraysia Australia Intervention Project. Int J Behav Dev 1992, 15:67–81.
27. Ziv Y: Attachment-Based Intervention Programs: Implications for Attachment Theory and Research. In Enhancing early attachments Theory, research, intervention, and policy. Edited by: Berlin LJ, Ziv Y, Amaya-Jackson L, Greenberg MT. New York: Guilford Press; 2005:61–78.
28. Cook CR, Little SG, Akin-Little A: Interventions based on attachment theory: A critical analysis. J Early Child Infant Psychol 2007, 3:61–73.
29. Brodén M: Psykoterapeutiska interventioner under spädbarnsperioden. [Psychotherapeutic interventions during infancy]. In Doctoral Dissertation. Lund University; 1992.
30. Wadsby M, Sydsjo G, Svedin CG: Evaluation of an intervention programme to support mothers and babies at psychosocial risk: assessment of mother/child interaction and mother's perceptions of benefit. Health Soc Care Community 2001, 9:125–133.
31. Heiervang E, Goodman A, Goodman R: The Nordic advantage in child mental health: Separating health differences from reporting style in a cross-cultural

comparison of psychopathology. J Child Psychol Psychiatry 2008, 49:678–685.

32. Kerns K: Attachment in middle childhood. In Handbook of attachment: theory, research, and clinical applications. 2nd edition. Edited by: Cassidy J, Shaver PR. New York: Guilford Press; 2008:366–382.
33. Ainsworth MDS: Some considerations regarding theory and assessment relevant to attachments beyond infancy. In Attachment in the preschool years: Theory, research, and intervention. Edited by: Greenberg MT, Cicchetti D, Cummings EM. Chicago: University of Chicago Press; 1990:463–488.
34. Antonovsky A: Unraveling the mystery of health: how people manage stress and stay well. San Francisco: Jossey-Bass; 1987.
35. Davison GC: Stepped care: Doing more with less? J Consult Clin Psychol 2000, 68:580–585.
36. Aarts M: Marte Meo: Basic manual. Harderwijk: Aarts Productions; 2000.
37. Hafstad R, Ovreeide H: Med barnet som fasit—Marte Meo-metoden. In Barn og unge i psykoterapi Vol 2 Terapeutiska fremgangsmåter og forandring [Children and young persons in psychotherapy]. Volume 2. Edited by: Haavind H, Ovreeide H. Oslo: Gyldendal; 2007:145–170.
38. Fonagy P, Target M: Attachment and reflective function: Their role in self-organization. Dev Psychopathol 1997, 9:679–700.
39. Cassidy J, Woodhouse SS, Cooper G, Hoffman K, Powell B, Rodenberg M: Examination of the Precursors of Infant Attachment Security: Implications for Early Intervention and Intervention Research. In Enhancing early attachments Theory, research, intervention, and policy. Edited by: Berlin LJ, Ziv Y, Amaya-Jackson L, Greenberg MT. New York: Guilford Press; 2005:34–60.
40. Fraiberg S, Adelson E, Shapiro V: Ghosts in the nursery: A psychoanalytic approach to the problems of impaired infant-mother relationships. J Am Acad Child Psychiatry 1975, 14:387–422.
41. Lieberman AF, Padron E, Van Horn P, Harris WW: Angels in the nursery: The intergenerational transmission of benevolent parental influences. Infant Ment Health J 2005, 26:504–520.
42. Ostberg M, Hagekull B, Wettergren S: A measure of parental stress in mothers with small children: dimensionality, stability and validity. Scand J Psychol 1997, 38:199–208.
43. Abidin RR: Parenting stress index (PSI)-manual. Odessa, FL: Psychological Assessment Resources Inc; 1990.

44. Bartholomew K, Horowitz LM: Attachment styles among young adults: a test of a four-category model. J Pers Soc Psychol 1991, 61:226–244.
45. Backstrom M, Holmes BM: Measuring adult attachment: A construct validation of two self-report instruments. Scand J Psychol 2001, 42:79–86.
46. Goldberg DP: General Health Questionnaire (GHQ). In Handbook of psychiatric measures. Edited by: Task Force for the Handbook of Psychiatric Measures. Washington, DC: American Psychiatric Association; 2000:75–79.
47. Lindstrom M, Moghaddassi M, Merlo J: Individual and contextual determinants of self-reported poor psychological health: a population-based multilevel analysis in southern Sweden. Scand J Public Health 2006, 34:397–405.
48. Cantril H: The patterns of human concerns. New Brunswick, NJ: Rutgers University Press; 1965.
49. Atkinson T: The stability and validity of quality of life measures. Soc Indic Res 1982, 10:113–132.
50. Henderson S, Byrne DG, Duncan-Jones P: Neurosis and the social environment. Sydney: Academic Press; 1981.
51. Unden AL, Orth-Gomer K: Development of a social support instrument for use in population surveys. Soc Sci Med 1989, 29:1387–1392.
52. Eklund M, Bengtsson-Tops A, Lindstedt H: Construct and discriminant validity and dimensionality of the Interview Schedule for Social Interaction (ISSI) in three psychiatric samples. Nord J Psychiatry 2007, 61:182–188.
53. Goodman R: Psychometric properties of the strengths and difficulties questionnaire. J Am Acad Child Adolesc Psychiatry 2001, 40:1337–1345.
54. Smedje H, Broman JE, Hetta J, von Knorring AL: Psychometric properties of a Swedish version of the "Strengths and Difficulties Questionnaire." Eur Child Adolesc Psychiatry 1999, 8:63–70.
55. Malmberg M, Rydell A-M, Smedje H: Validity of the Swedish version of the Strengths and Difficulties Questionnaire (SDQ-Swe). Nord J Psychiatry 2003, 57:357–363.
56. Cohen J: Statistical power analysis for the behavioral sciences. 2nd edition. Hillsdale, NJ: Lawrence Earlbaum Associates; 1988.
57. Ostberg M: Parental stress, psychosocial problems and responsiveness in help-seeking parents with small (2–45 months old) children. Acta Paediatr 1998, 87:69–76.
58. Olsson M: Unpublished work. In Faculty of Social Sciences. Göteborg University, Sweden; 2008.

59. Hälsa på lika villkor Resultat från nationella folkhälsoenkäten 2005. [Health on equal terms. Results from the national public health survey 2005] Swedish National Institute of Public Health; 2006.

60. Van Zeijl J, Mesman J, Van IMH, Bakermans-Kranenburg MJ, Juffer F, Stolk MN, Koot HM, Alink LR: Attachment-based intervention for enhancing sensitive discipline in mothers of 1- to 3-year-old children at risk for externalizing behavior problems: a randomized controlled trial. J Consult Clin Psychol 2006, 74:994–1005.

61. Axberg U, Hansson K, Broberg AG: Evaluation of the incredible years series. An open study of its effects when first introduced in Sweden. Nord J Psychiatry 2007, 61:143–151.

62. Malmberg M, Ingstedt Jarl B: Att mäta barns psykiska hälsa [Measuring children's health]. Uppsala University, Department of Psychology; 2000.

63. Berlin LJ: Interventions to Enhance Early Attachments. The State of the Field Today. In Enhancing early attachments: Theory, research, intervention, and policy. Edited by: Berlin LJ, Ziv Y, Amaya-Jackson L, Greenberg MT. New York: The Guilford Press; 2005:3–33.

64. Gray R, McCormick MC: Early Childhood Intervention Programs in the US: Recent Advances and Future Recommendations. J Prim Prev 2005, 26:259–275.

65. Kazdin AE: Dropping out of child psychotherapy: Issues for research and implications for practice. Clin Child Psychol Psychiatry 1996, 1:133–156.

66. Wierzbicki M, Pekarik G: A meta-analysis of psychotherapy dropout. Prof Psychol Res Pr 1993, 24:190–195.

67. Miller LM, Southam-Gerow MA, Allin RB Jr: Who stays in treatment? Child and family predictors of youth client retention in a public mental health agency. Child & Youth Care Forum Vol 2008, 37:153–170.

68. Staudt M: Treatment engagement with caregivers of at-risk children: Gaps in research and conceptualization. J Child Fam Stud 2007, 16:183–196.

69. Asay TP, Lambert MJ: The empirical case for the common factors in therapy: Quantitative findings. In The heart and soul of change What works in therapy. Edited by: Hubble MAE, Duncan BLE, Miller SDE. Washington, DC, US: American Psychological Association; 1999:23–55.

70. Kazdin AE, Wassell G: Predictors of barriers to treatment and therapeutic change in outpatient therapy for antisocial children and their families. Ment Health Serv Res 2000, 2:27–40.

71. Axberg U, Hansson K, Broberg AG, Wirtberg I: The Development of a Systemic School-Based Intervention: Marte Meo and Coordination Meetings. Fam Process 2006, 45:375–389.

72. Lipsey MW, Derzon JH: Predictors of violent or serious delinquency in adolescence and early adulthood: A synthesis of longitudinal research. In Serious & violent juvenile offenders Risk factors and successful interventions. Edited by: Loeber R, Farrington DP. Thousand Oaks, CA: Sage Publications; 1998:86–105.

73. Wilson SJ, Lipsey MW, Derzon JH: The effects of school-based intervention programs on aggressive behavior: A meta-analysis. J Consult Clin Psychol 2003, 71:136–149.

74. Greenberg MT: Attachment and psychopathology in childhood. In Handbook of attachment Theory, research, and clinical applications. Edited by: Cassidy J, Shaver PR. New York: Guilford Press; 1999:469–496.

75. Munson JA, McMahon RJ, Spieker SJ: Structure and variability in the developmental trajectory of children's externalizing problems: Impact of infant attachment, maternal depressive symptomatology, and child sex. Dev Psychopathol 2001, 13:277–296.

76. van Ijzendoorn MH, Bakermans-Kranenburg MJ: Intergenerational transmission of attachment: A move to the contextual level. In Attachment and psychopathology. Edited by: Atkinson L, Zucker KJ. New York: Guilford Press; 1997:135–170.

77. Anderson NE: The relationship between mothers' stress level and anxiety ratings of their children. Anderson, Nicole E.: U Arizona, US; 2007.

78. Green BL, Furrer C, McAllister C: How do relationships support parenting? Effects of attachment style and social support on parenting behavior in an at-risk population. Am J Community Psychol 2007, 40:96–108.

79. Neander K, Skott C: Bridging the Gap—the Co-creation of a Therapeutic Process. Reflections by Parents and Professionals on their Shared Experiences of Early Childhood Interventions. Qualitative Social Work 2008, 7:289–309.

80. Ford T, Collishaw S, Meltzer H, Goodman R: A prospective study of childhood psychopathology: Independent predictors of change over three years. Soc Psychiatry Psychiatr Epidemiol 2007, 42:953–961.

81. Zaslow MJ, Weinfield NS, Gallagher M, Hair EC, Ogawa JR, Egeland B, Tabors PO, De Temple JM: Longitudinal prediction of child outcomes from differing measures of parenting in a low-income sample. Dev Psychol 2006, 42:27–37.

Copyrights

Index

B

C

D

R

S

T

U